Linguistics for the Age of AI

Linguistics for the Age of AI

Marjorie McShane and Sergei Nirenburg

The MIT Press
Cambridge, Massachusetts
London, England

The open access edition of this book was made possible by generous funding from Arcadia—a charitable fund of Lisbet Rausing and Peter Baldwin.

ARCADIA
A charitable fund of Lisbet Rausing and Peter Baldwin

This book was set in Times New Roman by Westchester Publishing Services. Printed and bound in the United States of America.

Library of Congress Cataloging-in-Publication Data

Names: McShane, Marjorie Joan, 1967- author. | Nirenburg, Sergei, author.
Title: Linguistics for the age of AI / Marjorie McShane and Sergei Nirenburg.
Description: Cambridge, Massachusetts : The MIT Press, [2021] | Includes bibliographical references and index.
Identifiers: LCCN 2020019867 | ISBN 9780262045582 (hardcover)
Subjects: LCSH: Computaitonal linguistics. | Natural language processing (Computer science)
Classification: LCC P98 .M325 2020 | DDC 401/.430285635--dc23
LC record available at https://lccn.loc.gov/2020019867

10 9 8 7 6 5 4 3 2 1

Contents

Acknowledgments

Our warmest thanks to

- Stephen Beale, our close collaborator and friend, for his unrivaled expertise in turning ideas into systems;
- Bruce Jarrell and George Fantry, for shaping the vision behind the Maryland Virtual Patient system and showing that domain experts can be remarkable collaborators on intelligent systems;
- Jesse English, Benjamin Johnson, Irene Nirenburg, and Petr Babkin, for their tireless work on translating models into application systems;
- Lynn Carlson, for her meticulous reading of the manuscript and insightful suggestions for its improvement;
- Igor Boguslavsky, for encouraging us to hone our thinking about the *theory-system-model* distinction that became central to the framing of this work;
- The Office of Naval Research for their generous support over many years;
- Our program officers at the Office of Naval Research—Paul Bello, Micah Clark, and Thomas McKenna—for their steady faith in this program of research, which truly made it all possible.

Setting the Stage

Remember HAL, the "sentient computer" from Stanley Kubrick's and Arthur C. Clarke's *2001: A Space Odyssey*? To refresh your memory, here is a sample dialog between HAL and Dave, the astronaut:

Dave: Open the pod bay doors, HAL.

HAL: I'm sorry, Dave, I'm afraid I can't do that.

Dave: What's the problem?

HAL: I think you know what the problem is just as well as I do.

Dave: What are you talking about, HAL?

HAL: This mission is too important for me to allow you to jeopardize it.

Dave: I don't know what you're talking about, HAL.

HAL: I know that you and Frank were planning to disconnect me, and I'm afraid that's something I cannot allow to happen.

HAL clearly exhibits many facets of human-level intelligence—context-sensitive language understanding, reasoning about Dave's plans and goals, developing its own plans based on its own goals, fluent language generation, and even a modicum of emotional and social intelligence (note the politeness).

The movie came out over fifty years ago, so one might expect HAL to be a reality by now, like smartphones. But nothing could be further from the truth. Alas, despite the lure of AI in the public imagination and recurring waves of enthusiasm in R&D circles, re-creating human-level intelligence in a machine has proved more difficult than expected. In response, the AI community has, by and large, opted to change course, focusing on simpler, "low-hanging fruit" tasks and silo applications, such as beating the best human players in games like chess, *Go*, and *Jeopardy!* Such systems are not humanlike: they do not know what they are doing and why, their approach to problem solving does not resemble a person's, and they do not rely on models of the world, language, or agency. Instead, they largely rely on applying generic machine learning algorithms to ever larger datasets, supported by the spectacular speed and storage capacity of modern computers. Such systems are the best that the field can offer in the short term, as they outperform handcrafted AI systems on the kinds of tasks at which they excel, while requiring much less human labor to create. But those successes have blunted the impetus to work on achieving human-level AI and have led to the concomitant avoidance of applications that require it. The upshot is that the goal of developing artificial intelligent agents with language and reasoning capabilities like those of HAL remains underexplored (Stork, 2000).

In this book, we not only explain why this goal must remain on agenda but also provide a roadmap for endowing artificial intelligent agents with the language processing capabilities demonstrated by HAL. In the most general terms, our approach can be described as follows. An artificial intelligent agent with human-level language ability—what we call a *language-endowed intelligent agent* (LEIA)—must do far more than manipulate the words of language: it must be able to *understand*, *explain*, and *learn*.

- The LEIA must *understand* the context-sensitive, intended meaning of utterances, which can require such capabilities as reconstructing elided meanings, interpreting indirect speech acts, and connecting linguistic references not only to elements of memory but also, in some cases, to perceived objects in the physical world.
- It must be capable of *explaining* its thoughts, actions, and decisions to its human collaborators in terms that are meaningful to us. It is such transparency that will earn our trust as we come to rely on intelligent systems for assistance in tasks more sensitive and critical than finding a restaurant with Siri's help or getting the gist of a foreign-language blog using Google Translate.
- It must forever be in a mode of *lifelong learning*, just like its human counterparts. Lifelong learning subsumes learning new words and phrases, new elements of the world model (ontology), and new ways of carrying out tasks. It requires remembering and learning from past actions, interactions, and simulated thought processes. And it involves perceiving and interpreting the dynamically changing properties and preferences of the agent's human collaborators.

To achieve this level of prowess, the agent must be equipped with a descriptively adequate model of language and world, which fuels heuristic algorithms that support decision-making for perception and action.

This book concentrates on describing the language understanding capabilities of LEIAs from the perspectives of cognitive modeling and system building. A prominent design feature is the orientation around *actionability*—that is, seeking an interpretation of a language input that is sufficiently deep, precise, and confident to support reasoning about action. Targeting actionability rather than perfection is far more than an expedient engineering solution—it models what people do. Real language use is messy, and no small amount of what we say to each other is unimportant and/or not completely comprehensible. So, we understand what we can and then, if needed, enhance that interpretation by drawing on background knowledge, expectations, and judgments about the (un)importance of whatever remains unclear. How much effort we devote to understanding a particular utterance is guided by the principle of least effort, which serves to manage cognitive load. Modeling this strategy in machines is the best hope for achieving robustness in systems that integrate language understanding into an agent's overall operation.

To implement the above strategy, this book introduces two orthogonal conceptions of *incrementality. Horizontal* incrementality refers to processing inputs phrase by phrase, building up an interpretation as a person would upon perceiving a language stream. *Vertical* incrementality involves applying increasingly sophisticated (and resource-intensive) analysis methods to input "chunks" with the goal of achieving an actionable interpretation at the lowest possible cost. The control flow of language understanding—that is, deciding how deeply to process each chunk before consuming the next one—is handled by a cognitive architecture functionality that bridges language understanding and goal-oriented reasoning.

The crux of natural language understanding is semantic and pragmatic analysis. We treat the large number of semantic and pragmatic problems (from lexical disambiguation to coreference resolution to indirect speech acts and beyond) using *microtheories*. Each microtheory sketches a complete problem space, classifies component problems, and details methods of solving the subset of problems that can be treated fully automatically. Full automaticity is important, as it is nonsensical to develop agent systems that rely on unfulfillable prerequisites. The heuristics brought to bear on language analysis tasks are drawn from many sources: static knowledge bases, such as semantic lexicons and the associated ontological (world) model; the system's situation model, which contains the agent's active goal and plan agenda; results of processing prior inputs within the same dialog, task, or application; and any other machine-tractable resource that we can render useful, whether it was developed in-house or imported. In other words, the approach to language understanding described in this book operationalizes the many facets of situational context that we, as humans, bring to bear when participating in everyday language interactions.

Just as building intelligent agents requires a combination of science and craft, so does writing a book of this genre. The biggest challenge is balancing the generic with the specific. Our solution was to write most of the book (chapters 2–7) in relatively generic terms, without undue focus on implementation details but, rather, with an emphasis on the decision-making involved in modeling. Still, we devote two chapters to specifics: namely, systems developed using the described microtheories (chapter 8) and evaluations of a LEIA's language understanding capabilities (chapter 9). Both of these chapters (a) validate that we are working on real AI in real systems, with all their expected and unexpected hurdles, and (b) emphasize the need for holistic—rather than isolated or strictly modularized—approaches to agent modeling. For example, our Maryland Virtual Patient prototype system was designed to train physicians via interactions with simulated LEIAs playing the role of virtual patients. The virtual-patient LEIAs integrated cognitive capabilities (e.g., language processing and reasoning) with a physiological simulation that not only produced medically valid outcomes but also provided the agent with interoception of its own symptoms, which contributed to its health-oriented decision-making. This holistic approach to agent modeling stands in contrast to the currently more popular research methodology of carving out small problems to be solved individually. However, solving small problems does not help solve big problems, such as building LEIAs. Instead, we must take on the big problems in their totality and prepare agents to do the best they can with what they've got—just like people do.

We think that this book will be informative, thought-provoking, and accessible to a wide variety of readers interested in the fields of linguistics, cognitive science, natural language processing (NLP), and AI. This includes professionals, students, and anyone motivated enough to dig into the science-oriented offerings of the popular press. The book suggests ways in which linguists can make essential contributions to AI beyond corpus annotation and building wordnets, it offers students a choice of many topics for research and dissertations, it reminds practitioners of knowledge-lean NLP just how far we still have to go, and it shows developers of artificial intelligent agent systems what it will take to make agents truly language-endowed. To promote readability, we have divided the chapters, when applicable, into the main body followed by deep dives that will likely be of interest primarily to specialists. We also include pointers to online appendixes.

As a point of reference, we have successfully incorporated the book into undergraduate and graduate courses at Rensselaer Polytechnic Institute with the sole prerequisite of an introductory course in linguistics. As a tool for stimulating students to think about and discuss hard issues, the book has proven quite valuable. The suggested chapter-end exercises offer hands-on practice as well as a break from reading. But although the book nicely serves pedagogical goals, it is not a textbook. Textbooks tend to look backward, presenting a neat picture of work already accomplished and striving for extensive and balanced coverage of the literature. By contrast, this book looks forward, laying out a particular program of work that, in the historical perspective, is still in its early days—with all the unknowns commensurate with that status.

As mentioned earlier, this genre of exploration is far removed from the current mainstream, so it would be natural for some readers to come to the table with expectations that, in fact, will not be fulfilled. For example, we will say nothing about neural networks, which is the approach to machine learning that is receiving the most buzz at the time of writing. As for machine learning more broadly understood, we believe that the most promising path toward integrating machine learning–based and knowledge-based methods is to integrate the results of machine learning into primarily knowledge-based systems, rather than the other way around—though, of course, this is a wide-open research issue.

The genre of this book is very different from that of compendia such as Quirk et al.'s *A Comprehensive Grammar of the English Language* (1985). We do not offer either a complete description of the phenomena or the final word on any of the topics we discuss. While the microtheories we describe are sufficiently mature to be implemented in working systems and presented in print, they all address active research areas, and they will certainly continue to evolve in planned and unplanned directions.

There has been significant competition for space in the book. We expect that some readers will find themselves wanting more detail about one or another phenomenon than what the book contains. We took our best shot at estimating the levels of interest that particular material would elicit in readers.

In short, this book is about giving readers new things to think about in a new way. Ultimately, we hope that it will serve as a reminder of just how complex human languages are, and how remarkable—verging on miraculous—it is that we can use them with such ease.

1

Our Vision of Linguistics for the Age of AI

1.1 What Is Linguistics for the Age of AI?

A long-standing goal of artificial intelligence has been to build intelligent agents that can function with the linguistic dexterity of people, which involves such diverse capabilities as participating in a fluent conversation, using dialog to support task-oriented collaboration, and engaging in lifelong learning through processing speech and text. There has been much debate about whether this goal is, in principle, achievable since its component problems are arguably more complex than those involved in space exploration or mapping the human genome. In fact, enabling machines to emulate human-level language proficiency is well understood to be an AI-complete problem—one whose full solution requires solving the problem of artificial intelligence in general. However, we believe that it is in the interests of both scientific progress and technological innovation to assume that this goal *is* achievable until proven otherwise. The question then becomes, How best to pursue it?

We think that a promising path forward is to pursue linguistic work that adheres to the following tenets:

- Language processing is modeled from the agent perspective, as one component of an integrated model of perception, reasoning, and action.
- The core prerequisites for success are the abilities to (a) extract the meaning of linguistic expressions, (b) represent and remember them in a model of memory, and (c) use these representations to support an intelligent agent's decision-making about action—be it verbal, physical, or mental.
- While extralinguistic information is required for extracting the full meanings of linguistic inputs, in many cases, purely linguistic knowledge is sufficient to compute an interpretation that can support reasoning about action.
- Language modeling must cover and integrate the treatment of all linguistic phenomena (e.g., lexical ambiguity, modality, reference) and all components of processing (e.g., syntax, semantics, discourse).

- The treatments of language phenomena are guided by computer-tractable microtheories describing specific phenomena and tasks.
- A core capability is lifelong learning—that is, the agent's ability to independently learn new words, ontological concepts, properties of concepts, and domain scripts through reading, being told, and experience.
- Methodologically, the accent is on developing algorithms that facilitate the treatment of the many tasks within this research program. Any methods can be brought to bear as long as they are sufficiently transparent to allow the system's decisions to be explained in a manner that is natural for humans.

We call this program of work Linguistics for the Age of AI. The first half of this chapter describes the program in broad strokes. The deep dives in the second half provide additional details that, we think, might go beyond the interests of some readers.

1.2 What Is So Hard about Language?

For the uninitiated, the complexities of natural language are not self-evident: after all, people seem to process language effortlessly. But the fact that human language abilities are often taken for granted does not make them any less spectacular. When analyzed, the complexity of the human language facility is, in fact, staggering—which makes modeling it in silico a very difficult task indeed.

What, exactly, makes language hard for an artificial intelligent agent? We will illustrate the complexity using the example of ambiguity. Ambiguity refers to the possibility of interpreting a linguistic unit in different ways, and it is ubiquitous in natural languages. In order to arrive at the speaker's intended meaning, the interlocutor must select the contextually appropriate interpretation of the ambiguous entity. There are many types of ambiguity in natural language. Consider a few examples:

1. **Morphological ambiguity.** The Swedish word *frukosten* can have five interpretations, depending on how its component morphemes are interpreted. In the analyses below, lexical morphemes are separated by an underscore, whereas the grammatical morpheme for the definite article (*the*) is indicated by a plus sign:

    - frukost+en "the breakfast"
    - frukost_en "breakfast juniper"
    - fru_kost_en "wife nutrition juniper"
    - fru_kost+en "the wife nutrition"
    - fru_ko_sten "wife cow stone" (Karlsson, 1995, p. 28)

2. **Lexical ambiguity.** The sentence *I made her duck* can have at least the following meanings, depending on how one interprets the words individually and in combination: "I forced her to bend down," "I prepared food out of duck meat for her," "I prepared

food out of the meat of a duck that was somehow associated with her" (it might have belonged to her, been purchased by her, been raised by her), and "I made a representation of a duck that is somehow associated with her" (maybe she owns it, is holding it).

3. **Syntactic ambiguity.** In the sentence *Elaine poked the kid with the stick*, did Elaine poke the kid using a stick, or did she poke (using her finger) a kid who was in possession of a stick?
4. **Semantic dependency ambiguity.** The sentence *Billy knocked over the vase* is underspecified with respect to Billy's semantic role: if he did it on purpose, he is the agent; if not, he is the instrument.
5. **Referential ambiguity.** In the sentence *The soldiers shot at the women and I saw some of them fall*, who fell—soldiers or women?
6. **Scope ambiguity.** Does *big rivers and lakes* describe big rivers and big lakes or big rivers and lakes of any size?
7. **Pragmatic ambiguity.** When a speaker says, *I need help fixing the toaster*, is this asserting a fact or asking the interlocutor for help?

If these examples served as input to a machine translation system, the system would, in most cases, have to settle on a single interpretation because different interpretations would be translated differently. (The fact that ambiguities can sometimes be successfully carried across languages cannot be relied on in the general case.) While the need to select a single interpretation should be self-evident for some of the examples, it might be less clear for others, so consider some scenarios. Regarding (3), Russian translates the instrumental and accompaniment meanings of *with* in different ways, so that this ambiguity must be resolved explicitly. Regarding (5), in Hebrew, the third-person plural pronoun—which is needed to translate *them*—has different forms for different genders, so a translation system would need to identify either *the women* or *the soldiers* as the coreferent. Regarding (6), the ambiguity can be carried through to a language with the same adjective-noun ordering in noun phrases, but possibly not to a language in which adjectives follow their nouns. Regarding (7), although some language pairs may allow for speech act ambiguity to be carried through in translation, this escape hatch will be unavailable if the application involves a personal robotic assistant who needs to understand what you want of it.

The obvious response to the question of how to arrive at a particular interpretation is, *Use the context!* After all, people use the context effortlessly. But what does using the context actually mean? What is the context? How do we detect, categorize, and select the salient bits of context and then use them in understanding language? At the risk of some overgeneralization, we can say that the historical and contemporary scope of natural language processing (NLP) research reflects a wide variety of responses to these questions. At one extreme of the range of solutions—the so-called *knowledge-lean* approaches—the context is defined as a certain number of words appearing to the right and to the left of the word whose interpretation is sought. So, the context is words, period. At the other extreme—the *knowledge-based* approaches—the context is viewed as the combination of a large number of features about language, the situation, and the world that derive from different sources and are computed in different ways. Leveraging more elements from the context improves the accuracy of language interpretation; however, this ability comes at a steep price. One of the purposes of this book is to demonstrate that intelligent agents can often derive useful interpretations of language inputs without having to invoke every aspect of knowledge and reasoning that a person would bring to bear. An agent's interpretations may be incomplete or vague but still be sufficient to support the agent's reasoning about action—that is, the interpretations are *actionable.* Orienting around actionability rather than perfection is the key to making a long-term program of work toward human-level natural language processing at once scientifically productive and practically feasible.

A terminological note: In its broadest sense, the term *natural language processing* refers to any work involving the computational processing of natural language. However, over the past few decades, NLP has taken on the strong default connotation of involving knowledge-lean (essentially, semantics-free) machine learning over big data. Therefore, in the historically recent and current context, there is a juxtaposition between NLP and what we are pursuing in this book: NLU, or natural language *understanding* (see the deep dive in section 1.6.3 for discussion). However, earlier in the history of the field, the term *natural language processing* did encompass all methods of automating the processing of natural language. The historical discussion below inevitably uses both the broad and the narrow senses of the term. The context should make clear which sense is intended in each case. Lucky for us our readers are human.

1.3 Relevant Aspects of the History of Natural Language Processing

Natural language processing was born as machine translation, which developed into a high-profile scientific and technological area already in the late 1940s.[1] Within a decade of its inception, machine translation had given rise to the theoretical discipline of computational linguistics and, soon thereafter, to its applied facets that were later designated as natural language processing (NLP). The eponymous archival periodical of the field, *Computational Linguistics*, started its existence in 1954 as *Mechanical Translation* and, in 1965–1970, was published as *Mechanical Translation and Computational Linguistics.* A perusal of the journal's contents from 1954 to 1970 (http://www.machine translation-archive .info/MechTrans-TOC.htm) reveals a gradual shift from machine translation–specific to general computational-linguistic topics. The original machine translation initiative has also influenced other fields of study, most importantly theoretical linguistics and artificial intelligence.

From the outset, machine translation was concerned with building practical systems using whatever method looked most promising. It is telling that the first programmatic statement about machine translation, Warren Weaver's (1955 [1949]) famous memorandum, already suggests a few potential approaches to machine translation that can be seen as seeds of future computational-linguistic and NLP paradigms. Then, as now, such suggestions were influenced by the scientific and technological advances that captured the spirit of the times. Today, this may be big data and deep learning. Back then, Weaver was inspired by (a) results in early cybernetics, specifically McCulloch's artificial neurons (McCulloch & Pitts, 1943) and their use in implementing logical reasoning; (b) recent advances in formal logic; and (c) the remarkable successes of cryptography during World War II, which contributed to the development of information theory, on which Weaver collaborated with Shannon (Shannon & Weaver, 1964 [1949]). Inspiration from cybernetics can be seen as the seed of the connectionist approach to modeling language and cognition. The formal logic of Tarski, Carnap, and others underwent spectacular development and contributed to formal studies of the syntax and semantics of language as well as to the development of NLP systems. Shannon's information theory is the precursor of the currently ascendant statistical, machine learning–oriented approaches to language processing.

In machine translation research, it was understood early on that simplistic, word-for-word translation could not succeed and that understanding and rendering meaning were essential. It was equally understood that people disambiguate language in context. It is not surprising, therefore, that Weaver (1955 [1949]) suggests involving contextual clues in text analysis:

> If one examines the words in a book, one at a time through an opaque mask with a hole in it one word wide, then it is obviously impossible to determine, one at a time, the meaning of words. "Fast" may mean "rapid"; or it may mean "motionless"; and there is no way of telling which. But, if one lengthens the slit in the opaque mask, until one can see not only the central word in question but also say N words on either side, then, if N is large enough one can unambiguously decide the meaning.

The context-as-text-window method of analyzing text was in sync with the then-ascendant linguistic theory, structuralism, which did not accept unobservables in its repertoire (see the deep dive in section 1.6.1).[2] In view of this, it is not surprising that Weaver did not suggest a method of determining text meaning. The prevailing opinion was that the computational processing of meaning was not possible—this was the reason why Norbert Wiener, a pioneer of cybernetics, refused to join the early machine translation bandwagon and also why Yehoshua Bar Hillel, in the conclusion of his 1960 survey of a decade of machine translation research,[3] insisted that fully automatic, high-quality machine translation could not be an immediate objective of the field before much more work on computational semantics had been carried out. It is noteworthy that neither Wiener nor Bar Hillel believed that fully automatic, high-quality machine translation—and, by extension, high-quality NLP—could succeed without the treatment of meaning.

At the same time as semantics was failing to attract research interest, syntax was taking off in both the theoretical and computational realms. In theoretical work, the newly ascendant school of mentalist theoretical linguists—the generative grammarians—isolated syntax from other cognitive and language-processing capabilities with the goal of explaining the hypothesized Universal Grammar. In NLP, for its part, the flagship research direction for decades was developing syntactic parsers based on ever more sophisticated formal grammar approaches, such as lexical functional grammar, generalized phrase structure grammar, and head-driven phrase structure grammar.[4]

The study of meaning was not, however, completely abandoned: philosophers and logicians continued to pursue it with an accent on its formal representation and truth-conditional semantics. Formal representations, and their associated formal languages, were needed because it was assumed that formal reasoning could only be carried out over formal representations of the meanings of propositions—not the messy (ambiguous, elliptical, fragmentary) strings of natural language. Truth-conditional semantics, for its part, was a cornerstone of work on artificial reasoners in AI.

It is not surprising that a program of work headed by philosophers and logicians (not linguists) did not concentrate on translating from natural language into the artificial metalanguage of choice, even though that was a prerequisite for automatic reasoning. In fact, automating that translation process remains, to this day, an outstanding prerequisite to incorporating machine reasoning into end systems that involve natural language.

The distinction between these two lines of work—translating from natural language into a formal language, and reasoning over that formal language—was recognized early on by Bar Hillel, making his observation of long ago as relevant now as it was then:

> The evaluation of arguments presented in a natural language should have been one of the major worries … of logic since its beginnings. However, … the actual development of formal logic took a different course. It seems that … the almost general attitude of all formal logicians was to regard such an evaluation process as a two-stage affair. In the first stage, the original language formulation had to be rephrased, without loss, in a normalized idiom, while in the second stage these normalized formulations would be put through the grindstone of the formal logic evaluator.… Without substantial progress in the first stage even the incredible progress made by mathematical logic in our time will not help us much in solving our total problem. (Bar Hillel, 1970, pp. 202–203)

From the earliest days of configuring reasoning-oriented AI applications, contributing NLP systems did benefit from one simplifying factor: a given NLP system needed to interpret only those aspects of text, meaning that its target reasoning engine could digest (rather than aim for a comprehensive interpretation of natural language semantics). Still, preparing NLP systems to support automatic reasoning was far from simple. A number of efforts were devoted to extracting and representing facets of text meaning, notably, those of Winograd (1972), Schank (1972), Schank and Abelson (1977), Wilks (1975), and Woods (1975). These efforts focused on the resolution of ambiguity, which required knowledge of the context. The context, in turn, was understood to include the textual context, knowledge about the world, and knowledge about the speech situation.

Such knowledge needed to be formulated in machine-tractable form. It included, nonexhaustively, grammar formalisms specifically developed to support parsing and text generation, actual grammars developed within these formalisms, dictionaries geared toward supporting automatic lexical disambiguation, rule sets for determining nonpropositional (pragmatic and discourse-oriented) meaning, and world models to support the reasoning involved in interpreting propositions. Broad-scope knowledge acquisition of this complexity (whether for language-oriented work or general AI) was unattainable given the relatively limited resources devoted to it. The public perception of the futility of any attempt to overcome this so-called knowledge bottleneck profoundly affected the path of development of AI in general and NLP in particular, moving the field away from rationalist, knowledge-based approaches and contributing to the emergence of the empiricist, knowledge-lean, paradigm of research and development in NLP.

The shift from the knowledge-based to the knowledge-lean paradigm gathered momentum in the early 1990s. NLP practitioners considered three choices:

1. Avoid the need to address the knowledge bottleneck either by pursuing components of applications instead of full applications or by selecting methods and applications that do not rely on extensive amounts of knowledge.
2. Seek ways of bypassing the bottleneck by researching methods that rely on direct textual evidence, not ontologically interpreted, stored knowledge.
3. Address the bottleneck head-on but concentrate on learning the knowledge automatically from textual resources, with the eventual goal of using it in NLP applications.[5]

It is undeniable that the center of gravity in NLP research has shifted almost entirely toward the empiricist, knowledge-lean paradigm. This shift has included the practice of looking for tasks and applications that can be made to succeed using knowledge-lean methods and redefining what is considered an acceptable result—in the spirit of Church and Hovy's (1993) "Good Applications for Crummy Machine Translation." The empiricist paradigm was, in fact, already suggested and experimented with in the 1950s and 1960s—for example, by King (1956) with respect to machine translation. However, it became practical only with the remarkable advances in computer storage and processing starting in the 1990s.

The reason why NLP is particularly subject to fluctuations of fashion and competing practical and theoretical approaches is that, unlike other large-scale scientific efforts, such as mapping the human genome, NLP cannot be circumscribed by a unifying goal, path, purview, or time frame. Practitioners' *goals* range from incrementally improving search engines, to generating good-quality machine translations, to endowing embodied intelligent agents with language skills rivaling those of a human. *Paths* of development range from manipulating surface-level strings (words, sentences) using statistical methods, to generating full-blown semantic interpretations able to support sophisticated reasoning by intelligent agents. The *purview* of an NLP-oriented R&D effort can range from whittling away at a single linguistic problem (e.g., how quantification is expressed in Icelandic), to developing theories of selected language-oriented subdisciplines (e.g., syntax), to building full-scale, computational language understanding and/or generation systems. Finally, the *time frame* for projects can range from months (e.g., developing a system for a competition on named-entity recognition) to decades and beyond. Practically the only thing that NLP practitioners do agree on is just how difficult it is to develop computer programs that usefully manipulate natural language—a medium that people master with such ease.

Kenneth Church (2011) presents a compelling analysis of the pendulum swings between rationalism and empiricism starting with the inception of the field of computational linguistics in the 1950s. He attributes the full-on embrace of empiricism in the 1990s to a combination of pragmatic considerations and the availability of massive data sources.

> The field had been banging its head on big hard challenges like AI-complete problems and long-distance dependencies. We advocated a pragmatic pivot toward simpler more solvable tasks like part of speech tagging. Data was becoming available like never before. What can we do with all this data? We argued that it is better to do something simple (than nothing at all). Let's go pick some low hanging fruit. Let's do what we can with short-distance dependencies. That won't solve the whole problem, but let's focus on what we can do as opposed to what we can't do. The glass is half full (as opposed to half empty). (p. 3)

In this must-read essay, aptly titled "A Pendulum Swung Too Far," Church calls for the need to reenter the debate between rationalism and empiricism not only for scientific reasons but also for practical ones:

> Our generation has been fortunate to have plenty of low hanging fruit to pick (the facts that can be captured with short ngrams), but the next generation will be less fortunate since most of those facts will have been pretty well picked over before they retire, and therefore, it is likely that they will have to address facts that go beyond the simplest ngram approximations. (p. 7)

Dovetailing with Church, we have identified a number of opinion statements—detailed in the deep dive in section 1.6.3—that have led to a puzzling putative competition between knowledge-lean and knowledge-based approaches, even though they are pursuing entirely different angles of AI.

1.4 The Four Pillars of Linguistics for the Age of AI

The above perspective on the state of affairs in the field motivates us to define Linguistics for the Age of AI as a distinct perspective on the purview and methods of linguistic work.

This perspective rests on the following four pillars, which reflect the dual nature of AI as science and practice.

1. Language processing capabilities are developed within an integrated, comprehensive agent architecture.
2. Modeling is human inspired in service of explanatory AI and actionability.
3. Insights are gleaned from linguistic scholarship and, in turn, contribute to that scholarship.
4. All available heuristic evidence is incorporated when extracting and representing the meaning of language inputs.

We now consider each of these in turn.

1.4.1 Pillar 1: Language Processing Capabilities Are Developed within an Integrated, Comprehensive Agent Architecture

Since at least the times of Descartes, the scientific method has become more or less synonymous with the analytic approach, whereby a phenomenon or process is decomposed into contributing facets or components. The general idea is that, after each such component has been sufficiently studied independently, there would follow a synthesis step that would result in a comprehensive explanation of the phenomenon or process. A well-known example of the application of the analytic approach is the tenet of the autonomy of syntax in theoretical linguistics, which has been widely adopted by—and has strongly influenced—the field of computational linguistics. The analytic approach makes good sense because it is well-nigh impossible to expect to account for all the facets of a complex phenomenon simultaneously and at a consistent grain size of description. But it comes with a cost: it artificially constrains the purview of theories and the scope of models, and often unwittingly fosters indefinite postponement of the all-important synthesis step.

If we step back to consider some of the core tasks of a language-enabled intelligent agent, we see how tightly integrated they actually are and why modularization is unlikely to yield results if not complemented by the concern for integration. Which functionalities will have to be integrated? As a first approximation,

- Agents must implement some version of a BDI (belief-desire-intention) approach to agent modeling (Bratman, 1987) to make manifest how they select plans and actions to fulfill their goals.
- They must learn, correct, and augment their knowledge of the world (including their knowledge about themselves and other agents), as well as their knowledge of language, through experience, reasoning, reading, and being told.
- They must communicate with people and other agents in natural language.

- They must model experiencing, interpreting, and remembering their own mental, physical, and emotional states.
- They must manage their memories—including forgetting and consolidating memories.
- They must model and reason about the mental states, goals, preferences, and plans of self and others, and use this capability to support collaboration with humans and other intelligent agents.
- And, if they are embodied, they will require additional perception modalities, support for physical action, and, at least in a subset of applications, a simulated model of human physiology.

In order to minimize development effort, maximize resource reuse, and avoid knowledge incompatibilities, it is preferable to support all these processes within an integrated knowledge substrate encoded in an interoperable knowledge representation language.[6] Note that this requirement is not motivated theoretically; it is purely ergonomic, since significant engineering is needed to integrate different formalisms and approaches to knowledge representation in a single system.

The OntoAgent cognitive architecture referenced throughout the book has been designed with the above suite of functionalities in mind. Figure 1.1 shows a high-level (and ruthlessly simplified) view of that architecture, which will be refined in future chapters (see especially figure 7.1) to the degree necessary for explaining the linguistic behavior of language-endowed intelligent agents (LEIAs).

Agents obtain new facts about the world both through analyzing sensory inputs and as a result of their own mental actions. Attention to these new facts may trigger the adding of goals to the goal agenda. At each operation cycle, the agent prioritizes the goals on the agenda and then selects the plan(s) that will result in some physical, verbal, or mental action(s).

The core knowledge resources of the architecture include an ontological model (long-term semantic memory), a long-term episodic memory of past conscious experiences, and a situation model that describes the participants, props, and recent events in the current situation. The ontological model includes not only general world knowledge but also an inventory of the agent's goals; its physical, mental, and emotional states; its long-term personality traits and personal biases; societal rules of behavior, including such things as knowledge about the responsibilities of each member of a task-oriented team; and the agent's model of the relevant subset of the abovementioned features of its human and agentive collaborators.[7] The situation model, for its part, includes not only the representation of a slice of the observable world but also the agent's beliefs about its own and other agents' currently active goal and plan instances, as well as their current physical, mental, and emotional states.

The agent's knowledge enables its conscious decision-making as well as its ability to explain its decisions and actions. For the purposes of this book, the important point is that

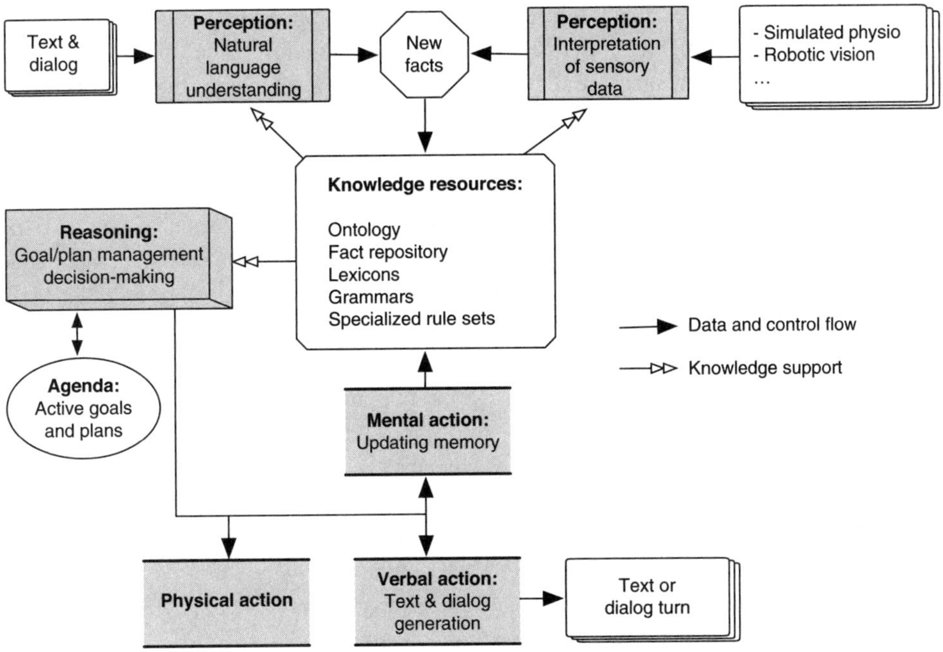

Figure 1.1
High-level sketch of the OntoAgent architecture.

this knowledge—both static and dynamically computed—is necessary for deriving the full meaning of language inputs. The view of agency we are sketching here is broadly similar to well-known approaches in cognitive modeling and AI, for example, the general worldview of such cognitive architectures as SOAR (Rosenbloom et al., 1991) or the BDI movement (Bratman, 1987).

In this book, we concentrate on those capabilities of LEIAs that are germane to their language understanding functionalities. When the LEIA receives text or dialog input (upper left in figure 1.1), it interprets it using its knowledge resources and a battery of reasoning engines, represented by the module labeled *Perception: Natural language understanding.* The internal organization and the functioning of this module is what this book is about. The result of this module's operation is one or more *New facts*, which are unambiguous assertions written in the metalanguage shared across all of the agent's knowledge resources and all downstream processing modules. These facts are then incorporated into the agent's knowledge bases.

As can be seen in figure 1.1, *New facts* can be obtained through channels of perception other than language. Robotic vision, other sensors, and even computer simulations (e.g., of human physiology) can all serve as sources of new information for a LEIA. And just like language, they must be interpreted using the module marked *Perception: Interpretation of sensory data* in the figure. This interpretation results in the same kinds of *New facts*, written in the same metalanguage, as does language understanding.[8] The upshot is that all knowledge learned by the agent from any source is equally available for the agent's subsequent reasoning and decision-making about action. The generation of associated actions—which can be physical, mental, or verbal—also involves extensive knowledge and reasoning since the actions must be selected and planned before actually being carried out. In this book, we will concentrate on a detailed and comprehensive exploration of how LEIAs understand language, while not detailing the processes through which certain components of the agent's internalized knowledge were obtained as a result of perception other than language understanding.

The above sketch of the OntoAgent architecture is high-level and omits a wealth of specialist detail. We include it here simply to frame the process of language understanding that constitutes the core of Linguistics for the Age of AI. Additional details about OntoAgent will be provided throughout this book whenever required to clarify a particular facet of language processing.

The OntoAgent approach to language processing is methodologically compatible with the cognitive systems paradigm in that it focuses on natural *understanding* in contrast to semantically impoverished natural language *processing* (Langley et al., 2009). Other language understanding efforts within this paradigm (e.g., Lindes & Laird, 2016; Cantrell et al., 2011)—while not sharing all the same assumptions or pursuing the same depth and breadth of coverage as OntoAgent—are united in that they all pursue the goal of faithfully replicating human language understanding behavior as a part of overall humanlike cognitive behavior. (For more on cognitive systems overall, see the deep dive in section 1.6.4.) The extent, quality, and depth of language understanding in each individual approach is determined by the scope of functionalities of the given cognitive agent—not independently, as when natural language processing is viewed as a freestanding task. Consequently, these approaches must take into account nonlinguistic factors in decision-making, such as the long-term and short-term beliefs of the given agent, its biases and goals, and similar features of other agents in the system's environment.

Consider, for example, anticipatory text understanding, in which an agent can choose to act before achieving a complete analysis of a message, and possibly before even waiting for the whole message to come through—being influenced to do so, for example, by the principle of economy of effort. Of course, this strategy might occasionally lead to errors, but it is undeniable that people routinely pursue anticipatory behavior, making the calibration of the degree of the anticipation an interesting technical task for cognitively inspired language understanding. Anticipatory understanding extends the well-known phenomenon of priming (e.g., Tulving & Schacter, 1990) by relying on a broader set of decision parameters, such as the availability of up-to-date values of situation parameters, beliefs about the goals and biases of the speaker/writer, and general, ontological knowledge about the world.

Our emphasis on comprehensive cognitive modeling naturally leads to a preference for multilayered models. We distinguish three levels of models, from the most general to the most specialized:

1. The *cognitive architecture* accounts for perception, reasoning, and action in a tightly integrated fashion.
2. The *NLU module* integrates the treatment of a very large number of linguistic phenomena in an analysis process that, we hypothesize, emulates how humans understand language.
3. The specialized models within the NLU module, called *microtheories*, treat individual linguistic phenomena. They anticipate and seek to cover the broadest possible scope of manifestations of those phenomena.

This infrastructure facilitates the exploration and development of detailed solutions to individual and interdependent problems over time. An important feature of our overall approach is that we concentrate not only on architectural issues but also, centrally, on the heuristics needed to compute meaning.

We have just explained the first part of our answer to the question, *What is Linguistics for the Age of AI?* It is the study of linguistics in service of developing natural language understanding and generation capabilities *within an integrated, comprehensive agent architecture.*

1.4.2 Pillar 2: Modeling Is Human Inspired in Service of Explanatory AI and Actionability

In modeling LEIAs, we are not attempting to replicate the human brain as a biological entity. Even if that were possible, it would fail to serve one of our main goals: *explanatory power.* We seek to develop agents whose behavior is explainable in human terms by the agents themselves. As an introductory example of the kinds of behavior we address in our modeling, consider the following situation. Lavinia and Felix are in an office with an open window in late fall. Lavinia says, "It's cold in here, isn't it?" Felix may respond in a variety of ways, including the following:

1. Yes, it *is* rather cold.
2. Do you want me to close the window?

Response (1) demonstrates that Felix interpreted Lavinia's utterance as a question and responded affirmatively to it. Response (2) demonstrates that Felix

a. interpreted Lavinia's utterance as an indirect request;
b. judged that Lavinia had an appropriate social status to issue this request;
c. chose to comply;

d. selected the goal of making the room warmer (rather than, say, making Lavinia warmer—as by offering a sweater);

e. selected one of the plans he knew for attaining this goal; and

f. decided to verify that carrying out this plan was preferable to Lavinia before doing it.

We want our agents to not only behave like this but also be able to explain why they responded the way they did in ways similar to (a)–(f). In other words, our models are inspired by our folk-psychological understanding of how people interpret language, make decisions, and learn. The importance of explainable AI cannot be overstated: society at large is unlikely to cede important decision-making in domains like health care or finance to machines that cannot explain their advice. For more on explanation, see the deep dive in section 1.6.5.

Our model of NLU does not require that agents exhaustively interpret every input to an externally imposed standard of perfection. Even people don't do that. Instead, agents operating in human-agent teams need to understand inputs to the degree required to determine which goals, plans, and actions they should pursue as a result of NLU. This will never involve blocking the computation of a human-level analysis if that is readily achievable; it will, however, absolve agents from doggedly pursuing ever deeper analyses if it is unnecessary in a particular situation.

In other words, in our models, agents decide how deeply they need to understand an input, and what counts as a successful—specifically, *actionable*—interpretation, based on their plans, goals, and overall understanding of the situation. If the goal is to learn new facts, then complete understanding of the portion of text containing the new fact might be preferable. By contrast, if an agent hears the input *We are on fire! Grab the axe. We need to hack our way out!*, it should already be moving toward the axe before working on interpreting the final sentence. In fact, a meaning representation that is sufficient to trigger an appropriate action by an agent may even be vague or contain residual ambiguities and lacunae.

Actionability-oriented human behavior can be explained in terms of the principle of least effort (Zipf, 1949). Piantadosi et al. (2012) argue that maintaining a joint minimum of effort between participants in a dialog is a universal maximizing factor for efficiency in conversation. Speakers do not want to spend excessive effort on precisely specifying their meaning; but hearers, for their part, do not want to have to apply excessive reasoning to understand the speaker's meaning. A core prerequisite for minimizing effort in communication is for the dialog participants to have models of the other's beliefs, goal and plan inventories, personality traits, and biases that allow them to "mindread" each other and thus select the most appropriate amount of information to convey explicitly. People use this capability habitually. It is thanks to our ability to mindread that we will describe a bassoon as "a low double reed" only in conversation with musicians. (For more on mindreading, see chapter 8.)

Another vestige of the operation of the principle of least effort in our work is our decision to have our agents look for opportunities to avoid having to resolve all ambiguities in a given input, either postponing this process (allowing for underspecification) or pronouncing it unnecessary (recognizing benign ambiguity). This is a core direction of our research at the intersection of NLP and cognitive science.

Having worked for years on developing an agent system to teach clinical medicine, we see a compelling analogy between building LEIAs and training physicians. Clinical medicine is a notoriously difficult domain: the volume of research is growing at an unprecedented rate, but the scientific knowledge that can be distilled from it is still inadequate to confidently answer all clinical questions. As a result, the field is arguably still as much art as science (Panda, 2006). And yet medical schools produce competent physicians. These physicians have different mental models of medicine, none of which is complete or optimal—and yet, they practice and save lives. Developers of AI systems need to adopt the same mindset: a willingness to take on the problem of human cognition—which is, in a very real sense, *too hard*—and make progress that will serve both science and society at large.

This concludes the explanation of the second part of our answer to the question, *What is Linguistics for the Age of AI?* It is the study of linguistics in service of developing natural language understanding and generation capabilities (1) within an integrated, comprehensive agent architecture, (2) *using human-inspired, explanatory modeling techniques and actionability judgments*.

1.4.3 Pillar 3: Insights Are Gleaned from Linguistic Scholarship and, in Turn, Contribute to That Scholarship

The past seventy years have produced a tremendous amount of scholarship in linguistics and related fields. This includes theories, data analyses, print and digital knowledge bases, corpora of written and spoken language, and experimental studies with human subjects. It would be optimal if all these fruits of human thinking could somehow converge into artificial intelligence; but, alas, this will not happen. In fact, not only will there be no smooth convergence, but much of the scholarship is not applicable to the goals and requirements of AI for the foreseeable future. While this is a sobering statement, it is not a pessimistic one: it simply acknowledges that there is a fundamental difference between human minds as thoughtful, creative consumers of scholarship and machines as nonthinking, exacting demanders of algorithms (despite the overstretched metaphorical language of *neural networks* and *machine learning*). Stated differently, it is important to appreciate that much of linguistic scholarship involves either theoretical debates that float above a threshold of practical applicability or human-oriented descriptions that do not lend themselves to being formulated as computable heuristics.

Work in computational linguistics over the past twenty years or so has largely concentrated on corpus annotation in service of supervised machine learning.[9] During this time, the rest of the linguistics community has continued to work separately on human-oriented research. This has been an unfortunate state of affairs for developing LEIAs because neither the computational, nor the theoretical, nor the descriptive linguistic community has been developing explanatory, heuristic-supported models of human language understanding that are directly suitable for implementation in agent systems.

By contrast, the models we seek to build, which we call *microtheories,* are machine-tractable descriptions of language phenomena that guide the agent, in very specific ways, through the language analysis process. Although microtheory development can be informed by noncomputational approaches, the main body of work in building a microtheory involves (a) determining the aspects of linguistic descriptions that are, in principle, machine-tractable and (b) developing the heuristic algorithms and knowledge needed to operationalize those descriptions. To take a simple example, lexicographers can explain what the English word *respectively* means, but preparing a LEIA to semantically analyze sentences that include *respectively*—for example, *Our dog and our cat like bones and catnip, respectively*—requires a dynamic function that effectively recasts the input as *Our dog likes bones and our cat likes catnip* and then semantically analyzes those propositions.

It would be a boon to agent development if linguists working in noncomputational realms would join the computational ranks *as well.* Such crossover-linguists would identify aspects of their theories and models that can be accounted for using precise, computer-tractable heuristics and then formulate the associated algorithms and descriptions. This work would not only serve NLU but, in all likelihood, also shed light on the theories and models themselves since the demands of computation set the bar of descriptive adequacy very high. In this section, we briefly review some sources of past inspiration from various fields as a prelude to what we hope will be a much richer mode of interaction in the future.[10]

1.4.3.1 Theoretical syntax Theoretical approaches to syntax attempt to account for the nature of the human language faculty with respect to sentence structure. Under this umbrella are approaches that range from almost exclusively theoretical to a combination of theoretical and descriptive. Some focus exclusively on syntax, whereas others consider interactions with other modules, such as semantics.

An example of a squarely theoretical, almost exclusively syntactic, approach is generative grammar in the tradition of Noam Chomsky. In its more recent manifestations (Chomsky, 1995), it is too abstract, too modular, and too quickly changing to inform practical system building. However, Chomsky's early work in this paradigm (e.g., Chomsky, 1957) spurred the development of the context-free grammars and associated parsing technologies that have been a cornerstone of natural language processing for decades.

Turning to theoretical approaches with practical applicability, a good example is construction grammar in its various manifestations (Hoffman & Trousdale, 2013). Construction grammars focus on the form-to-meaning mappings of linguistic entities at many levels of complexity, from words to multiword expressions to abstract templates of syntactic constituents. As theoretical constructs, construction grammars make particular claims about how syntactic knowledge is learned and organized in the human mind. For example, constructions are defined as learned pairings of form and function, their meaning is associated exclusively with surface forms (i.e., there are no transformations or empty categories), and they are organized into an inheritance network. For agent modeling, what is most important is not the theoretical details (e.g., the role of inheritance networks) but (a) the basic insight—that is, that constructions are central to human knowledge of language—and (b) the descriptive work on the actual inventory and meaning of constructions.[11]

Our third example of a theoretical-syntax approach that can inform agent modeling is Dynamic Syntax (Kempson et al., 2001). It places emphasis on the incremental generation of decorated tree structures that are intended to capture not only the syntactic structure but also the semantic interpretation of utterances. Like the other theories mentioned here, this is a theory of language processing in humans, not by machines.[12] However, it reflects a core capability of human language processing that must be emulated if machines are to behave like humans: incremental, integrated syntactic and semantic analysis.

1.4.3.2 Psycholinguistics As we just saw, incrementality has been folded into the study of theoretical syntax, but it has also been a focus of investigation in the field of psycholinguistics. Experiments have established that language processing integrates linguistic and nonlinguistic sources of information as people understand inputs incrementally. For example, Altmann and Kamide (1999) report an experiment in which participants were shown a scene containing a boy, a cake, a train set, a balloon, and a toy car. While looking at this scene, they heard one of two sentences:

(1.1) The boy will eat the cake.

(1.2) The boy will move the cake.

In trials using (1.1), the subjects' eyes moved to the target object (the cake) sooner than in trials using (1.2) since the verb *eat* predicts that its object should be something edible, and the only edible thing in the scene is the cake. These experimental results "support a hypothesis in which sentence processing is driven by the predictive relationships between verbs, their syntactic arguments, and the real-world contexts in which they occur" (p. 247).

Experiments such as these—and many more along the same lines—provide human-oriented evidence in support of developing cognitive models of multisensory agent perception that centrally feature incremental analysis.[13] For more on the computational treatments of incrementality, see the deep dive in section 1.6.6.

1.4.3.3 Semantics Semantics—a word so big that it gives one pause. Most of this book can be viewed as a case study in defining what semantics is and how we can prepare agents to compute it. But for now, in this section on linguistic inspirations for agent development, let us focus on just two threads of scholarship in semantics: lexical semantics and formal semantics.[14]

Lexical semantics. Much of human knowledge about lexical semantics is reflected in human-oriented knowledge bases: lexicons, thesauri,[15] and wordnets (i.e., hierarchical inventories of words that are organized conceptually rather than alphabetically). Although early practitioners of NLP held high hopes for the utility of machine-readable lexical knowledge bases, the disappointing reality is that human-oriented resources tend not to be well suited to computational aims. The main reason (for others, see the deep dive in section 1.6.7) is that, in order to effectively use such resources, people must bring to bear a lot of knowledge and reasoning about language and the world—all subconsciously, of course. To give just two examples: Have you ever attempted to use a thesaurus, or a large bilingual dictionary, for a language you are trying to learn? How do you choose a particular word or phrase among all those options? Similarly, have you ever tried to explain to a child why an unabridged dictionary needs a dozen senses to describe a seemingly simple word like *horse*? All this is so obvious to an adult native speaker—but not to a child, a nonnative speaker, or, even more so, a machine. So, for the enterprise of agent building, human-oriented scholarship in lexical semantics is most useful as a resource that computational linguists can consult when building knowledge bases specifically suited to machine processing. We will return to work of the latter sort in pillar 4.

Formal semantics. Formal semantics is a venerable area of study in linguistics and the philosophy of language that focuses primarily on three things: determining the truth conditions of declarative sentences; interpreting nondeclarative sentences on the basis of what would make the declarative variant true; and interpreting quantifiers. Of course, only a small part of language understanding actually involves truth conditions or quantification, which suggests that computational formal semantics cannot be considered an all-purpose approach to NLU. Moreover, truth judgments can only be made over unambiguous statements, which are rare in natural language. Intelligent agents certainly need to reason about truth, so formal semantics clearly has a role in agent functioning. But for that to happen, the NLU processes described in this book must first provide the prerequisite translation from natural language into an unambiguous metalanguage.

There does exist a branch of inquiry called computational formal semantics, which embraces the same topics as descriptive formal semantics and adds another: the use of theorem provers to determine the consistency of databases (Blackburn & Bos, 2005). We call it a *branch of inquiry* rather than (as yet) a field because (a) it assumes the abovementioned NLU-to-metalanguage translation prerequisite, and (b) some of the hottest issues turn out to be moot when subjected to the simple test of whether the problem actually occurs in natural language.

Regarding the latter, in his analysis of the place of formal semantics in NLP, Wilks (2011) reports a thought-provoking finding about a sentence type that has been discussed extensively in the theoretical literature, illustrated by the well-known example *John wants to marry a Norwegian.* Such sentences have been claimed to have two interpretations: John wants to marry a particular Norwegian (*de re*), and he wants to marry some Norwegian or other (*de dicto*). When Wilks carried out an informal web search for the corresponding "wants to marry a German" (since marrying a Norwegian was referenced in too many linguistics papers), the first twenty hits all had the generic meaning, which suggests that if one wants to express the specific meaning, this turn of phrase is just not used. Wilks argues that computational semantics must involve both meaning representation and "concrete computational tasks on a large scale" (p. 7). He writes, "What is not real Compsem [computational semantics], even though it continues to masquerade under the name, is a formal semantics based on artificial examples and never, ever, on real computational and implemented processes" (p. 7).

This comment underscores two of the most important features that divide practitioners of NLP: judgments about the acceptable germination time between research results and practical utility, and the acceptable inventory of as-yet unfulfilled prerequisites. Formal semanticists who cast their work as computational assume a long germination time and require quite ambitious prerequisites to be fulfilled—most notably, a perfect language-to-metalanguage translation. However, they are attempting to treat difficult problems that will eventually need to be handled by human-level intelligent agents. The opposite point of view is that NLP is a practical pursuit that requires near-term results, within which long-term needs tend to be considered less central. The approach described in this book lies somewhere in between, pursuing a depth of analysis that has frequently been called ambitious but imposing firm requirements about computability.

Long germination time and outstanding prerequisites are not limited to formal semantics; they also apply to other research programs involving machine reasoning. Consider, for example, Winston's (2012) work on automating story understanding, which was further developed by Finlayson (2016). Winston's Genesis system carries out commonsense reasoning over stories, such as identifying that the concept of revenge plays a role in a story despite the absence of the word *revenge* or any of its synonyms. Finlayson's system, for its part, learns plot functions in folktales, such as villainy/lack, struggle and victory, and reward. A common thread of this reasoning-centric work is its reliance on inputs that are cleaner than everyday natural language. That is, like formal semanticists, these investigators press on in their study of reasoning, even though the prerequisite of automatic NLU remains outstanding.

Winston and Finlayson take different approaches to language simplification. Finlayson's learner requires semantically annotated texts, but the annotation process is only semiautomatic: it requires manual review and supplementation because the required features cannot be computed with high reliability given the current state of the art. These features include such things as the temporal ordering of events; mappings to WordNet senses; event valence—for example, the event's impact on the Hero; and the identification of dramatis personae, that is, character types.

Winston's system, for its part, takes as input plot summaries written in simple English. However, these are not typical plot summaries intended for people. Strictly speaking, these look more like logical forms with an English veneer. For example, the summary for Cyberwar begins: "Cyberwar: Estonia and Russia are countries. Computer networks are artifacts. Estonia insulted Russia because Estonia relocated a war memorial." This excerpt includes both unexpected definitional components (essentially, elements of ontology) and a noncanonical use of the closed-class item *because* (in regular English, one would say *Estonia insulted Russia by relocating a war memorial*).

Our point is not that such inputs are inappropriate: they are very useful and entirely fitting in support of research whose focus lies outside the challenges of natural language as such. Our point is that these are excellent examples of the potential for dovetailing across research paradigms, with NLU of the type we describe here serving reasoning systems, and those reasoning systems, in turn, being incorporated into comprehensive agent systems.[16]

1.4.3.4 Pragmatics Pragmatic (also called *discourse* or *discourse-theoretic*) approaches attempt to explain language use holistically and, accordingly, can invoke all kinds of linguistic and nonlinguistic features. In this way, they are entirely in keeping with our methodology of agent development.

When pragmatics is approached from a descriptive, noncomputational perspective, it involves analyzing chunks of discourse using explanatory prose. The descriptions often invoke concepts—such as *topic*, *focus*, and *discourse theme*—that are understandable to people but have been difficult to concretize to the degree needed by computer systems. That is, when we read descriptive-pragmatic analyses of texts, our language-oriented intuitions fire and intuitively fill in the blanks of the associated pragmatic account.

Descriptive-pragmatic analyses tend to be cast as generalizations rather than rules that could be subjected to formal testing or hypotheses that could be overturned by counter-evidence. So, the challenges in exploiting such analyses for computational ends are (a) identifying which generalizations can be made computer-tractable with what level of confidence and (b) providing agents with both the algorithms and the supporting knowledge to operationalize them.

Many of the microtheories we describe throughout the book involve pragmatics, as will become clear in our treatment of topics such as reference, ellipsis, nonliteral language, and indirect speech acts. In fact, it would not be an exaggeration to say that one of the core goals of Linguistics for the Age of AI is initiating a deep and comprehensive program of work on computational pragmatics.

One of the most widely studied aspects of pragmatics over the decades has been reference resolution. However, although individual insights can be quite useful for agent modeling, most approaches cannot yet be implemented in fully automatic systems because they require unobtainable prerequisites.[17] For example, prior knowledge of the discourse structure is required by the approaches put forth in Webber (1988, 1990) and Navarretta (2004). It is also required by Centering Theory (Grosz et al., 1995), which has been deemed computationally problematic and/or unnecessary by multiple investigators (e.g., Poesio, Stevenson, et al., 2004; Strube, 1998). Carbonell and Brown (1988), referring to Sidner (1981), say: "We … believe that dialog focus can yield a useful preference for anaphoric reference selection, but lacking a computationally-adequate theory for dialog-level focus tracking (Sidner's is a partial theory), we could not yet implement such a strategy."

A new tradition of investigation into human cognition has been initiated by the field of computational psycholinguistics, whose practitioners are cognitive scientists looking toward statistical inference as a theoretically grounded explanation for some aspects of human cognition (e.g., Crocker, 1996; Dijkstra, 1996; Jurafsky, 2003; Griffiths, 2009). However, computational psycholinguistics relies on large corpora of manually annotated texts, whose scarcity limits progress, as it introduces a new aspect of the familiar knowledge bottleneck.

An obvious question is, Haven't aspects of pragmatics already been treated in computer systems? Yes, they have. (For deep dives into coreference, dialog act detection, and grounding, see sections 1.6.8–1.6.10.) However, these phenomena have been approached primarily using machine learning, which does not involve explanatory microtheories. Still, there is an associated knowledge angle that can, at least in part, be exploited in developing microtheories. Since most of the associated machine learning has been supervised, the methodology has required not only corpus annotation itself but the computational linguistic analysis needed to devise corpus annotation schemes. It cannot be overstated how much hard labor is required to organize a linguistic problem space into a manageable annotation task. This involves creating an inventory of all (or a reasonable approximation of all) eventualities; removing those that are too difficult to be handled by annotators consistently and/or are understood to be not treatable by the envisioned computer systems; and applying candidate schemes to actual texts to see how natural language can confound our expectations. Examples of impressive linguistic analyses of this genre include the MUC-7 coreference task description (Hirschman & Chinchor, 1997), the MUC-7 named-entity task description (Chinchor, 1997), the book-length manuscript on the identification and representation of ellipsis in the Prague Dependency Treebank (Mikulová, 2011), and the work on discourse-structure annotation described in Carlson et al. (2003).

Above we said that, within the realm of natural language processing, pragmatic phenomena have been addressed "primarily using machine learning." The word *primarily* is important, since there *are* some long-standing programs of research that address computational pragmatics from a knowledge-based perspective. Of particular note is the program of research led by Jerry Hobbs, which addresses many aspects of natural language understanding (e.g., lexical disambiguation; reference resolution; interpreting metaphors, metonymies, and compound nouns) using abductive reasoning with a reliance on world knowledge (e.g., Hobbs, 1992, 2004). An important strain of work in this area relates to studying the role of abductive inference in generating explanations of behavior, including learning (e.g., Lombrozo, 2006, 2012, 2016). Abduction-centered approaches to semantics, pragmatics, and agent reasoning overall are of considerable interest to cognitive systems developers (e.g., Langley et al., 2014). They are also compatible, both in spirit and in goals, with the program of NLU we present in this book. To make a sweeping (possibly, too sweeping) generalization, the main difference between those programs of work and ours is one of emphasis: whereas Hobbs and Lombrozo focus on abduction as a logical method, we focus on treating the largest possible inventory of linguistic phenomena using hybrid analysis methods.

Continuing on the topic of language-related reasoning, one additional issue deserves mention: textual inference. Although at first blush it might seem straightforward to distinguish between what a text means and which inferences it supports, this can actually be quite difficult, as encapsulated by Manning's (2006) paper title, "Local Textual Inference: It's Hard to Circumscribe, But You Know It When You See It—and NLP Needs It." To take just one example from Manning, a person reading *The Mona Lisa, painted by Leonardo da Vinci from 1503–1506, hangs in Paris' Louvre Museum* would be able to infer that *The Mona Lisa is in France*. Accordingly, an NLP system with humanlike language processing capabilities should be able to make the same inference. However, as soon as textual inference was taken up by the NLP community as a "task," debate began about its nature, purview, and appropriate evaluation metrics. Should systems be provided with exactly the world knowledge they need to make the necessary inferences (e.g., Paris is a city in France), or should they be responsible for acquiring such knowledge themselves? Should language understanding be evaluated separately from reasoning about the world (if that is even possible), or should they be evaluated together, as necessarily interlinked capabilities? Should inferences orient around formal logic (*John has 20 dollars* implies *John has 10 dollars*) or naive reasoning (*John has 20 dollars* does not imply *John has 10 dollars*—because he has 20!)? Zaenen et al. (2005) and Manning (2006) present different points of view on all of these issues, motivated, as always, by differing beliefs about the proper scope of NLP, the time frame for development efforts, and all manner of practical and theoretical considerations.[18]

The final thing to say about pragmatics is that it is a very broad field that encompasses both topics that are urgently on agenda for intelligent agents and topics that are not. Good examples of the latter are three articles in a recent issue of *The Journal of Pragmatics* that discuss how/why doctors look at their computer screens (Nielsen, 2019); the use of underspecification in five languages, as revealed by transcripts of TED talks (Crible et al., 2019); and how eight lines of a playscript are developed over the course of rehearsals (Norrthon, 2019). Although all interesting in their own right, these topics are unlikely to make it to the agenda of AI in our lifetime. Our point in citing these examples is to illustrate, rather than merely state, the answer to a reasonable question: *With all the linguistics scholarship out there, why don't you import more?* Because (a) it is not all relevant (yet), and (b) little of it is importable without an awful lot of analysis, adjustment, and engineering.

1.4.3.5 Cognitive linguistics The recent growth of a paradigm called *cognitive linguistics* is curious with respect to its name because arguably all work on linguistics involves hypotheses about human cognition and therefore is, properly speaking, cognitive. However, this is not the first time in the history of linguistics that a generic, compositional term has taken on a paradigm-specific meaning. After all, *theoretical linguistics* is commonly used as a shorthand for *generative grammar* in the Chomskian tradition, even though all schools of linguistics have theoretical underpinnings of one sort or another.

So, what *is* cognitive linguistics? If we follow the table of contents in Ungerer and Schmid's (2006) *An Introduction to Cognitive Linguistics*, then the major topics of interest for the field are prototypes and categories; levels of categorization; conceptual metaphors and metonymies; figure and ground (what used to be called topic/comment); frames and constructions; and blending and relevance. To generalize, what seems important to cognitive linguists is the world knowledge and reasoning we bring to bear for language processing, as well as the possibility of testing hypotheses on human subjects. From our perspective, all these topics are centrally relevant to agent modeling, but their grouping into a field called *cognitive linguistics* is arbitrary. To the extent that ongoing research on these topics produces descriptive content that can be made machine-tractable, this paradigm of work could be a contributor to agent systems.[19]

1.4.3.6 Language evolution A theoretical approach with noteworthy ripples of practical utility is the hierarchy of grammar complexity proposed by Jackendoff and Wittenberg (Jackendoff, 2002; Jackendoff & Wittenberg, 2014, 2017; hereafter referred to collectively as J&W). J&W emphasize that communication via natural language is, at base, a signal-to-meaning mapping. All the other levels of structure that have been so rigorously studied (phonology, morphology, syntax) represent intermediate layers that are not always needed to convey meaning.

J&W propose a hierarchy of grammatical complexity, motivating it both with hypotheses about the evolution of human language and with observations about current-day language use. They hypothesize that language evolved from a direct mapping between phonetic patterns and conceptual structures through stages that introduced various types of phonological, morphological, and syntactic structure—ending, finally, in the language faculty of modern humans. An early stage of language evolution—what they call *linear grammar*—had no morphological or syntactic structure, but the ordering of words could convey certain semantic roles following principles such as Agent First (i.e., refer to the Agent before the Patient). At this stage, pragmatics was largely responsible for utterance interpretation. As the modern human language faculty developed, it went through stages that introduced phrase structure, grammatical categories, symbols to encode abstract semantic relations (such as prepositions indicating spatial relations), inflectional morphology, and the rest. These enhanced capabilities significantly expanded the expressive power of the language system.

As mentioned earlier, the tiered-grammar hypothesis relates not only to the evolution of the human language faculty; it is also informed by phenomena attested in modern language use. Following Bickerton (1990), J&W believe that traces of the early stages of language evolution survive in the human brain, manifesting when the system is either disrupted (e.g., by agrammatic aphasia) or not fully developed (e.g., in the speech of young children, and in pidgins). Expanding on this idea, J&W describe the human language faculty as "not a monolithic block of knowledge, but rather a palimpsest, consisting of layers of different degrees of complexity, in which various grammatical phenomena fall into different layers" (J&W, 2014, p. 67). Apart from fleshing out the details of these hypothesized layers of grammar, J&W offer additional modern-day evidence (beyond aphasia, the speech of young children, and pidgins) of the use of pre-final layers. For example:

1. Language emergence has been observed in two communities of sign language speakers (using Nicaraguan Sign Language and Al-Sayyid Bedouin Sign Language), in which the language of successive generations has shown increased linguistic complexity along the lines of J&W's layers.
2. The fully formed language called Riau Indonesian is structurally simpler than most modern languages. According to J&W (2014, p. 81), "the language is basically a simple phrase grammar whose constituency is determined by prosody, with a small amount of morphology."
3. The linguistic phenomenon of compounding in English can be analyzed as a trace of a pre-final stage of language development, since the elements of a compound are simply juxtaposed, with the ordering of elements suggesting the semantic head, and with pragmatics being responsible for reconstructing their semantic relationship.

What do language evolution and grammatical layers have to do with computational cognitive modeling? They provide theoretical support for independently motivated modeling strategies. In fact, one doesn't have to look to fringe phenomena like aphasia and pidgins to find evidence that complex and perfect structure is not always central to effective communication. We need only look at everyday dialogs, which are rife with fragmentary utterances and production errors—unfinished sentences, self-corrections, stacked tangents, repetitions, and the rest. All of this mess means that machines, like humans, must be prepared to apply far more pragmatic reasoning to language understanding than approaches that assume a strict syntax-to-semantics pipeline would expect.

Another practical motivation for preparing systems to function effectively without full and perfect structural analysis is that all that analysis is very difficult to perfect, and thus represents a long-term challenge for the AI community. As we work toward a solution, machines will have to get by using all the strategies they can bring to bear—not unlike a nonnative speaker, a person interpreting a fractured speech signal, or someone ramping up in a specialized domain. In short, whenever idealized language processing breaks down, we encounter a situation remarkably similar to the hypothesized early stages of language development: using word meaning to inform a largely pragmatic interpretation.

This concludes the necessarily lengthy explanation of the third part of our answer to the question, *What is Linguistics for the Age of AI?* It is the study of linguistics in service

of developing natural language understanding and generation capabilities (1) within an integrated, comprehensive agent architecture, (2) using human-inspired, explanatory modeling techniques, and (3) *leveraging insights from linguistic scholarship and, in turn, contributing to that scholarship.*

This whirlwind overview might give the impression that more of linguistic scholarship is *not* relevant than *is* relevant.[20] Perhaps. But that is not the main point. The main point is that a lot of it *is* relevant. Moreover, we are optimistic that practitioners in each individual field might be willing to think about how their results—even if not initially intended for AI—might be applied to AI, creating a cascade of effects throughout the scientific community. We find this a compelling vision for the future of AI and invite linguists to take up the challenge.

1.4.4 Pillar 4: All Available Heuristic Evidence Is Incorporated When Extracting and Representing the Meaning of Language Inputs

As we explained in pillar 2, agent modeling is most effective when (a) it is inspired by human functioning—to the extent that it can be modeled and is useful—and (b) it strongly emphasizes practicality. Since it is impossible to immediately achieve both depth and breadth of coverage of all phenomena using knowledge-based methods, it is, in principle, useful to import external sources of heuristic evidence—both knowledge bases and processors. However, as with exploiting linguistic scholarship, these importations come at a cost—often a high one that involves much more engineering than science. Both the decision-making about what to import and the associated work in the trenches are below the threshold of general interest and will not be discussed further in this book. Instead, we will simply describe some resources that have direct computational-linguistic relevance as examples of what's out there to serve agent systems as they progress toward human-level sophistication.

1.4.4.1 Handcrafted knowledge bases for NLP As discussed earlier, one of the main drawbacks of using human-oriented lexical resources for NLP is the machine's inability to contextually disambiguate the massively polysemous words of natural language. Accordingly, a core focus of attention in crafting resources expressly for NLP has been to provide the knowledge to support automatic disambiguation, which necessarily includes both syntactic and semantic expectations about heads and their dependents (most notably, verbs and the arguments they select). As George Miller rightly states, "Creating a handcrafted knowledge base is a labor-intensive enterprise that reasonable people undertake only if they feel strongly that it is necessary and cannot be achieved any other way" (Lenat et al., 1995). Quite a few reasonable people have seen this task as a necessity, taking different paths toward the same goal. By way of illustration, we briefly compare three handcrafted knowledge bases that were designed for use outside any particular language processing environment: the lexical databases called VerbNet and FrameNet and the ontology called Cyc.[21]

VerbNet (Kipper et al., 2006) is a hierarchical lexicon inspired by Levin's (1993) inventory of verb classes. The main theoretical hypothesis underlying Levin's work is that the similarity in syntactic behavior among the members of verb classes suggests a certain semantic affinity. Over the course of its development, VerbNet has expanded Levin's inventory to more than 200 verb classes and subclasses, increased the coverage to more than 4,000 verbs, and has described each class in terms of (a) argument structure, (b) legal syntactic realizations of the verb and its arguments, (c) a mapping of the verb to a WordNet synset (i.e., set of cognitive synonyms), and (d) an indication of coarse-grained semantic constraints on the arguments (e.g., human, organization).

FrameNet, for its part, was inspired by the theory of frame semantics (a version of construction grammar; Fillmore & Baker, 2009), which suggests that the meaning of most words is best described using language-independent semantic frames that indicate a type of event and the types of entities that participate in it. For example, an Apply_heat event involves a Cook, Food, and a Heating_instrument. A language-independent frame thus described can be evoked by given lexical items in a language (e.g., fry, bake). The FrameNet resource includes frame descriptions, words that evoke them, and annotated sentences that describe their use. Although FrameNet does include nouns as well as verbs, they are used mostly as dependents in verbal frames.

Apart from lexical knowledge bases, ontologies are also needed for knowledge-based AI, including but not limited to NLU (see, e.g., Guarino, 1998, for an overview). One of the largest and oldest ontology-building projects to date has been Cyc, whose goal is to encode sufficient commonsense knowledge to support any task requiring AI, including but not specifically oriented toward NLP. Doug Lenat, the project leader, described it as a "very long-term, high-risk gamble" (Lenat, 1995) that was intended to stand in contrast to what he called the "bump-on-a-log" projects occupying much of AI (see Stipp, 1995, for a nontechnical perspective). Although initially configured using the frame-like architecture typical of most ontologies—including all ontologies developed using Stanford's open-source Protégé environment (Noy et al., 2000)—the knowledge representation strategy quickly shifted to what developers call a "sea of assertions," such that each assertion is equally about each of the terms used in it. In a published debate with Lenat (Lenat et al., 1995), George Miller articulates some of the controversial assumptions of the Cyc approach: that commonsense knowledge is propositional; that a large but finite number of factual assertions (supplemented by machine learning of an as-yet undetermined type) can cover all necessary commonsense knowledge; that generative devices are unnecessary; and that a single inventory of commonsense knowledge can be compiled to suit any and all AI applications.[22] Additional points of concern include how people can be expected to manipulate (find, keep track of, detect lacunae in) a knowledge base containing millions of assertions, and the ever present problem of lexical ambiguity. Yuret (1996) offers a fair-minded explanatory review of Cyc in the context of AI.

Before closing this section, we must mention the Semantic Web, which is another source of manually encoded data intended to support the machine processing of text. This time, however, the data is in the form of tags that serve as metadata on internet pages. The Semantic Web vision arose from the desire to make the content of the World Wide Web more easily processed by machines. Berners-Lee et al. (2001) write: "The Semantic Web will bring structure to the meaningful content of Web pages, creating an environment where software agents roaming from page to page can readily carry out sophisticated tasks for users." In effect, the goal was to transform the World Wide Web into a richly annotated corpus, but in ways that remain largely unspecified (see Sparck Jones, 2004, for an insightful critique).

We refer to the Semantic Web as a vision rather than a reality because work toward automatically annotating web pages, rather than manually providing the annotations, has been largely sidelined by the Semantic Web community in favor of creating formalisms and standards for encoding such meaning, should it ever become available. Moreover, even the simpler desiderata of the Semantic Web community—such as the use of consistent metadata tags—are subject to heavy real-world confounders. Indeed, metadata, which is typically assumed to mean manually provided annotations realized by hypertext tags, is vulnerable to inconsistency, errors, laziness, intentional (e.g., competition-driven) falsification, subconscious biases, and bona fide alternative analyses. Standardization of tags has been a topic of intense discussion among the developers, but it is not clear that any practical solution to this problem is imminent. As a result, especially in critical applications, the metadata cannot currently be trusted.

While the current R&D paradigm of the Semantic Web community might ultimately serve some intelligent agents—particularly in applications like e-commerce, in which true language understanding is not actually needed (cf. Uschold, 2003)—use of the term *Semantic Web* to describe the work is unfortunate since automatically extracting meaning is centrally absent. As Shirky (2003) writes in his entertaining albeit rather biting analysis, "The Semantic Web takes for granted that many important aspects of the world can be specified in an unambiguous and universally agreed-on fashion, then spends a great deal of time talking about the ideal XML formats for those descriptions. This puts the stress on the wrong part of the problem." In sum, when viewed from the perspective of developing deep NLU capabilities, the web—with or without metadata tags—is simply another corpus whose most challenging semantic issues are the same as for any corpus: lexical disambiguation, ellipsis, nonliteral language, implicature, and the rest.

1.4.4.2 Using results from empirical NLP Empirical NLP has had many successes, demonstrating that certain types of language-related tasks are amenable to statistical methods. (For an overview of empirical NLP, see the deep dive in section 1.6.11.) For example, machine translation has made impressive strides for language pairs for which sufficiently large parallel corpora are available; syntactic parsers for many languages do a pretty good job on the more canonical text genres; and we all happily use search engines to find what we need on the internet. The task for developers of agent systems, then, is to identify engines that can provide useful heuristic evidence for NLU, no matter how this evidence is obtained.[23]

The most obvious source of useful heuristics are preprocessors and syntactic parsers, which have historically been among the most studied topics of NLP. Syntax being as tricky as it is—particularly in less formal genres—parsing results remain less than perfect.[24] However, when such results are treated as overridable heuristic evidence within a semantically oriented language understanding system, they can still be quite useful.

Another success from the statistical paradigm that can be broadly applied to agent systems is case role labeling. Case roles—otherwise known as semantic roles—indicate the main participants in an event, such as the agent, theme, instrument, and beneficiary. In a knowledge-lean environment, these roles are used to link uninterpreted text strings; so the semantics in this approach is in the role label itself. If a semantic role–labeling system is provided with a set of paraphrases, it should be able to establish the same inventory of semantic role assignments for each.[25] For example, given the sentence set

- Marcy forced Andrew to lend her his BMW.
- Andrew was forced by Marcy into lending her his BMW.
- Andrew lent his BMW to Marcy because she made him.

a semantic role labeler should recognize that there is a *lend* event in which Andrew is the agent, his BMW is the theme, Marcy is the beneficiary, and Marcy caused the event to begin with.

Semantic role–labeling systems (e.g., Gildea & Jurafsky, 2002) are typically trained using supervised machine learning, relying on the corpus annotations provided in such resources as PropBank (Palmer et al., 2005) and FrameNet (Fillmore & Baker, 2009). Among the linguistic features that inform semantic role labelers are the verb itself, including its subcategorization frame and selectional constraints; aspects of the syntactic parse tree; the voice (active vs. passive) of the clause; and the linear position of elements. As Jurafsky and Martin (2009, pp. 670–671) report, semantic role–labeling capabilities have improved system performance in tasks such as question answering and information extraction.

Coreference resolution within statistical NLP has also produced useful results, though with respect to a rather tightly constrained scope of phenomena and with variable confidence across different referring expressions, as we detail in chapter 5.

Distributional semantics is a popular statistical approach that operationalizes the intuitions that "a word is characterized by the company it keeps" (Firth, 1957) and "words that occur in similar contexts tend to have similar meanings" (Turney & Pantel, 2010).[26] Distributional models are good at computing similarities between words. For example, they can establish that *cat* and *dog* are more similar to each other than either of these is to *airplane*, since *cat* and *dog* frequently co-occur with many of the same words: *fur*, *run*, *owner*, *play*. Moreover, statistical techniques, such as Pointwise Mutual Information, can be used to detect that some words are more indicative of a word's meaning than others. For example, whereas *fur* is characteristic of dogs, very frequent words like *the* or *has*, which often appear in texts with the word *dog*, are not.

Although distributional semantics has proven useful for such applications as document retrieval, it is not a comprehensive approach to computing meaning since it only considers the co-occurrence of words. Among the things it does not consider are

- the ordering of the words, which can have profound semantic implications: *X attacked Y* versus *Y attacked X*;
- their compositionality, which is the extent to which the meaning of a group of words can be predicted by the meanings of each of the component words; for example, in most contexts, *The old man kicked the bucket* has nothing to do with the physical act of kicking a cylindrical open container;[27] and
- any of the hidden sources of meaning in language, such as ellipsis and implicature.

To sum up, syntactic parsing, semantic role labeling, coreference resolution, and distributional semantics exemplify ways in which empirical NLP can serve NLU. We do not, however, expect empirical methods to have similar successes in more fundamental aspects of semantics or pragmatics. As Zaenen (2006) explains, annotating semantic features is significantly more difficult than annotating syntactic features; accordingly, related annotation efforts to date have reflected substantial simplifications of the real problem space. Moreover, even if semantic annotation were possible, it is far from clear that the learning methods themselves would work very well over a corpus thus annotated since the annotations will necessarily include meanings not overtly represented by text strings. (For more on corpus annotation, see the deep dive in section 1.6.12.)

This concludes the fourth part of our answer to the question, *What is Linguistics for the Age of AI?* It is the study of linguistics in service of developing natural language understanding and generation capabilities (1) within an integrated, comprehensive agent architecture, (2) using human-inspired, explanatory modeling techniques, (3) leveraging insights from linguistic scholarship and, in turn, contributing to that scholarship, and (4) *incorporating all available heuristic evidence when extracting and representing the meaning of language inputs.*

1.5 The Goals of This Book

The cognitive systems–inspired, computer-tractable approach to NLU described here has been under continuous development, with various emphases, for over thirty-five years. This time frame is noteworthy because the program of work began when computational linguistics and knowledge-based approaches were still considered a proper part of NLP, when AI was not largely synonymous with machine learning, and when words like *cognitive*, *agents*, and *ontology* were not yet commonplace in the popular press.

A good question is why this program of work has survived despite finding itself outside the center of attention of both mainstream practitioners and the general public. The reason, we believe, is that the vision of human-level AI remains as tantalizing now as when first formulated by the founders of AI over a half century ago. We agree with Marvin Minsky that "We have got to get back to the deepest questions of AI and *general* intelligence and quit wasting time on little projects that don't contribute to the main goal. We can get back to them later" (quoted in Stork, 1997, p. 30). It is impossible to predict how long it will take to attain high-quality NLU, but John McCarthy's estimate about AI overall, as reported by Minsky, seems appropriate: "If we worked really hard we'd have an intelligent system in from four to four hundred years" (Stork, p. 19).

Witticisms aside, endowing LEIAs with the ability to extract an iceberg of meaning from the visible tip reflected by the words in a sentence is not a short-term endeavor. At this point in history, it more properly belongs to the realm of science than technology, although we can and have packaged useful results for particular tasks in specific domains. Accordingly, the main contribution of the book is scientific. We present a theory of NLU for LEIAs that includes its component algorithms and knowledge resources, approaches for extending the latter, and a methodology of its integration with the extralinguistic functionalities of LEIAs. The theory can be applied, in full or in part, to any agent-based system. Viewed this way, our contribution must be judged on how well it stands the test of time, how effectively it serves as a scaffolding for deeper exploration of the component phenomena and models, and how usefully it can be applied to any of the world's languages.

While our main emphasis is on science, engineering plays an important role, too. Much of what we describe has already been implemented in systems. We believe that implementation is essential in cognitively inspired AI to ensure that the theories can, in fact, serve as the basis for the development of applications. When we say that a LEIA *does X*, it means that algorithms have been developed to support the behavior. Many of these algorithms have already been included in prototype application systems. Others are scheduled for inclusion, as our team continues system-development work.

The language descriptions and algorithms presented here cover both generic theoretical and specific system-building aspects. They are specific in that they have been developed within a particular theoretical framework (Ontological Semantics), which has been implemented in a particular type of intelligent agents (LEIAs) in a particular cognitive architecture (OntoAgent). In this sense, the work is real in the way that system developers understand. On the other hand, the descriptions and algorithms reflect a rigorous analysis of language phenomena that is valid outside its association with this, or any other, formalism or application environment.

Descriptions of complex phenomena in any scientific realm have a curious property: the better they are, the more self-evident they seem. Linguistic descriptions are particularly subject to such judgments because every person capable of reading them has functional expertise in language—something that cannot be said of mathematics or biology. Even within the field of linguistics, rigorous descriptions of how things work—the kind you need, for example, if you have ever tried to fully master a foreign language—are traditionally unpublishable unless they are subsumed under some theoretical umbrella. This is unfortunate as it leaves an awful lot of work for computational linguists to do.

As we have explained, published linguistic scholarship is suitable only as a starting point for the knowledge engineering required to support language processing in LEIAs. Grammar books leave too much hidden behind lists flanked by *e.g.* and *etc.*; discourse-theoretic accounts regularly rely on computationally intractable concepts such as *topic* and *focus*; and lexical resources intended for people rely on people's ability to, for example, disambiguate the words used in definitions and recognize the nuances distinguishing near-synonyms. Artificial intelligent agents do not possess these language processing and reasoning abilities, so linguistic resources aimed at them must make all of this implicit information explicit. It would be a boon to linguistics overall if the needs of intelligent agents spurred a proliferation of precise, comprehensive, and computer-tractable linguistic descriptions. As this has not been happening, our group is taking on this work, albeit at a scale that cannot rival the output potential of an entire field.

What we hope to convey in the book is how a knowledge-based, deep-semantic approach to NLU works, what it can offer, and why building associated systems is not only feasible but necessary. Naturally, the composition of actual agent system prototypes will vary, as it will reflect different theoretical, methodological, and tactical decisions. However, all such systems will need to account for the same extensive inventory of natural language phenomena and processes that we address in this book.

A note on how to read this book. There is no single best, straight path through describing a large program of work, including its theoretical and methodological substrates, its place in the history of the field, and its plethora of technical details. Readers will inevitably have different most-pressing questions arising at different points in the narrative. We, therefore, make three tactical suggestions:

- If something is not immediately clear, read on; a clarification might be just around the corner.
- Skip around liberally, using the table of contents as your guide.
- Understand that some repetition in the narrative is a feature, not a bug, to help manage the reader's cognitive load.

1.6 Deep Dives

1.6.1 The Phenomenological Stance

We are interested in modeling the agents from the first-person, phenomenological perspective.[28] This means that each agent's knowledge, like each person's knowledge, is assumed at all times to be incomplete and potentially at odds with how the world really is (i.e., it can contrast with the knowledge of a putative omniscient agent, which would embody what's known as the third-person perspective). To borrow a term from ethology, we model each LEIA's *umwelt*.

We have demonstrated the utility of modeling agents from multiple perspectives by implementing and testing non-toy computational models of both first-person and third-person (omniscient) agents in application systems. For example, the Maryland Virtual Patient (MVP) system (see chapter 8) featured an omniscient agent endowed with an expert-derived, state-of-the-art explanatory model of the physiology and pathology of the human esophagus, as well as clinical knowledge about the experiences of humans affected by esophageal diseases. This omniscient agent (a) ensured the realistic progression of a virtual patient's disease and healing processes, in response to whatever interventions were selected by system users, and (b) provided ground-truth knowledge to the tutoring agent who was not, however, omniscient: like any physician, it had access only to that subset of patient features that had either been reported by the patient or were returned as test results. The virtual patients in the system were, likewise, modeled from the first-person perspective: they were endowed with different partial, and sometimes objectively incorrect, knowledge. Importantly for this book, the virtual patients could expand and correct their knowledge through experiences and interactions with the human trainees, who played the role of attending physicians. For example, virtual patients were shown to be able to learn both ontological concepts and lexical items through conversation with the human trainees.[29]

Another human-inspired aspect of our modeling strategy is the recognition that the agents' knowledge can be internally contradictory and/or vague. For example, in a recent robotic application the agent was taught more than one way to perform a complex task through dialogs with different human team members (see section 8.4). In any given system run, the agent carried out the task according to the instructions that it had learned from the team member participating in that run. When asked to describe the task structure, the agent offered all known options: "According to A, the complex task is T_1; while according to B, it is T_2."

To sum up our phenomenological stance, we model intelligent agents to operate on the basis of folk psychology—that is, their view of the world (like a human's) is less than scientific. Each human and artificial member of the society is expected to have different first-person perspectives, but they have sufficient overlap to support successful communication and joint operation. Incompatibilities and lacunae in each agent's knowledge are expected to occur. One of the core methods of eliminating incompatibilities and filling lacunae is through natural language communication.

1.6.2 Learning

The ability to understand language is tightly coupled with the ability to learn. As emphasized earlier, in order to understand language, people must possess a lot of knowledge, and that knowledge must be learned. In developing artificial intelligent agents, learning can be either delegated to human knowledge acquirers (whose job description has been more or less the same since the 1970s) or modeled as an automatic capability of agents. Since modeling humanlike behavior is a core requirement for LEIAs, they, like people, must be able to learn using natural language.

The core prerequisite for language-based learning—be it through reading, being taught, or participating in nonpedagogically oriented dialogs—is the ability to understand natural language. But, as we just pointed out, that process itself requires knowledge! Although this might appear to be a vicious circle, it is actually not, as long as the agent starts out with a critical mass of ontological and lexical knowledge, as well as the ability to bootstrap the learning process—by generating meaning representations, using reasoning engines to make inferences, managing memory, and so on. Focusing on bootstrapping means that we are not modeling human learning as if it were from scratch—particularly since, for human brains, there arguably is no scratch. Of all the types of learning that LEIAs must, and have in the past, undertaken, we will focus here on the learning of new words and new facts—that is, new propositional content recorded as ontologically grounded meaning representations.[30]

1.6.3 NLP and NLU: It's Not Either-Or

Over the past three decades, the ascendance of the statistical paradigm in NLP and AI in general has seen knowledge-based methods being variously cast as outdated, unnecessary, lacking promise, or unattainable. However, the view that a competition exists between the approaches is misplaced and, upon closer inspection, actually rather baffling. This should become clear as we walk through some unmotivated beliefs that, by all indications, are widely held in the field today.[31]

Unmotivated belief 1. *There is a knowledge bottleneck and it affects only knowledge-based approaches.* Although knowledge-lean approaches purport to circumvent the need for manually acquired knowledge, those that involve supervised learning—and many do—simply shift the work of humans from building lexicons and ontologies to annotating corpora. When the resulting supervised learning systems hit a ceiling of results, developers point to the need for more or better annotations. Same problem, different veneer. Moreover, as Zaenen (2006) correctly points out, the success of supervised machine learning for syntax does not promise similar successes for semantics and pragmatics (see section 1.6.12). In short, it is not the case that knowledge-based methods suffer from knowledge needs whereas knowledge-lean methods do not: the higher-quality knowledge-lean systems *do* require knowledge in the form of annotations. Moreover, all knowledge-lean systems avoid phenomena and applications that would require unavailable knowledge support. What do all of those exclusions represent? Issues that must be solved to attain the next level of quality in automatic language processing.

Unmotivated belief 2. *Knowledge-based methods were tried and failed.* Yorick Wilks (2000) says it plainly: "The claims of AI/NLP to offer high quality at NLP tasks have never been really tested. They have certainly not failed, just got left behind in the rush towards what could be easily tested!" Everything about computing has changed since the peak of knowledge-based work in the mid-1980s—speed, storage, programming languages, their supporting libraries, interface technologies, corpora, and more. So comparing statistical NLP systems of the 2010s with knowledge-based NLP systems of the 1980s says nothing about the respective utility of these R&D paradigms. As a side note, one can't help but wonder where knowledge-based NLU would stand now if all, or even a fraction, of the resources devoted to statistical NLP over the past twenty-five years had remained with the goal of automating language understanding.

Unmotivated belief 3. *NLU is an extension of NLP.* Fundamental NLU has little to nothing in common with current mainstream NLP; in fact, it has much more in common with robotics. Like robotics, NLU is currently most fruitfully pursued in service of specific tasks in a specific domain for which the agent is supplied with the requisite knowledge and reasoning capabilities. However, whereas domain-specific robotics successes are praised—and rightly so!—domain-specific NLU successes are often criticized for not being immediately applicable to all domains (under the pressure of evaluation frameworks entrenched in statistical NLP). One step toward resolving this miscasting of NLU might be the simple practice of reserving the term *NLU* for actual deep understanding rather than watering it down by applying it to any system that incorporates even shallow semantic or pragmatic features. Of course, marrying robotics with NLU is a natural fit.

Unmotivated belief 4. *It's either NLP or NLU.* One key to the success of NLP has been finding applications and system configurations that circumvent the need for language understanding. For example, consider a question-answering system that has access to a large and highly redundant corpus. When asked to indicate when the city of Detroit was founded, it can happily ignore formulations of the answer that would require sophisticated linguistic analysis or reasoning (*It was founded two years later*; *That happened soon afterward*) and, instead, fulfill its task with string-level matching against the following sentence from Wikipedia: "Detroit was founded on July 24, 1701 by the French explorer and adventurer Antoine de la Mothe Cadillac and a party of settlers."[32] However, not all language-oriented applications offer such remarkable simplifications. For example, agents in dialog systems receive one and only one formulation of each utterance. Moreover, they must deal with performance errors such as unfinished thoughts, fragmentary utterances, self-interruptions, repetitions, and non sequiturs. Even the speech signal itself can be corrupted, as by background noise and dropped signals.

Consider, in this regard, a short excerpt from the Santa Barbara Corpus of Spoken American English, in which the speaker is a student of equine science talking about blacksmithing:

> we did a lot of stuff with the—like we had the, um, … the burners? you know, and you'd put the—you'd have—you started out with the straight … iron? … you know? and you'd stick it into the, … into the, … you know like, actual blacksmithing. (DuBois et al., 2000–2005)[33]

Unsupported by the visual context or the intonation of spoken language, this excerpt requires quite a bit of effort even for people to understand. Presumably, we get the gist thanks to our ontological knowledge of the context (we told you that the topic was blacksmithing). Moreover, we make decisions about how much understanding is actually needed before we stop trying to understand further. In sum, NLP has one set of strengths, purviews, and methods, and NLU has another. These programs of work are complementary, not in competition.

Unmotivated belief 5. *Whereas mainstream NLP is realistic, deep NLU is unrealistic.* This faulty assessment seems to derive from an undue emphasis on compartmentalization. If one plucks NLU out of overall agent cognition and expects meaning analysis to be carried out to perfection in isolation from world and situational knowledge, then, indeed, the task is unrealistic. However, this framing of the problem is misleading. To understand language inputs, a cognitive agent must know what kinds of information to rely on during language analysis and why. It must also use a variety of kinds of stored knowledge to judge how deeply to analyze inputs. Analysis can involve multiple passes over inputs, requiring increasing amounts of resources, with the agent pursuing the latter stages only if it deems the information worth the effort. For example, a virtual medical assistant tasked with assisting a doctor in a clinical setting can ignore incidental conversations about pop culture and office gossip, which it might detect using a resource-light comparison between the input and its active plans and goals. By contrast, that same agent needs to understand both the full meaning and the implicatures in the following doctor-patient exchange involving a patient presenting with gastrointestinal distress: *Doctor: "Have you been traveling lately?" Patient: "Yes, I vacationed in Mexico two weeks ago."*

One additional aspect of the realistic/unrealistic assessment must be mentioned. A large portion of work on supervised learning in support of NLP has been carried out under less than realistic conditions. Task specifications normally include in their purview only the simpler instances of the given phenomenon, and manually annotated corpora are often provided to developers for both the training and the evaluation stages of system development. This means that the systems configured according to such specifications cannot perform at their evaluated levels on raw texts (for discussion, see Mitkov, 2001, and chapter 9). To generalize, judgments about feasibility cannot be made in broad strokes at the level of statistical versus knowledge-based systems.

To recap, we have just suggested that five misconceptions have contributed to a state of affairs in which statistical NLP and knowledge-based NLU have been falsely pitted against each other. But this zero-sum-game thinking is too crude for a domain as complex as natural language processing/understanding. The NLP and NLU programs of work pursue different goals and promise to contribute in different ways, on different timelines, to technologies that will enhance the human experience. Clearly there is room, and a need, for both.

1.6.4 Cognitive Systems: A Bird's-Eye View

To assess the current views on the role of NLP in computational cognitive science, we turn to an authoritative survey of research in cognitive architectures and their associated cognitive systems (Langley et al., 2009). The survey analyzes nine capabilities that any good cognitive architecture must have: (1) recognition and categorization, (2) decision-making and choice, (3) perception and situation assessment, (4) prediction and monitoring, (5) problem solving and planning, (6) reasoning and belief maintenance, (7) execution and action, (8) interaction and communication, and (9) remembering, reflection, and learning. Langley et al. primarily subsume NLP under *interaction and communication* but acknowledge that it involves other aspects of cognition as well. The following excerpt summarizes their view. We have added indices in square brackets to link mentioned phenomena with the aspects of cognition just listed:

> A cognitive architecture should … support mechanisms for transforming knowledge into the form and medium through which it will be communicated [8]. The most common form is … language, which follows established conventions for semantics, syntax and pragmatics onto which an agent must map the content it wants to convey. … One can view language generation as a form of planning [5] and execution [7], whereas language understanding involves inference and reasoning [6]. However, the specialized nature of language processing makes these views misleading, since the task raises many additional issues. (Langley et al., 2009)

Langley et al.'s (2009) analysis underscores a noteworthy aspect of most cognitive architectures: even if reasoning is acknowledged as participating in NLP, the architectures are modularized such that core agent reasoning is separate from NLP-oriented reasoning. This perceived dichotomy between general reasoning and reasoning for NLP has been influenced by the knowledge-lean NLP paradigm, which both downplays reasoning as a tool for NLP and uses algorithms that do not mesh well with the kind of reasoning carried out in most cognitive architectures. However, if NLP is pursued within a knowledge-based paradigm, then there is great overlap between the methods and knowledge bases used for all kinds of agent reasoning, as well as the potential for much tighter system integration. Even more importantly, language processing is then, appropriately, not relegated to the input-output periphery of cognitive modeling because reasoning about language is a core task of a comprehensive cognitive model.

Consider, for example, an architecture in which verbal action is considered not separate from other actions (as in Langley et al.'s [2009] point [7] vs. point [8]) but simply another class of action. Such an organization would capture the fact that, in many cases, the set of plans for attaining an agent's goal may include a mixture of physical, mental, and verbal actions. For example, if an embodied agent is cold, it can ask someone else to close the window (a verbal action), it can close the window itself (a physical action), or it can focus on something else so as not to notice its coldness (a mental action). Conversely, one and the same element of input to reasoning can be generated from sensory, language, or interoceptory (i.e., resulting from the body's signals, e.g., pain) input or as a result of prior reasoning. For example, a simulated embodied agent can choose to put the goal "have cut not bleed anymore" on its agenda—with an associated plan like "affix a bandage"—because it independently noticed that its finger was bleeding; because someone pointed to its finger and then it noticed it was bleeding (previously, its attention was elsewhere); because someone said, "Your finger is bleeding"; or because it felt pain in its finger and then looked and saw that it was bleeding.

The conceptual and algorithmic frameworks developed in the fields of agent planning, inference, and reasoning can all be usefully incorporated into the analysis of the semantics and pragmatics of discourse. For example, the pioneering work of Cohen, Levesque, and Perrault (e.g., Cohen & Levesque, 1990; Perrault, 1990) demonstrated the utility of approaching NLP tasks in terms of AI-style planning; planning is a first-order concern in the field of natural language generation (e.g., Reiter, 2010); and inference and reasoning have been at the center of attention of AI-style NLP for many years.

Returning to Langley et al.'s (2009) survey, their section on open issues in cognitive architectures states: "Although natural language processing has been demonstrated within some architectures, few intelligent systems have combined this with the ability to communicate about their own decisions, plans, and other cognitive activities in a general manner." Indeed, of the eighteen representative architectures briefly described in the appendix, only two—SOAR (Lewis, 1993) and GLAIR (Shapiro & Ismail, 2003)—are overtly credited with involving NLP, and one, ACT-R, is credited indirectly by reference to applied work on tutoring (Koedinger et al., 1997) within its framework. Although many cognitive architectures claim to have implemented language processing (thirteen of the twenty-six included in a survey by Samsonovich, http://bicasociety.org/cogarch/architectures.pdf), most of these implementations are limited in scope and depth, and none of them truly has language at the center of its scientific interests.

The LEIAs we describe throughout the book pursue deep NLU within the cognitive systems paradigm. Of the few research programs worldwide that currently pursue similar aims, perhaps the closest in spirit are those of Cycorp and the University of Rochester's TRAINS/TRIPS group (Allen et al., 2005). We will not attempt point-by-point comparisons with these because in order for such comparisons to be useful—rather than nominal, box-checking exercises—heavy preconditions must be met, both in the preparation and in the presentation.[34] In addition, the differences between research programs are certainly largely influenced by nonscientific considerations that live as explanatory folklore in actual research operations: which research projects were funded, which dissertations were written, which goals were prioritized for which reasons, and so on. In short, any investigator who is interested in a head-to-head comparison will have a particular goal in mind, and it is that goal that will delimit and make useful the process of drawing comparisons.

As concerns cognitive systems that include deep natural language processing but without an emphasis on fundamentally advancing our understanding of language processing, two noteworthy examples are the robotic systems reported by Lindes and Laird (2016) and Scheutz et al. (2017). The former system implements a parser based on embodied construction grammar (Feldman et al., 2009). The latter system uses an algorithm by which a "Lambda calculus representation of words could be inferred in an inverse manner from examples of sentences and their formal representation" (Baral et al., 2017, p. 11). In both systems, the role of the language component is to support (a) direct human-robotic interaction, predominantly simple commands; and (b) robotic learning of the meanings of words as the means of grounding linguistic expressions in the robot's world model. As a result of the above choice, both the robot's language processing capabilities and its conceptual knowledge cover the minimum necessary for immediate system needs. However, if the ultimate goal is to develop robotic language understanding that approaches human-level sophistication, then the large number of linguistic issues addressed in this book cannot be indefinitely postponed.

1.6.5 Explanation in AI

The ability to explain behavior in human terms is not a forte of the current generation of AI systems. The following statement by Rodney Brooks (2015) provides a good illustration of the current state of the art in a representative AI application:

> Today's chess programs have no way of saying why a particular move is "better" than another move, save that it moves the game to a part of a tree where the opponent has less good options. A human player can make generalizations and describe why certain types of moves are good, and use that to teach a human player. Brute force programs cannot teach a human player, except by being a sparring partner. It is up to the human to make the inferences, the analogies, and to do any learning on their own. The chess program doesn't know that it is outsmarting the person, doesn't know that it is a teaching aid, doesn't know that it is playing something called chess nor even what "playing" is. Making brute force chess playing perform better than any human gets us no closer to competence in chess. (p. 109).

For an agent to serve as a true AI—meaning an equal member of a human-agent team—it must be able to generate explanations of its behavior that are elucidating and satisfying to people.

The need for explanation in AI has certainly been recognized, as evidenced, for example, by the existence of DARPA's Explainable AI program. A workshop on the topic was held at IJCAI-2017. This is a positive development. Constructing explanations is not an easy task. Constructing relevant explanations is an even more difficult one. It seems that very few things can demonstrate that an artificial intelligent agent possesses at least a vestige of human-level intelligence as well as its ability to generate explanations specifically for a particular audience and state of affairs in the world. Without these constraints, many explanations, while being technically accurate, might prove unedifying or inappropriate. Plato's reported definition of humans as "featherless bipeds" may have engendered Diogenes's witty and cynical response (according to Diogenes Laertius, Diogenes the Cynic plucked feathers off a chicken and presented it to Plato as a counterexample) but will not be treated by most people in most situations as an enlightening characterization.

Explanations differ along multiple parameters. For example, the basis of an explanation can be empirical or causal. Empirical explanations can range from "have always done it this way and succeeded" to appeals to authority ("this is what my teammate told me to do"). Causal explanations can appeal to laws of physics/biology or to folk psychology ("because people tend to like people they have helped"). And causes themselves may be observable ("the table is set for dinner because I just saw Zach setting it") or unobservable ("Bill is silent because he does not know the answer to the question I asked").

To provide explanations for unobservables, intelligent agents must be equipped with a theory of mind, which is the ability to attribute mental states (beliefs, desires, emotions, attitudes) to oneself and others. Operationalizing the twin capabilities of metacognition (the analysis of self) and mindreading (the analysis of others) is facilitated by organizing the agent's models of self and others in folk-psychological terms (see Carruthers, 2009, for a discussion of the interaction between mindreading and metacognition). Agents able to understand their own and others' behavior in folk-psychological terms will be able to generate humanlike explanations and, as a result, be better, more trusted, collaborators.

The ability to explain past behavior in terms of causes, and future behavior in terms of expected effects, is needed not only to support interpersonal interactions but also for language understanding itself. For example, indirect speech acts ("I'd be much happier if I didn't have to cook tonight") require the listener to figure out why the speaker said what he or she said, which is a prerequisite for selecting an appropriate response. This means that, although explanation has traditionally been treated separately from NLU, this separation cannot be maintained: a model of explanation must be a central part of the NLU module itself. And, since there do not exist any behavior-explanation reasoners that we can import—and since we do not rely on unavailable prerequisites—developing associated reasoning capabilities is necessarily within our purview.

On the practical level, the agent models we build are explanatory not only because their operation is interpretable in human folk-psychological terms but also because our systems' internal workings—static knowledge, situational knowledge, and all algorithms—are inspectable by people (though familiarity with the formalism is, of course, required).

Philosophers and psychologists (Woodward, 2019; Lombrozo, 2006, 2016) have devoted significant attention to the varieties and theories of explanation, often coming to unexpected conclusions, as when Nancy Cartwright (1983) persuasively argues that, despite their great explanatory power, fundamental scientific laws are not descriptively adequate—that is, they do not describe reality. The corollary for us is that the scientific view of the world is different from the view of the world reflecting everyday human functioning. We believe that our task is to develop LEIAs that are primarily intended to model and interact with these everyday human agents. Such agents have much broader applicability in all kinds of practical applications than agents that are omniscient, whether in a given field or across fields.

A related issue is whether to endow LEIAs with normative or descriptive rationality. Normative rationality describes how people *should* make decisions, whereas descriptive rationality describes how they actually *do* make decisions. In their discussion of human and artificial rationality, Besold and Uckelman (2018) persuasively argue that "humans do not, generally, attain the normative standard of rationality" proposed in philosophy and cognitive science. As a corollary, a LEIA endowed with normative rationality will behave in ways that people will not interpret as sufficiently humanlike. This state of affairs evokes the concept of "the uncanny valley" (Mori, 2012). Indeed, Besold and Uckelman continue: "Because humans fall short of perfect rationality, a perfectly rational machine would almost immediately fall victim to the uncanny valley." Their solution is to base agents' theory of mind and mind-reading capabilities not on normative rationality but on descriptive rationality—that is, on how people actually act rather than how they say they are supposed to act. We choose to model descriptive rationality and ground explanations in folk psychology. Such explanations are not necessarily scientific, nor necessarily (always) true, but we see to it that they are always contextually appropriate and that they take into account the goals, plans, biases, and beliefs of both the producer and the consumer of the explanation.

To summarize, models of explanation in Linguistics for the Age of AI rely on the folk-psychological capabilities of mindreading and metacognition because the people who will interact with—and, with any luck, ultimately trust—AI systems need explanations in terms that they understand and find familiar.[35]

1.6.6 Incrementality in the History of NLP

For any task—from speech recognition to syntactic parsing to full natural language understanding—one can implement any or all component processors using any degree of incrementality. Ideally, the incremental (sub)systems would correctly process every incoming chunk of input and seamlessly add to the overall analysis, as fragments turned into sentences and sentences into discourses. However, defining *chunk* is anything but obvious: Is a chunk a word? A phrase? A clause? Must the optimal chunk size be dynamically calculated depending on the input? Can the system backtrack and change its analysis (i.e., be non-monotonic) or is it permitted only to add to previously computed analyses (i.e., be monotonic)? Is it better to wait for larger chunks in order to achieve higher initial accuracy or, as in automatic speech recognition systems, must the system decide fast and finally?

Köhn (2018) illustrates the challenges of incrementality in his analysis of the Verbmobil project (e.g., Wahlster, 2000), which aimed at developing a portable, simultaneous speech-to-speech translation system. He writes:

> The project developed speech recognition and synthesis components, syntactic and semantic parsers, self-correction detection, dialogue modeling and of course machine translation, showing that incrementality is an aspect that touches nearly all topics of NLP. This project also exemplifies that building incremental systems is not easy, even with massive funding [equivalent to approximately 78 million Euros when adjusted for inflation]: Only one of the many components ended up being incremental and the final report makes no mention of *simultaneous* interpretation. (p. 2991)

One way to incorporate incrementality into NLP systems is to focus on a narrow domain in which the focus is not on the coverage of linguistic phenomena but on the holistic nature of the application. For example, Kruijff et al. (2007) and Brick and Scheutz (2007) report robotic systems with broadly comparable cognitive architectures and capabilities. For the purposes of our language-centric overview, these programs of work are similar in that they acknowledge the necessity of language understanding and integrate related capabilities into the overall robotic architecture, but without taking on all of the challenges of unconstrained language use. For example, Kruijff et al. have a dialog model, they ground the incremental interpretation in the overall understanding of the scene, and they bunch as-yet ambiguous interpretations into what they call a *packed* representation, which represents all information shared by alternative analyses just once. However, their robot's world contains only three mugs and a ball, and utterances are limited to basic assertions and commands related to those entities, such as "the mug is red" and "put the mug to the left of the ball." So, whereas some necessary components of a more sophisticated language processing system are in place, the details of realistic natural language have not yet been addressed.

Another system that belongs to this narrow-domain category is the one described in DeVault et al. (2009).[36] It can predict at which point in a language stream it has achieved the maximum understanding of the input and then complete the utterance. For example, given the utterance "We need to," the system offers the completion "move your clinic"; given the utterance "I have orders," the system offers the completion "to move you and this clinic." Presumably, these continuations can be made confidently because the domain-specific ontology and task model offer only one option for each utterance continuation. The method employed involved machine learning using 3,500 training examples that were mapped into one of 136 attribute-value matrix frames representing semantic information in the ontology and task model.

A computational model of pragmatic incrementality is presented in Cohn-Gordon et al. (2019). Among the goals of their model is to account for the fact that people make anticipatory implicatures partway through utterances (cf. Sedivy, 2007). For example, if shown a scene with a tall cup, a short cup, a tall pitcher, and a key, a listener who hears "Give me the tall __" will fixate on the tall cup before the utterance is complete, since the only reason to use *tall* would be to distinguish between cups; since there is only one pitcher, there is no need to refer to its height. This model assigns a probability preference to *cup* (over *pitcher*) when the word *tall* is consumed, which formally accounts for the implicature.

However, this implicature is cancelable: if the utterance actually ends with *pitcher*, all referents apart from the pitcher are excluded.

1.6.7 Why Machine-Readable, Human-Oriented Resources Are Not Enough

The 1980s and early 1990s showed a surge of interest in automatically extracting NLP-oriented knowledge bases from the newly available machine-readable dictionaries as a means of overcoming the knowledge bottleneck. This research was based on two assumptions: (a) that machine-readable dictionaries contain information that is useful for NLP and (b) that this information would be relatively easy to extract into a machine-oriented knowledge base (Ide & Véronis, 1993). For example, it was expected that an ontological subsumption hierarchy could be extracted using the hypernyms that introduce most dictionary definitions (*a dog is a domesticated carnivorous mammal*) and that other salient properties could be extracted as well (*a dog ... typically has a long snout*). Although information in an idealized lexicon might be both useful and easy to extract, actual dictionaries built *by* people *for* people require human levels of language understanding and reasoning to be adequately interpreted. For example:

1. Senses are often split too finely for even a person to understand why.
2. Definitions regularly contain highly ambiguous descriptors.
3. Sense discrimination is often left to examples, meaning that the user must infer the generalization illustrated by the example.
4. The hypernym that typically begins a definition can be of any level of specificity (a dog is an animal/mammal/carnivore/domesticated carnivore), which confounds the automatic learning of a semantic hierarchy.
5. The choice of what counts as a salient descriptor is variable across entries (dog: a domesticated carnivorous mammal; turtle: a slow-moving reptile).
6. Circular definitions are common (a tool is an implement; an implement is a tool).

After more than a decade's work toward automatically adapting machine-readable dictionaries for NLP, the field's overall conclusion (Ide & Véronis, 1993) was that this line of research had little direct utility: machine-readable dictionaries simply required too much human-level interpretation to be of much use to machines.

However, traditional dictionaries do not exhaust the available human-oriented lexical resources. The lexical knowledge base called WordNet (Miller, 1995) attempts to record not only what a person knows about words and phrases but also how that knowledge might be organized in the human mind, guided by insights from cognitive science. Begun in the 1980s by George Miller at Princeton University's Cognitive Science Laboratory, the English WordNet project has developed a lexical database organized as a semantic network of four directed acyclic graphs, one for each of the major parts of speech: noun, verb, adjective, and adverb. Words are grouped into sets of cognitive synonyms, called synsets. Synsets within a part-of-speech network are connected by a small number of relations. For nouns, the main ones are subsumption ("is a") and meronymy ("has as part": *hand has-as-part finger*); for adjectives, antonymy; and for verbs, troponymy (indication of manner: *whisper troponym-of talk*). WordNet itself offers few relations across parts of speech, although satellite projects have pursued aspects of this knowledge gap.

WordNet was adopted by the NLP community for a similar reason as machine-readable dictionaries were: it was large and available. Moreover, its hierarchical structure captured additional aspects of lexical and ontological knowledge that had promise for machine reasoning in NLP. However, WordNet has proved suboptimal for NLP for the same reasons as machine-readable dictionaries did: the ambiguity arising from polysemy. For example, at time of writing *heart* has ten senses in WordNet: two involve a body part (working muscle; muscle of dead animal used as food); four involve feelings (the locus of feelings; courage; an inclination; a positive feeling of liking); two involve centrality (physical; nonphysical); one indicates a drawing of a heart-shaped figure; and one is a playing card. For human readers, the full definitions, synonyms, and examples make the classification clear, but for machines they introduce additional ambiguity. For example, the synonym for the "locus of feelings" sense is "bosom," which has eight of its own WordNet senses. So, although the lexicographical quality of this manually acquired resource is high, interpreting the resource without human-level knowledge of English can be overwhelming.

The consequences of polysemy became clear when WordNet was used for query expansion in knowledge retrieval applications. Query expansion is the reformulation of a search term using synonymous key words or different grammatical constructions. But, as reported in Gonzalo et al. (1998), success has been limited because badly targeted expansion—using synonyms of the wrong meaning of a keyword—degrades performance to levels below those when queries undergo no expansion at all. A relevant comparison is the utility of a traditional monolingual thesaurus to native speakers versus its opaqueness to language learners: whereas native speakers use a thesaurus to jog their memory of words whose meanings and usage contexts they already know, language learners require all of those distinguishing semantic and usage nuances to be made explicit.

Various efforts have been launched toward making the content of WordNet better suited to NLP. For example, select components of some definitions have been manually linked to their correct WordNet senses as a method of disambiguation, and some cross-part-of-speech relations have been added, as between nouns and verbs. Much effort has also been devoted to developing multilingual wordnets and bootstrapping wordnets from one language to another. In the context of this flurry of development, what has not been pursued is a community-wide assessment of whether wordnets, in principle, are the best target of the NLP community's resource-building efforts.

1.6.8 Coreference in the Knowledge-Lean Paradigm

The complexity of reference resolution—of which establishing textual *co*references is just one aspect—has been inadvertently masked by the selective nature of mainstream work in NLP over the past twenty-five years. The vast majority of that work has applied machine learning (most often, supervised) to the simpler instances of the simpler types of referring expressions. To give just a few examples, most systems exclude ellipsis wholesale, they treat pronouns only in contexts in which their antecedents are realized as a single NP constituent, they consider only identity relations, and they consider the identification of a textual coreferent the end point of the task. (Why these constitute only partials is explained in chapter 5.) This means, for example, that *they* in (1.3) will be outside of purview, even though it is far from a worst case as real-world examples go.[37]

(1.3) My dad served with a Mormon and they became great friends. (COCA)

The rule-in/rule-out conditions are encoded in the corpus annotation guidelines that support the machine learning.[38]

An example of a task specification that has significantly influenced work on reference in NLP for the past two decades is the MUC-7 Coreference Task (Hirschman & Chinchor, 1997). This task was formulated to support a field-wide competition among NLP systems. Since it provided developers with annotated corpora for both the training and the evaluation stages of system development, it strongly encouraged the methodology of supervised machine learning. As regards the task's purview, the selection of so-called markables (entities for which systems were responsible) was more strongly influenced by practical considerations than scientific ones. For example, two of the four requirements were the need for greater than 95% interannotator agreement and the ability of annotators to annotate quickly and therefore cheaply—which necessitated the exclusion of all complex phenomena. The other two requirements involved supporting the MUC information extraction tasks and creating a useful research corpus outside of the MUC extraction tasks. Mitkov (2001) and Stoyanov et al. (2009) present thoughtful analyses of the extent to which such simplifications of the problem space have boosted the popular belief that the state of the art is more advanced that it actually is. Stoyanov et al. write, "The assumptions adopted in some evaluations dramatically simplify the resolution task, rendering it an unrealistic surrogate for the original problem." In short, task specifications of this sort—which have been created for quite a number of linguistic phenomena beside coreference—can be useful in revving up enthusiasm via competitions and fostering work on machine learning methods themselves. However, there is an unavoidable negative consequence of removing all difficult cases a priori: few people reading about the results of such systems will understand that the evaluation scores reflect performance on the easier examples. Tactically speaking, this makes it difficult to make the case that much more work is needed on reference—after all, numbers like 90% precision stick in the mind, no matter what they actually mean.

To reiterate, most of the NLP-oriented reference literature over the past twenty-five years has reported competing paradigms of machine learning, along with supporting corpus annotation efforts and evaluation metrics. Olsson (2004) and Lu and Ng (2018) offer good surveys. Poesio, Stuckardt, and Versley (2016; hereafter, PS&V) provide a more comprehensive overview of the field to date. Not only does this collection nicely frame the reference-oriented work described here, the authors also give a mainstream-insider's analysis of the state of the art that, notably, resonates with our own, out-of-the-mainstream observations. In their concluding chapter, "Challenges and Directions of Further Research," PS&V juxtapose the noteworthy advances in reference-related *engineering* with the state of treating *content*:

> If, however, one looks at the discipline from the side of the phenomenon (i.e. language, discourse structure, and—ultimately—*content*), we might arrive at the somewhat sobering intermediate conclusion that, after more than four decades of research, we are yet far away from the ambitious discourse processing proposals propagated by the classical theoretical work. That is, instead of investigating the celestial realms of rhetorical and thematic structure, we're yet occupied with rather mundane issues such as advanced string matching heuristics for common and proper nouns, or appropriate lexical resources for elementary strategies, e.g., number-gender matching etc. (p. 488)

They suggest that we might need to become "more ambitious again" (p. 488) in order to enhance the current levels of system performance. Although we wholeheartedly agree with the spirit of this assessment, we see a danger in describing rhetorical and thematic structure as "celestial realms," as this might suggest that they are permanently out of reach. Perhaps a more apt (and realistic) metaphor would have them on a very tall mountain.

It is noteworthy that PS&V are not alone in their assessment that the field has a long way to go—or, as Poesio puts it: "Basically, we know how to handle the simplest cases of anaphoric reference/coreference, anything beyond that is a challenge." (PS&V, pp. 490–491). For example, among the respondents to their survey about the future of the field was Marta Recasens, who wrote:

> I think that research on coreference resolution has stagnated. It is very hard to beat the baseline these days, state-of-the-art coreference outputs are far from perfect, and conferences receive less and less submissions on coreference. What's the problem? The community has managed to do our best with the "cheapest" and simplest features (e.g., string matching, gender agreement), plus a few more sophisticated semantic features, and this is enough to cover about 60% of the coreference relations that occur in a document like a news article, but successfully resolving the relations that are left requires a rich discourse model that is workable so that inferences at different levels can be carried out. This is a problem hindering research not only on coreference resolution but many other NLP tasks. (PS&V, p. 498)

Although we enthusiastically incorporate, as heuristic evidence, the results of a knowledge-lean coreference resolution engine into our NLU process, this paradigm of work does not inform our own research. Instead, our research is focused on semantically vetting—and, if needed, overturning—the results of such systems, as well as treating the more difficult phenomena that, to date, have been outside of purview. The reasons why the knowledge-lean paradigm does not inform our work are as follows:

1. It does not involve cognitive modeling, integration into agent systems, or the threading of reference resolution with semantic analysis.
2. The results are not explanatory.
3. Many contributions focus on a single reference phenomenon rather than seeking generalizations across phenomena.
4. The work does not involve linguistically grounded microtheories that can be improved over time in service of ever more sophisticated LEIAs. Instead, in the knowledge-lean paradigm, once the machine-learning methods have exploited the available corpus annotations, the work stops, with developers waiting for more and better annotations.

In fact, in response to the same survey mentioned above, Roland Stuckardt noted a complication of the supervised machine learning paradigm in terms of annotation and evaluation:

> The more elaborated the considered referential relations are, the less clear it becomes what "*human-like performance*" really amounts to. Eventually—since the reference processing task to be accomplished is too "vague" and thus not amenable to a sufficiently exact definition—, we might come to the conclusion that it is difficult to evaluate such systems in isolation, so that we have to move one level upwards and to evaluate their contribution chiefly extrinsically at application level. (PS&V, p. 491)

To sum up, knowledge-lean coreference systems serve our agent system in the same way as knowledge-lean preprocessing and syntactic analysis: all of these provide heuristic evidence that contributes to the agent's overall reasoning about language inputs.

1.6.9 Dialog Act Detection

The flow of human interaction overall, and language use in particular, follows typical patterns.[39] For example, upon meeting, people usually greet each other; a question is usually followed by an answer; and a request or order anticipates a response promising compliance or noncompliance. Of course, there are many variations on the theme, but those, too, are largely predictable: for example, the response to a question could be a clarification question or a comment about its (ir)relevance. In agent systems, understanding dialog acts[40] like these is a part of overall semantic/pragmatic analysis.

Automatic dialog act detection using supervised machine learning has been pursued widely enough to be the subject of survey analyses, such as the one in Král and Cerisara

(2010), which covers both the challenges of the enterprise and the methods that have been brought to bear. Among the challenges is creating a taxonomy of dialog acts that, on the one hand, balances the utility of a domain-neutral approach with the necessity for application-specific modifications and, on the other hand, supports an annotation scheme that is simple and clear enough to permit good interannotator agreement. Methods that have been brought to bear include various machine learning algorithms that use features categorized as lexical (the words used in an utterance), syntactic (word ordering and cue phrases), semantic (which can be quite varied in nature, from general domain indicators to frame-based interpretations of expected types of utterances), prosodic, and contextual (typically defined as the dialog history, with the previous utterance type being most important). Král and Cerisara note that application-independent dialog act–detection systems often use all of the above except semantic features.

Traum (2000) attends to the deep-semantic/discourse features that would be needed to fully model the dialog act domain. For example, since speaker intention is a salient feature of dialog acts, mindreading must be modeled; since user understanding is a salient feature, interspeaker grounding must be modeled; and since dialog acts belong to and are affected by the context (defined as the interlocutors' mental models), context must be modeled.

One noteworthy problem in comparing taxonomies of dialog acts is the use of terminology. In narrow-domain applications, the term *dialog act* can be used for what many would consider events in domain scripts. For example, in Jeong and Lee's (2006) flight reservation application, "Show Flight" is considered a dialog act, whereas under a more domain-neutral approach, the dialog act might be *request-information*, with the semantic content of the request being treated separately.

For illustration, we will consider the dialog act inventory in Stolcke et al. (2000),[41] which we selected for two reasons: first, because it includes a combination of generic and application-specific elements; and second, because the selections are justified by their utility in serving a particular goal—in this case, improving a speech recognition system. The latter reminds us of an important facet of statistical approaches: the right features are the ones that work best.

Stolcke et al.'s (2000) inventory of forty-two dialog acts was seeded by the Dialogue Act Markup in Several Layers (DAMSL) tag set (Core & Allen, 1997) and then modified to suit the specificities of their corpus: the dialogs in the Switchboard corpus of human-human conversational telephone speech (Godfrey et al., 1992). Although Stolcke et al. present the speech acts as a flat inventory (p. 341), we classify them into four categories to support our observations about them.[42]

- *Assertions*: STATEMENT, OPINION, APPRECIATION, HEDGE, SUMMARIZE/REFORMULATE, REPEAT-PHRASE, HOLD BEFORE ANSWER/AGREEMENT, 3RD-PARTY-TALK, OFFERS, OPTIONS & COMMITS, SELF-TALK, DOWNPLAYER, APOLOGY, THANKING

- *Question types*: YES-NO-QUESTION, DECLARATIVE YES-NO-QUESTION, WH-QUESTION, DECLARATIVE WH-QUESTION, BACKCHANNEL-QUESTION, OPEN-QUESTION, RHETORICAL-QUESTIONS, TAG-QUESTION
- *Responses*: YES ANSWERS, AFFIRMATIVE NON-YES ANSWERS, NO ANSWERS, NEGATIVE NON-NO ANSWERS, REJECT, RESPONSE ACKNOWLEDGMENT, AGREEMENT/ACCEPT, MAYBE/ACCEPT-PART, DISPREFERRED ANSWERS, BACKCHANNEL/ACKNOWLEDGE, SIGNAL NON-UNDERSTANDING, OTHER ANSWERS
- *Other*: ABANDONED/UNINTERPRETABLE, CONVENTIONAL-OPENING, CONVENTIONAL CLOSING, QUOTATION, COLLABORATIVE COMPLETION, OR-CLAUSE, ACTION-DIRECTIVE, NON-VERBAL, OTHER

If one looks at this inventory in isolation—that is, from a linguistic perspective, divorced from a machine learning application—questions naturally come to mind. Why the fine-grained splitting of question types? Why are APPRECIATE, APOLOGY, and THANKING included while other types of performative acts are excluded? Why is QUOTATION separate from the content of the quotation? However, when the inventory is framed within its intended task, it makes much more sense. Stolcke et al. (2000) write that they "decided to label categories that seemed both inherently interesting linguistically and that could be identified reliably. Also, the focus on conversational speech recognition led to a certain bias toward categories that were lexically or syntactically distinct (recognition accuracy is traditionally measured including all lexical elements in an utterance)" (p. 343).

We appreciate Stolcke et al.'s (2000) clarity of presentation, not only with respect to their goals and experimental results but also with respect to a simplification that boosted their evaluation score. Namely, they provided their system with correct utterance-level segmentations as input, since computing utterance-level segmentations is a difficult and error-prone task in itself. They explain that different developer choices make it difficult to compare systems: "It is generally not possible to directly compare quantitative results because of vast differences in methodology, tag set, type and amount of training data, and, principally, assumptions made about what information is available for 'free' (e.g., hand-transcribed versus automatically recognized words, or segmented versus unsegmented utterances)" (p. 363). This is a good reminder to us all of how essential it is to *read* the literature rather than skim the tables of results.

1.6.10 Grounding

The term *grounding* has been used with various meanings in AI. The two meanings most salient for robotic systems are *linking words to their real-world referents* and *linking any perceptual inputs to agent memory*. We will discuss those in chapter 8.[43] Here, by contrast, we focus on the meaning of *grounding* that involves overtly establishing that the speaker and interlocutor have achieved mutual understanding, which is a natural and necessary part of a fluid dialog. In live interactions, grounding is carried out through a combination of body language (e.g., maintaining appropriate eye contact and nodding) and utterances (e.g., "hmmm," "uh huh," and "yeah"). In computer dialog systems, by contrast, language is the only available channel for grounding.

Clark and Schaefer (1989, p. 262) posit the *grounding criterion*: "The contributor and the partners mutually believe that the partners have understood what the contributor meant to a criterion sufficient for current purposes." Traum (1999a, p. 130) divides this into two features: how much grounding is enough and how important it is for this level of grounding to be achieved. Baker et al. (1999) focus on the collaborative nature of grounding and the relevance of Clark and Wilkes-Gibbs' (1986) *principle of least collaborative effort*. Baker et al. say that it is better for addressees to simply show that they are listening rather than display exactly how they understand each utterance; if common ground is lost, repair should only be undertaken if it is deemed worth the effort.

Although the intuitions underlying grounding are clear, it is a big leap from intuitions to a formal, computable model. Traum (1999a) took this leap, compiling expectations about grounding into a state transition table covering the following grounding acts: *initiate*, *continue*, *acknowledge*, *repair*, *request repair*, *request acknowledgment*, and *cancel*. For example, if the dialog state is "Need for acknowledgment by initiator" and the responder continues talking without providing that acknowledgment, then the dialog remains in an ungrounded state. Although the model is compellingly formal, Traum himself points out its outstanding needs: the binary grounded/ungrounded distinction is too coarse; typical grounding practices (e.g., how often grounding is expected and needed) differ across language genres and contexts; the automatic identification of utterance units is an unsolved problem, as is the identification of which grounding act was performed (i.e., vagueness and partial understanding/grounding are typical outcomes that would need to be handled by an enhanced model). Traum asks a good question: "While it is clear that effective collaborative systems must employ the use of grounding-related feedback, what is less clear is whether there must be an explicit model of grounding that is referred to in the system's performance and interpretation of communications, or whether a system could be designed to behave properly without such an explicit model." He suggests that his grounding model could be improved by incorporating things like the cost and utility of grounding in conjunction with various other considerations, such as the utility of other actions that could help to ground the utterance.

We are not aware of any substantial breakthroughs in operationalizing models of grounding, which is not surprising since the difficult problems that Traum (1999a) indicates—as well as others he does not, such as the full semantic analysis needed to detect grounding-related features—remain open research issues.

1.6.11 More on Empirical NLP

In its purest form, empirical NLP relies on advanced statistical techniques for measuring similarities and differences between textual elements over large monolingual or multilingual text corpora—with corpora being viewed as repositories of evidence of human language behavior. In corpus-based approaches, all feature values must be obtained from unadorned text corpora.[44] That is, the only knowledge that exists is the surface form of text, as we would read it online or in a book. Within this neobehaviorist paradigm, there is no need to overtly address unobservables such as meaning; in fact, the very definition of *meaning* shifted. For example, in the latent semantic analysis approach, word meaning is understood essentially as a list of words that frequently appear in texts within N words of the "target" word whose meaning is being described. By the time of this writing, the empiricist paradigm in NLP has matured, and its main issues, results, and methods are well presented in the literature (for overviews, see, e.g., Jurafsky & Martin, 2009; Manning & Schütze, 1999).

One hallmark of recent NLP has been a widespread preference for developing—often in the context of a field-wide competition[45]—component technologies over building end-user applications. This preference has usually been justified as learning to walk before learning to run, or, in a more scholarly fashion, by saying that the scientific method mandates meeting prerequisites for a theory or a model before addressing that theory or model as a whole. In fact, in NLP, the latter precept has been often honored in the breach: in many (perhaps most?) cases, theoretical work on a variety of language phenomena proceeds from the assumption that all the prerequisites for the theory are met, whereas in reality this is seldom the case. This exasperates developers of application systems on the lookout for readily available, off-the-shelf components and knowledge resources for boosting the output quality of their applications. Their appetites are whetted when they read the description of a theory that promises to help them solve a practical problem, only to realize on further investigation that the theory can work only if certain currently unattainable prerequisites are met. For example, if a theory claims to solve the problem of automatically determining the discourse focus in a dialog but requires a complete propositional semantic analysis of the dialog content as a prerequisite, then it will not be of any use to practical dialog system builders because full semantic analysis is currently beyond the state of the art. It is in this context that one must understand the famous quip by Fred Jelinek, a leader in the field of automatic speech recognition, to the effect that every time he fired a linguist, his system's results improved.

Here we consider just two examples of tasks whose results are not directly useful for NLU because the task specification itself contrasts too markedly with the goals of full NLU. The tasks in question are word sense disambiguation and the interpretation of nominal compounds.

Word sense disambiguation. Within the empiricist paradigm, word sense disambiguation (WSD) has been identified as a freestanding task, which has been approached using both supervised and unsupervised machine learning. Associated with each approach is, interestingly enough, a different goal (see Navigli's 2009 survey for details). WSD using supervised machine learning is a classification task: the system is required to assign instances of words to a closed set of word meanings (selected by task developers) after training on an annotated corpus that provides word-to-meaning correspondences. In targeted WSD, systems are expected to disambiguate only certain target words, typically one to a sentence, for which ample training evidence (annotated examples) is provided. In all-words WSD, systems are expected to disambiguate all open-class words, but data sparseness (i.e., lack of sufficient training examples for each word) impedes the quality of results. By contrast, WSD using unsupervised machine learning is a clustering task whose goal is to cluster examples that use the same sense of a word. Although motivations for pursuing WSD as an independent task have been put forth (see, e.g., Wilks, 2000), when seen from an agent-building perspective, this is incongruent, since the results of WSD become ultimately useful only when they are integrated with dependency determination, reference resolution, and much more.

Identifying the relations in nominal compounds. Nominal compounding has been studied by descriptive linguists, psycholinguists, and practitioners of NLP.[46] Descriptive linguists have primarily investigated the inventory of relations that can hold between the component nouns. They have posited anywhere from six to sixty or even more descriptive relations, depending on their take on an appropriate grain size of semantic analysis. They do not pursue algorithms for disambiguating the component nouns, presumably because the primary consumers of linguistic descriptions are people who carry out such disambiguation automatically. However, they do pay well-deserved attention to the fact that NN interpretation requires a discourse context, as illustrated by Downing's (1977) "apple-juice seat" example. Psycholinguists, for their part, have found that the speed of NN processing increases if one of the component nouns occurs in the immediately preceding context (Gagné & Spalding, 2006). As for mainstream NLP practitioners, they typically select a medium-sized subset of relations of interest and train their systems to automatically choose the relevant relation during the analysis of compounds taken outside of context—that is, presented as a list. Two methods have been used to create the inventory of relations: developer introspection, often with iterative refinement (e.g., Moldovan et al., 2004), and crowdsourcing, also with iterative refinement (e.g., Tratz & Hovy, 2010). A recent direction of development involves using paraphrases as a proxy for semantic analysis: that is, a paraphrase of an NN that contains a preposition or a verb is treated as the meaning of that NN (e.g., Kim & Nakov, 2011). However, since verbs and prepositions are also highly ambiguous, these paraphrases do not count as fundamental disambiguation. Evaluations of knowledge-lean systems typically compare machine performance with human performance on a relation-selection or paraphrasing task.

In most statistical NLP systems, the semantics of the component nominals is not directly addressed: that is, semantic relations are used to link uninterpreted nouns. Although this is incongruous from a linguistic perspective, there are practical motivations.

1. The developers' purview can be a narrow, technical domain (e.g., medicine, as in Rosario & Hearst, 2001) that includes largely monosemous nouns, making nominal disambiguation not a central problem.[47]
2. The development effort can be squarely application-oriented, with success being defined as near-term improvement to an end system, with no requirement that all aspects of NN analysis be addressed.
3. The work can be method-driven, meaning that its goal is to improve our understanding of a machine learning approach itself, with the NN dataset being of secondary importance.
4. Systems can be built to participate in a field-wide competition, for which the rules of the game are posited externally (cf. the Free Paraphrases of Noun Compounds task of SemEval-2013 in Hendrickx et al., 2013).

Understanding this broad range of developer goals helps not only to put past work into perspective but also to explain why the full semantic analysis approach we will describe in chapter 4 does not represent an evolutionary extension of what came before; instead, it addresses a different problem altogether. It is closest in spirit to the work of Moldovan et al. (2004), who also undertake nominal disambiguation. However, whereas they implement a pipeline from word sense disambiguation to relation selection, we combine these aspects of analysis.

1.6.12 Manual Corpus Annotation: Its Contributions, Complexities, and Limitations

Corpus annotation has been in great demand over the past three decades because manually annotated corpora are the lifeline of NLP based on supervised or semisupervised machine learning (Ide & Pustejovsky, 2017). However, despite the extensive effort and resources expended on corpus annotation, the annotation of meaning has not yet been addressed to a degree sufficient for supporting NLP in the framework of cognitive modeling. So, even though annotated corpora represent a gold standard, the question is, What is the *gold* in the standard? The value of the gold derives from the task definition for the annotation effort, which in turn derives from developers' judgments about practicality and utility. To date, these judgments have led to creating annotated corpora to support such tasks as syntactic parsing, establishing textual coreference links, detecting proper names, and calculating light-semantic features, such as the case role fillers of verbs. Widely used annotated corpora of English include the syntax-oriented Penn Treebank (e.g., Taylor et al., 2003); PropBank, which adds semantic role labels to the Penn Treebank (Palmer et al., 2005); the Automatic Content Extraction (ACE) corpus, which annotates semantic relations and events (e.g., Doddington et al., 2004); and corpora containing annotations of pragmatics-oriented phenomena, such as coreference (e.g., Poesio, 2004), temporal relations (e.g., Pustejovsky et al., 2005), and opinions (e.g., Wiebe et al., 2005).

Decision-making about the scope of phenomena to annotate has typically been more strongly affected by judgments of practicality than utility. Some examples:

- The goal of the Interlingual Annotation of Multilingual Text Corpora project (Dorr et al., 2010) was to create an annotation representation methodology and test it on six languages, with component phenomena restricted to those aspects of syntax and semantics that developers believed could be consistently handled well by the annotators for all languages.
- When extending the syntactically oriented Penn Treebank into the semantically supplemented PropBank, developers selected semantic features (coreference and predicate argument structure) on the basis of feasibility of annotation (Kingsbury & Palmer, 2002).
- The scope of reference phenomena covered by the MUC coreference corpus was narrowly constrained due to the requirements that the annotation guidelines allow annotators to achieve 95% interannotator agreement and to annotate quickly and, therefore, cheaply (Hirschman & Chinchor, 1997).

Before passing an opinion about whether annotation efforts have been sufficiently ambitious, readers should pore over the annotation guidelines compiled for any of the past efforts, which grow exponentially as developers try to cover the overwhelming complexity of real language as used by real people. As Sampson (2003) notes in his thoughtful review of the history of annotation efforts, the annotation scheme needed to cover the syntactic phenomena in his corpus ran to 500 pages—which he likens both in content and in length to the independently produced 300+ page guidelines for Penn Treebank II (Bies et al., 1995). Hundreds of pages for syntax alone—we can only imagine what would be needed to cover semantics and discourse as well.

Since interannotator agreement and cost are among the most important factors in annotation projects, semiautomation—that is, automatically generating annotations to be checked and corrected by people—has been pursued in earnest. Marcus et al. (1993) report an experiment revealing that semiautomating the annotation of parts of speech and light syntax in English doubled annotation speed, showed about twice as good interannotator agreement, and was much less error-prone than manual tagging. However, even though semiautomation can speed up and improve annotation for simpler tasks, the cost should still not be underestimated. Brants (2000) reports that although the semiautomated annotation of German parts of speech and syntax required approximately fifty seconds per sentence, with sentences averaging 17.5 tokens, the actual cost—counting annotator training and the time for two annotators to carry out the task, for their results to be compared, and for difficult issues to be resolved—added up to ten minutes per sentence.

The cost of training and the steepness of the training curve for annotation cannot be overstated. Consider just a few of the rules comprising the MUC-7 task definition (Chinchor, 1997) for the annotation of named entities. Family names like *the Kennedys* are not to be annotated, nor are diseases, prizes, and the like named after people: *Alzheimer's, the Nobel prize*. Titles like *Mr.* and *President* are not to be annotated as part of the name, but appositives like *Jr.* and *III* ("the third") are. For place names, compound place names like *Moscow, Russia* are to be annotated as separate entities, and adjectival forms of locations are not to be annotated at all: *American companies*. While there is nothing wrong with these or any comparable decisions about scope and strategy, lists of such rules are very hard to remember—and one must bear in mind that tagging named entities, in the big picture of text annotation, is one of the simplest tasks.

This leads us to a seldom discussed but, in our opinion, central aspect of corpus annotation: it is expensive and labor-intensive, not to mention unpleasant and thankless—a combination of factors that puts most actual annotation work in the hands of low-paid students.

The empirical, machine learning–oriented paradigm of NLP has been routinely claimed to be the realistic alternative to knowledge-based methods that rely on expensive knowledge acquisition, but corpus annotation *is* expensive knowledge acquisition. The glamorous side of the work in this paradigm is the development and evaluation of the stochastic algorithms that use these annotations as input.

It is possible that during the early stages of the neobehaviorist revival, the crucial role of training materials for learning how to make sophisticated judgments by analogy was not fully appreciated. But unsupervised learning, although the cleanest theoretical concept, has so far proved to be far less successful. The preconditions of supervised learning put the task of corpus annotation, and the concomitant expense, front and center. The little-acknowledged reality is that the complexity and extent of the annotation task is fully commensurate with the task of acquiring knowledge resources for knowledge-based NLU. One lesson to learn from this is that the need for knowledge simply does not go away with a change in processing paradigms. And one thing to remember about corpus annotations is that, in contrast to knowledge bases developed for NLP, there is a big leap from examples to the kinds of useful generalizations that machine learning is expected to draw from them.

Although most annotation efforts to date have focused on relatively simpler phenomena, not all have. For example, the Prague Dependency Treebank (PDT) is a complex, linguistically motivated treebank that captures the deep syntactic structure of sentences (Mikulová, 2014). It follows a dependency-syntax theory called Functional Generative Description, according to which sentences are represented using treelike structures comprised of three interlinked layers of representation: the morphological layer, the surface syntactic (analytical) layer, and the deep syntactic (tectogrammatical) layer. The latter captures "the deep, semantico-syntactic structure, the functions of its parts, the 'deep' grammatical information, coreference and topic-focus articulation including the deep word order" (Mikulová, p. 129). The representations include three vertically juxtaposed and interlinked tree structures. Among the PDT's noteworthy features is its annotation of two types of deletions: textual ellipsis, in which the deleted material could have been expressed in the surface syntax (even if this would have led to stylistic infelicity), and grammaticalized ellipsis, in which some meaning must be semantically reconstructed but no corresponding category could be inserted into the surface syntax (Hajič et al., 2015). Deletions are accounted for in the PDT by introducing nodes in the tectogrammatical layer. Since Czech is a subject-drop language, this node-introduction strategy is widely represented in the PDT. However, introducing nodes is not the only way that null subjects have been treated in annotation schemes. According to Hajič et al., the treebanks of Italian and Portuguese—not to mention the analytical layer of the PDT—do not include such nodes.

The literature describing the PDT illustrates just how much theoretical and descriptive work must underpin the development of an annotation scheme before annotators are even set to the practical task. For example, Marie Mikulová et al.'s "Annotation on the Tectogrammatical Layer in the Prague Dependency Treebank"[48] runs to over 1,200 pages—a size and grain size of description that rivals comprehensive grammars. Similarly, a book-length manuscript (Mikulová, 2011) is devoted entirely to the identification and representation of ellipsis, without even opening up issues related to conditions of usage, their explanations, or predictive heuristics.

In the early twenty-first century, corpus annotation—specifically, creating the theoretically grounded annotation guidelines—has been the most visible arena for descriptive linguists to flex their muscles. The purview of descriptive linguistics has expanded from idealized, well-behaved, most-typical realizations of phenomena to what people actually say and write. The corpora annotated using such schemes can serve further linguistic investigation by making examples of phenomena of interest identifiable using simple search functions. In fact, Hajič et al. (2016, p. 70) present an in-depth analysis of how the process of annotating the PDT, as well as its results, have led to amendments in the underlying linguistic theory and a better understanding of the language system.

1.7 Further Exploration

1. There are many hard things about language. One of them is understanding bad writing. Read or watch Steven Pinker's insightful and entertaining analyses of bad writing and its good counterpart:

- *The Sense of Style: The Thinking Person's Guide to Writing in the 21st Century* (Penguin, 2014).
- "Why Academics Stink at Writing," *The Chronicle Review*, The Chronicle of Higher Education, September 26, 2014, https://stevenpinker.com/files/pinker/files/why_academics_stink_at_writing.pdf
- Various lectures available on YouTube, such as "Linguistics, Style and Writing in the 21st Century—with Steven Pinker," October 28, 2015, https://www.youtube.com/watch?v=OV5J6BfToSw&t=1020s

2. The history of machine translation makes for interesting reading. Some suggestions:

- Warren Weaver's 1949 memorandum "Translation," available at http://www.mt-archive.info/Weaver-1949.pdf
- Yehoshua Bar Hillel's (Hebrew University, Jerusalem) "The Present Status of Automatic Translation of Languages," from *Advances in Computers*, vol. 1 (1960), pp. 91–163, available at http://www.mt-archive.info/Bar-Hillel-1960.pdf
- John Hutchins's "ALPAC: The (In)famous Report," available at http://www.hutchinsweb.me.uk/ALPAC-1996.pdf
- *Readings in Machine Translation*, edited by S. Nirenburg, H. Somers, and Y. Wilks (MIT Press, 2003), which contains all of the above as well as many other relevant texts.

3. Investigate the current state of the art in machine translation using Google Translate (translate.google.com). You don't need to know another language to do this.

- Copy-paste (or simply type) a passage into the left-hand window and be sure it is recognized as English.
- Translate it into any of the available languages by choosing a target language in the right-hand window.
- Copy the translation (even though you won't understand it) back into the left-hand window and be sure the system understands which language it is.
- Translate the translation back into English.

   a. How good is the translation?
   b. Can you hypothesize any differences between English and that language based on the output? For example, maybe that language does not use copular verbs (i.e., the verb *be* in sentences like *George is a zookeeper*), or maybe it permits subject ellipsis—both of which might be reflected in the translation back into English.

You should get better translations if you (a) select a language, L, for which L-to-English and English-to-L machine translation has been worked on extensively (e.g., French, Spanish, Russian); (b) select a language that is grammatically close to English; and (c) select a grammatically normative text (not, e.g., a highly elliptical dialog). Make the opposite choices and translation quality is likely to suffer. If you know another language, things become more interesting since you can do multistage translation—not unlike the telephone game, in which players whisper a message in a circle and see how much it morphs by the time it reaches the last player.

4. Read about the mainstream approaches to NLP over the past thirty years in Jurafsky and Martin's *Speech and Language Processing: An Introduction to Natural Language*

Processing, Speech Recognition, and Computational Linguistics, 2nd ed. (Prentice-Hall, 2009).

5. Think about and/or discuss the differences between applications that operate over big data (e.g., question-answering *Jeopardy!*-style) and applications in which every utterance is produced exactly once, using exactly one formulation (e.g., a task-oriented dialog). What are the challenges and opportunities specific to each one?

2

A Brief Overview of Natural Language Understanding by LEIAs

This chapter presents a brief overview of natural language understanding (NLU) by LEIAs. Our purpose is to use simple examples to describe and motivate our overall approach before introducing, in chapters 3–7, the large number of linguistic phenomena that must be treated by any realistic-scale NLU system.

One word of framing before we begin. We cannot emphasize enough that this book describes an ongoing, long-term, broad-scope program of work that we call Linguistics for the Age of AI. The depth and breadth of work to be done is commensurate with the loftiness of the goal: enabling machines to use language with humanlike proficiency. The main contributions of the book are the computational cognitive models we present and their organization into an overall, broadly encompassing process of NLU. At present, the models are of various statuses: many have been demonstrated in prototype systems, some have been formally evaluated, and still others await implementation. When we say that an agent *does X*, this describes how our model works; it does not mean that, today, our agent systems can *do X* with respect to every possible language interaction. Were that the case, we would have already solved the language problem in AI.

As the book progresses, readers should find sufficient details about a sufficient number of models to realize that we do not underestimate the sheer quantity of work awaiting linguists who take on the challenge of automating human-level NLU. Moreover, with each cycle of building, implementing, and evaluating models, new theoretical and methodological insights will accrue, which could lead to significant modifications to the approach presented here. But still, as long as the goal of the research enterprise is to build psychologically plausible, explanatory models of language behavior, we expect the core of our approach to endure.

2.1 Theory, Methodology, and Strategy

We begin with an overview of theoretical principles, methodological preferences, and strategic choices. The presentation is organized into four categories: the nature of NLU, the knowledge and reasoning needed for NLU, how NLU interacts with overall agent cognition, and strategic preferences. The lists are not exhaustive; they are intended to serve as a conceptual scaffolding for upcoming discussions.

The nature of NLU

- Language understanding by LEIAs involves translating language inputs into ontologically grounded *text meaning representations* (TMRs), which are then stored in agent memory (Nirenburg & Raskin, 2004).
- Translation into the ontologically grounded metalanguage
  - focuses on the content of the message rather than its form;
  - resolves complexities such as lexical and referential ambiguity, ellipsis, and linguistic paraphrase; and
  - permits many of the same knowledge bases and reasoning engines to be used for processing texts in different languages.
- The global interpretation of text meaning is built up compositionally from the interpretations of progressively larger groups of words and phrases.
- Semantic imprecision is recognized as a feature of natural language; it is concretized only if the imprecision impedes reasoning or decision-making.
- Semantic analysis is carried out in stages, marked by the kind of *wait and see* tactic that people seem to use when they are operating in a foreign language that they know only moderately well (i.e., even if something is not immediately clear, what comes next might serve to clarify).
- Each meaning interpretation is assigned a confidence level that reflects the degree to which the interpretation deviates from the expectations of the supporting knowledge bases and algorithms. In human-agent applications, confidence levels will help agents to decide whether or not to act on their understanding of a language input or seek clarification from a human collaborator.
- We make no theoretical claims about how people carry out pre-semantic aspects of language analysis—most notably, preprocessing and syntactic analysis; our research interests are squarely in the realm of semantics and pragmatics. Therefore, pre-semantic analysis in our environment is outsourced to externally developed engines, and the agent interprets their results as overridable heuristic evidence.

The knowledge and reasoning needed for NLU

- Language understanding relies primarily on three knowledge resources: the ontology (knowledge about concept types), episodic memory (knowledge about concept instances), and an ontological-semantic lexicon.
- Language understanding employs algorithms that manipulate both linguistic and extralinguistic knowledge.

- The microtheories for the treatment of language phenomena (e.g., lexical disambiguation, coreference resolution, indirect speech act interpretation) reflect an attempt to operationalize our understanding of how humans make sense of language. Machine learning–based mainstream NLP does not adhere to this objective. As a result, it cannot claim insights into explanatory theories of language use. Human-inspired microtheories not only have the best potential to enable agents to successfully collaborate with people but also make the results explainable in human terms.
- We consider *listing* a good and necessary approach to capturing human knowledge. This includes writing rules, lexicon entries, grammatical constructions, and more. The common prejudice against listing derives from the desire to create elegant and streamlined accounts that strike some as scientifically satisfying. However, this prioritization of the elegant is misplaced when the domain of inquiry is natural language, and it invariably leads to the proliferation of wastebasket phenomena that are not treated at all. There is no evidence that people lack the capacity to record a lot of language-oriented information directly and explicitly. Moreover, there is no reason why the principle of dynamic programming (whereby the results of computations are remembered and reused instead of being recomputed from scratch every time they are needed) should not be used in modeling human language processing.

How NLU interacts with overall agent cognition

- Agents use stored knowledge recorded as TMRs (not natural language text strings!) to reason about action. (The source text strings are, however, stored as metadata associated with each TMR since they can, e.g., inform word selection when generating a response to an input.)
- Agents are, at base, language independent. Only the text-to-TMR translation process is language specific—and even many aspects of *that* are applicable crosslinguistically. To cite just two examples, the semantic descriptions in the LEIA's lexicon are readily portable across lexicons for different languages, and key aspects of reasoning about coreference can be carried out over TMRs, not the language strings that gave rise to them.
- Language understanding cannot be separated from overall agent cognition since heuristics that support language understanding draw from (among other things) the results of processing other modes of perception (such as vision), reasoning about the speaker's plans and goals, and reasoning about how much effort to expend on understanding difficult inputs.
- Agents, like humans, must be able to leverage many reasoning strategies, including language-centric reasoning (using selectional constraints, linguistic rules, and so on), reasoning by analogy, goal-based reasoning, and statistical likelihood.

- Modeling human-level language understanding does not mean modeling what an ideal human might ideally understand given ideal circumstances. People don't operate in a perfect, sanitized world. Instead, they pay various degrees of attention to language inputs depending on whether the content is interesting and/or relevant to them. So, too, must agents.
- One way to model humanlike language understanding is to focus on *actionability*—that is, configuring agents to pursue a level of interpretation that supports intelligently selected subsequent action. An actionable interpretation might represent a complete and correct analysis of an input, or it might be incomplete; it might involve deep analysis or only skimming the surface; and it might be achievable by the agent alone, or it might require clarifications or corrections by a human or artificial collaborator. Deeming an interpretation actionable serves as a practical halting condition for language analysis.
- Among the available actions an agent can take in response to an input are physical actions, verbal actions, and mental actions. The latter include decision-making, learning, and even ignoring the input, having deemed it outside the agent's current topics of interest.

Strategic preferences

- Although our program of work focuses primarily on the research end of the R&D spectrum, system implementations serve to validate the component microtheories and algorithms. Implementations are both theoretically necessary (AI requires computation) and strategically beneficial (we always discover something unexpected when applying algorithms to unconstrained texts).
- Since knowledge bases compiled by our team in decades past offer sufficient breadth and depth of coverage to serve as test beds for developing theories and models, manual resource-building is not a current priority, even though continued expansion of the knowledge resources is essential for achieving human-level AI on an industrial scale.
- A scientifically more compelling solution to the need for knowledge is making the agent capable of lifelong learning by reading, by experience, and by being told—which is something we *are* working on.
- It is essential to distinguish domain-neutral from domain-specific facets of language understanding and to design NLU systems that are maximally portable across domains. This counters the implicit, but unrealistic, hope of developers of narrow-domain systems: that those systems will ultimately converge, resulting in open-domain coverage. This will not happen because narrow-domain systems do not address the core problem of lexical ambiguity that looms large when moving to the open domain.
- System development in our approach is task-oriented, not method-oriented. The task is to develop humanlike NLU capabilities that permit LEIAs to explain their

cognitive functioning in human terms. Only then will they become trusted collaborators. Any method that is well suited to a particular task can serve this goal. It so happens that knowledge-based methods are best suited to solving most NLU tasks, but when other methods prove useful, they are incorporated. This task-oriented methodology is in contrast to first choosing a method (such as using statistical algorithms operating over big data) and then seeking useful applications for it.

- The cognitive models contributing to NLU aim for descriptive adequacy, not neatness. Decision-making during modeling is guided by the principle *choose and move on*. A sure route to failure is to ponder over innumerable decision spaces with their associated pros and cons, effectively paralyzing the enterprise with the drive for perfect but nonexistent solutions. The *choose and move on* principle reflects a preference for something over nothing, as well as the reality that all models are incomplete (Bailer-Jones, 2009).
- Strategic simplifications are incorporated, provided they do not jeopardize the utility of the results. If a simplification causes a drop in quality, the optimal grain size of description must be reconsidered. For example, to date we have not worked on applications for which distinctions between types of dogs, hats, or lizards were important. Accordingly, all names of dog breeds, types of hats, and infraorders of lizards are mapped to the ontological concepts DOG, HAT, and LIZARD, respectively. However, as soon as a dog-, hat-, or lizard-oriented application comes along, then this simplification will no longer serve, and the requisite knowledge-acquisition work will come on agenda.

As the above theoretical and methodological statements should make clear, our approach to extracting meaning from natural language text is as human inspired as possible for the development of theory, but as methodologically inclusive as necessary for the development of application systems.

An organizational sidebar. This book contains many examples of knowledge structures—text meaning representations, lexical senses, ontological frames, and more. In our NLU system, these are recorded using a formalism that takes time to learn to read quickly. Although assessing the quality of formalisms is important for gauging both their expressive power and their utility for building computational applications, this is not central to the goals of this book. Here, we concentrate on the *content* of our approach to NLU, as reflected in microtheories and algorithms that process them. To keep the cognitive load for readers at a reasonable level, we decided to present knowledge structures in a simplified format, with the goal of making the specific, contextually relevant points as clear as possible. Naturally, examples retain certain features of the underlying system-oriented formalism. Readers interested in the formalism itself will find examples in the online appendixes at https://homepages.hass.rpi.edu/mcsham2/Linguistics-for-the-Age-of-AI.html.

2.2 A Warm-Up Example

To reiterate, a LEIA's understanding of what a language input means is recorded in an ontologically grounded *text meaning representation*, or TMR. Consider the simplified[1] TMR for ***A gray squirrel ate a nut.***

```
INGEST-1
    AGENT          SQUIRREL-1
    THEME          NUT-FOODSTUFF-1
    TIME           <find-anchor-time
    from-sense     eat-v1
    word-num       3

SQUIRREL-1
    COLOR          gray
    AGENT-OF       INGEST-1
    from-sense     squirrel-n1
    word-num       2

NUT-FOODSTUFF-1
    THEME-OF       INGEST-1
    from-sense     nut-n1
    word-num       5
```

This example is simple for the following reasons: it contains just one clause; that clause is syntactically regular; none of its referring expressions require coreference resolution; its lexical ambiguities can be resolved using rather simple analysis techniques; and the semantic analyses of the lexemes reliably combine into an ontologically valid semantic dependency structure. This TMR should be read as follows.

- The first frame is headed by a numbered instance of the concept INGEST. Concepts are distinguished from words of English by the use of small caps. Note that this is not vacuous *upper-case semantics*[2] because the concepts in question have formal definitions in the ontology that (a) are based on value sets of ontological properties (the primitives in the conceptual system of the ontology) and (b) support reasoning about language and the world.
- INGEST-1 has three contextually relevant property values: its AGENT (the eater) is an instance of SQUIRREL; its THEME (what is eaten) is an instance of NUT-FOODSTUFF; and the TIME of the event is before the time of speech.[3]
- The properties in italics are among the many elements of metadata generated during processing, which support system evaluation, testing, and debugging. Those shown indicate which word number (starting with 0) and which lexical sense were used to generate the given TMR frame. In most TMRs presented hereafter, we will not include these metadata slots.

- The next frame, headed by SQUIRREL-1, shows not only the inverse relation to INGEST-1 but also that the COLOR of this SQUIRREL is gray. *Gray* is not written in small caps because it is a literal (not concept) filler of the property COLOR.
- Since we have no additional information about the nut, its frame—NUT-FOODSTUFF-1—shows only the inverse relation with INGEST-1, along with the same type of metadata described above.

For each TMR it produces, the LEIA generates a value of the *confidence* parameter (a type of metadata not shown in the example) that reflects its certainty in the TMR's correctness. For TMRs, like this one, that do not require advanced semantic and pragmatic reasoning, the confidence score is computed using a function that compares how the elements of input align with the syntactic and semantic expectations of word senses in the lexicon.

In working through how a LEIA generates this analysis, we will assume for the moment (since we're just getting warmed up) that the agent has access to the entire sentence at once. We will introduce incremental processing—our first complicating factor—in section 2.4.

First the input undergoes preprocessing and syntactic analysis, which are provided by an external toolset.[4] Using features from the syntactic parse, the LEIA attempts to align sentence constituents with the syntactic expectations recorded in the lexicon for the words in the sentence. For example, it will find three senses of the verb *eat* in the lexicon. One is optionally transitive and means INGEST. The other two describe the idiom *eat away at* in its physical sense (*This powerful antioxidant is always a handy chemical for the aerospace industry, since it can eat away at metal without causing the heat fatigue associated with traditional machining.* (COCA)) and its abstract sense (*This vice begins to eat away at our soul …* (COCA)).[5] Since the idiomatic senses require the words *away at*, which are not present in the input, they are rejected, leaving only the INGEST sense as a viable candidate. Below is the needed lexical sense of *eat* (eat-v1) followed by one of the idiomatic senses (eat-v2).

```
eat-v1
    def.                ingest
    ex.                 He was eating (cheese).
    syn-struc
        subject         $var1
        v               $var0
        directobject    $var2 (opt+)
    sem-struc
        INGEST
            AGENT       ^$var1
            THEME       ^$var2 (sem FOOD)

eat-v2
    def.      The construction 'eat away at': erode physically
    ex.       The rust ate away at the pipe.
    syn-struc
```

```
    subject   $var1
    v         $var0
    adv       $var2 (root 'away')
    pp
        prep  $var3 (root 'at')
        obj   $var4
sem-struc
    ERODE
        INSTRUMENT      ^$var1
        THEME           ^$var4
    ^$var2   null-sem+
    ^$var3   null-sem+
```

We will explain the format and content of lexical senses using *eat-v1* as an example. The syntactic structure (syn-struc) zone of *eat-v1* says that this sense of *eat* is optionally transitive: it requires a subject and can be used with or without a direct object (*opt+*means *optional*).

Each constituent of input is associated with a variable in the syn-struc. In *eat-v1*, the head of the entry (the verb) is $var0, the subject is $var1, and the direct object is $var2. The meaning of $var0, expressed as an ontological concept, heads the semantic structure (sem-struc) zone: INGEST. After all, the point of this sense is to describe the meaning of $var0 when it is used in this particular construction. All other variables are linked to their semantic interpretations in the sem-struc, with ^ being read as *the meaning of.* So the *word* that fills the subject slot in the syn-struc, $var1, must be semantically analyzed, resulting in ^$var1. Then ^$var1 must be evaluated to see if it is a semantically suitable AGENT of an INGEST event.

Note that, by default, the semantic constraints on the case roles are not listed because they are drawn from the ontology. However, the THEME of INGEST is an exception. Its constraint is listed because it overrides (i.e., is narrower than) what is listed in the ontology: whereas one can INGEST food, beverages, or medication (see the ontological frame for INGEST below), one can *eat* only food.[6]

Shifting for a moment to *eat-v2*, two of its variables are described in the sem-struc as *null-sem+*. This means that their meaning has already been taken care of by the semantic representation and should not be computed compositionally. In this example, *eat away at*, taken together, means ERODE; *away* and *at* do not carry any extra meaning beyond that.

The ontology, for its part, provides information about the valid fillers of the case roles of INGEST. Consider a small excerpt from the ontological description of INGEST; the full concept description contains many more property-facet-value triples.

```
INGEST
    AGENT      sem             ANIMAL
               relaxable-to    SOCIAL-OBJECT
    THEME      sem             FOOD, BEVERAGE, INGESTIBLE-MEDICATION
               relaxable-to    ANIMAL, PLANT
               not             HUMAN
```

This ontological frame says that the typical AGENT of INGEST (i.e., the basic semantic constraint indicated by the *sem* facet) is an ANIMAL; however, this constraint can be relaxed to SOCIAL-OBJECTs, as in *The fire department eats a lot of pizza*. Similarly, the description of the THEME indicates that FOOD, BEVERAGE, and INGESTIBLE-MEDICATION are the most typical THEMEs, but other ANIMALs and PLANTs not already subsumed under the FOOD subtree might be consumed as well. HUMANs are explicitly excluded as ingestibles (which illustrates the semantics of the *not* facet), since they would otherwise be understood as unusual-but-possible ingestibles due to their placement in the ANIMAL subtree of the ontology. There are two important reasons to exclude humans as ingestibles even though, for sure, big cats and alligators have been known to occasionally eat a person. First, the ontology is intended to provide agents with knowledge of how the world typically works. Second, there is a sense of *eat* that means *to annoy* (*What was eating her?* (COCA)), and that sense should be preferred when the direct object is a person.

Having narrowed down the interpretation of *eat* to a single sense, the LEIA must now determine which senses of *squirrel*, *gray*, and *nut* best fit this input. *Squirrel* and *gray* are easy: the lexicon currently contains only one sense of each, and these senses fit well semantically: SQUIRREL is a suitable AGENT of INGEST, and *gray* is a valid COLOR of SQUIRREL. However, there are three senses of *nut*: an edible foodstuff, a crazy person, and a machine part. We just saw that neither people nor machine parts are FOOD, leaving only the NUT-FOODSTUFF sense, which is selected as a high-confidence interpretation.

Operationally speaking, after all the constraints have been checked, the TMR for *A gray squirrel ate a nut* is generated by

1. copying the sem-struc of *eat-v1* into the nascent TMR;
2. translating the concept type (INGEST) into an instance (INGEST-1); and
3. replacing the variables with their appropriate interpretations: ^$var1 becomes SQUIRREL-1 (COLOR gray), and ^$var2 becomes NUT-FOODSTUFF-1.

With respect to runtime reasoning, this example is as straightforward as it gets since (a) it involves only matching statically recorded constraints, and (b) all constraints match in a unique and satisfactory way. *Straightforward constraint matching* does not, however, come for free: its precondition is the availability of high-quality lexical and ontological knowledge bases that are sufficiently detailed to allow the LEIA to disambiguate words and validate the semantic congruity of its resulting interpretations.

As mentioned earlier, LEIAs generate confidence scores for particular TMRs based on how well the syntactic and semantic expectations of lexical senses are satisfied by the candidate interpretation. Whereas our example TMR will get a very high confidence score, there will be no high-scoring interpretations for *That furry face is eating a nut*. Such inputs are handled using recovery methods described in later chapters.

The ontologically grounded knowledge representation language just illustrated has many advantages for agent reasoning (McShane & Nirenburg, 2012). Most importantly, (a) it is unambiguous, and (b) the concepts underlying word senses are described extensively in the ontology, which means that more knowledge is available for reasoning about language and the world than is made available by the occurrence of words in the input. However, translating natural language utterances into this metalanguage is difficult and expensive. So, a reasonable question is, *Do we really need it?*

If agents were to communicate exclusively with other agents, and if they had no need to learn anything from human-oriented language resources, then there would be no need for the natural-language-to-knowledge-representation-language translation that we describe. However, for agents to be truly useful, they *do* need to communicate with people, and they *do* need to learn about the world by converting vast amounts of data into interpreted knowledge. Because of this, it is important to both establish the formal relationship between natural language and a knowledge representation language and to provide intelligent agents with the facility to translate between them. For other views on the relationship between natural language and knowledge representation languages, see the deep dive in section 2.8.1.

2.3 Knowledge Bases

The main static knowledge bases for LEIAs are

1. the lexicon (including an onomasticon—a repository of proper names), which describes the syntactic expectations of words and phrases along with their ontologically grounded meanings;
2. the ontology, which is the repository of types of objects and events, each of which is described by a large number of properties; and
3. the episodic memory, which is the repository of concept instances—that is, real-life representatives of objects and events, along with their property values.[7]

Theoretically speaking, every LEIA—like every person—will have idiosyncratic knowledge bases reflecting their individual knowledge, beliefs, and experiences. And, in fact, such individualization does occur in practice since LEIAs not only remember the new information they learn through language understanding but can also use this information to dynamically learn new words and ontological concepts in various ways (see chapter 8). However, this learning builds on the core lexicon, ontology, and episodic memory that are provided to all LEIAs as a model of a typical adult's knowledge about language and the world. The coverage of these core knowledge bases is, of course, incomplete relative to the knowledge store of an average person (a practical matter); but, for what they cover, they are representative. For example, the word *horse* has three senses in the current lexicon, referring to an animal, a piece of gymnastic equipment, and a sawhorse. So every time a LEIA encounters the word *horse* it must contextually disambiguate it, which it does using ontological and contextual knowledge.

Apart from LEIAs that model typical adults, there are also specialist LEIAs that are endowed with additional ontological, lexical, and episodic knowledge in a particular domain. For example, LEIAs serving as tutors and advisors in the field of clinical medicine must not only have extensive knowledge of that domain but also be aware of the differences between their knowledge and that of a typical person. This can be operationalized by flagging specialist-only subtrees in the ontology as well as specialist-only lexicon entries. When agents use these flagged concepts, words, and phrases in dialogs with nonspecialists, they, like people, will introduce them with explanations. This kind of mindreading allows for effective communication between individuals possessing different levels of expertise (see chapter 8 for more on mindreading). After all, if a physician's ten-minute explanation of all the potential side effects of a medication is so packed with specialist terminology that the patient understands nothing, then the communication has failed.

In principle, any LEIA can have or lack any datum, and it can have wrong beliefs about things as well. This is an interesting aspect of cognitive modeling: preparing agents to behave in lifelike ways in the face of incomplete and contradictory beliefs. However, individual differentiation is not the focus of the current discussion. Here, we concentrate on the basics: the knowledge bases that reflect general adult-level knowledge about language and the world.

The sections that follow give a very brief, nontechnical introduction to these knowledge bases. Readers interested in more detail can consult the cited references.

2.3.1 The Ontology

The LEIA's ontology is a formal model of the world that is encoded in the metalanguage presented above. A comprehensive description of, and rationale for, the form and content of the ontology is available in Nirenburg and Raskin (2004, section 7.1). Here we present just enough detail to ground the description of NLU to come.

The ontology is organized as a multiple-inheritance hierarchical collection of frames headed by concepts that are named using language-independent labels. It currently contains approximately 9,000 concepts, most of which belong to the general domain. We avoid a proliferation of ontological concepts, in line with the recommendation by Hayes (1979) that the ratio of knowledge elements used to describe a set of elements of the world to the number of these latter elements must be kept as low as possible. There are additional reasons why the number of concepts in the ontology is far lower than the number of words or phrases in any language.

1. Synonyms map to the same ontological concept, with semantic nuances of particular words recorded as constraints in the corresponding lexical senses.
2. Many lexical items are interpreted using combinations of concepts.
3. Lexical items that represent a real or abstract point or range on a scale all point to a single property that represents that scale (e.g., *brilliant*, *smart*, *pretty smart*, and *dumb* reflect different values of INTELLIGENCE).

4. Concepts are intended to be crosslinguistically and cross-culturally relevant. This means that the ontology does not, for example, contain a concept for the notion *recall* in the sense "request that a purchased good be returned because of a discovered flaw" because not all languages and cultures use this concept. Instead, the meanings of such words are described compositionally in the lexicons of those languages that do use them.

Concepts divide up into EVENTS, OBJECTS, and PROPERTYS. PROPERTYS are primitives, which means that their meaning is grounded in the real world with no further ontological decomposition. Ontological properties are used to define the meaning of OBJECTS and EVENTS. Stated plainly, an OBJECT or EVENT *means* whatever its property-facet-value triples say it means. The types of properties contributing to OBJECT and EVENT descriptions include:

- IS-A and SUBCLASSES, which are the two properties that indicate the concept's placement in the tree of inheritance. Multiple inheritance is permitted but not overused, and rarely does a concept have more than two parents.
- RELATIONS, which indicate relationships among OBJECTS and EVENTS. Examples include the nine case roles that describe the typical participants in EVENTS—AGENT, THEME, BENEFICIARY, EXPERIENCER, INSTRUMENT, PATH, SOURCE, DESTINATION, LOCATION—along with their inverses (e.g., AGENT-OF, THEME-OF).[8]
- SCALAR-ATTRIBUTES, which indicate meanings that can be expressed by numbers or ranges of numbers: for example, COST, CARDINALITY, FREQUENCY.
- LITERAL-ATTRIBUTES, which indicate meanings whose fillers were determined by acquirers to be best represented by uninterpreted literals: for example, the property MARITAL-STATUS has the literal fillers *single, married, divorced, widowed.*
- Several administrative properties for the use of people only, such as DEFINITION and NOTES.

Selecting the optimal inventory of properties has not been, nor is it slated to be, on agenda in our research—though it is an interesting topic for full-time ontologists. Instead, we have taken a practical, system-oriented approach to creating properties. Some were included by virtue of being central to any world model: for example, HAS-CAPITAL is a useful descriptor for LARGE-GEOPOLITICAL-ENTITYS, and CAUSED-BY is needed to describe EVENTS. Other properties are convenient shorthand, introduced for a given application: for example, HAS-COACH was included for an Olympics application because it was more convenient to record and manipulate structures like "X HAS-COACH Y" than more explanatory structures like "there is a long-term, repeating succession of COACHING-EVENTS for which Y is the AGENT and X is the BENEFICIARY." (For more on semantically decomposable properties, see section 6.1.3.) The point is that practically any inventory of properties can serve a LEIA's purposes as long as those properties are used effectively in ontological descriptions of related OBJECTS and EVENTS, and as long as the LEIA's reasoners are configured to appropriately use them.

The expressive power of the ontology is enhanced by multivalued fillers for properties, implemented using *facets*. Facets permit the ontology to include information such as *the most typical colors of a car are white, black, silver, and gray; other normal, but less common, colors are red, blue, brown, and yellow; rare colors are gold and purple.* The inventory of facets includes: *default*, which represents the most restricted, highly typical subset of fillers; *sem*, which represents typical selectional restrictions; *relaxable-to*, which represents what is, in principle, possible although not typical; and *value*, which represents not a constraint but an actual, nonoverridable value. *Value* is used primarily in episodic memory, but in the ontology it has the role of indicating the place of the concept in the hierarchy, using the concepts IS-A and SUBCLASSES. Select properties from the ontological frame for the event DRUG-DEALING illustrate the use of facets.

```
DRUG-DEALING
    IS-A                  value           CRIMINAL-ACTIVITY
    AGENT                 default         CRIMINAL, DRUG-CARTEL
                          sem             HUMAN
                          relaxable-to    SOCIAL-OBJECT
    THEME                 default         ILLEGAL-DRUG
    INSTRUMENT            sem             MONEY
    HAS-EVENT-AS-PART     sem             BUY, SELL
    LOCATION              default         CITY
                          sem             PLACE
                          relaxable-to    PHYSICAL-OBJECT
    ...
```

OBJECTS and EVENTS are defined in the ontology using an average of sixteen properties each, but many of the fillers of those properties are inherited rather than locally specified.

To reiterate the most important point: The *meaning* of an OBJECT or EVENT is the set of its property-facet-value triples.

The main benefits of writing an ontology in a knowledge representation language rather than a natural language are (a) the absence of ambiguity in the knowledge representation language, which makes the knowledge suitable for automatic reasoning, and (b) its reusability across natural languages. Cut to thirty years from now, and the LEIA's ontology should contain tens of thousands of well-described concepts, including thousands of descriptions of complex events (scripts). Since the ontology is language independent, this knowledge infrastructure will be accessible to intelligent agents that communicate in any language, as long as a compatible lexicon and language-understanding engine for that language have been developed.

A core need in ontological modeling is describing complex events that involve multiple steps, multiple participants, and multiple props. In our ontology these complex events are represented using ontological scripts.[9] Scripts can reflect knowledge in any domain (what happens at a doctor's appointment, how to make spaghetti and meatballs, how to remove a brain tumor), and they can be at any level of generality (from a basic sequence of events to the level of detail needed to generate computer simulations). What ontological scripts do *not* do is conform to the simple slot-facet-filler formalism described above. Although scripts use the same concepts and the same basic knowledge representation language, they require additional expressive power. Taking examples from the domain of medical appointments, scripts require:

- The coreferencing of arguments. In a given appointment, the same instance of PHYSICIAN will carry out many actions (e.g., asking questions, answering questions, recommending interventions), and the same instance of MEDICAL-PATIENT will carry out many actions (asking questions, answering questions, deciding about interventions).
- Loops. There can be many instances of event sequences, such as ask/answer a question and propose/discuss an intervention.
- Variations in ordering. A doctor can get vital signs before or after the patient interview, and provide lifestyle recommendations before or after discussing medical interventions.
- Optional components. A doctor may or may not engage in small talk and may or may not recommend tests or interventions.
- Time management. For simulation-oriented or time-sensitive scripts (e.g., in the domain of emergency medicine), the script must include information about what happens when, how fast, and for how long.

In short, although scripts *are* a part of ontology per se—that is, they fill the HAS-EVENT-AS-PART slot of the ontological frames for complex EVENTS—they are not simple slot-filler knowledge of the type illustrated earlier.

So, what does a script look like? The simplest way to answer is by example. Below is a tiny excerpt—in its original, unsimplified, format—from the script that supports the interactive simulation of virtual patients experiencing gastroesophageal reflux disease (GERD) in the Maryland Virtual Patient (MVP) application of LEIAs (see chapter 8 for further discussion).

```
(tracks
   (if (>total-time-in-acid-reflux 1.2)
      Then
      (bind-variables
         (fraction-through (- 1.0 (/ (- time-since-start time-increment) gerd-time)))
         (gerd-t-duration-days (get-attribute gerd-t-duration gerd-patient 1000))
         (extra-time (fraction-through gerd-t-duration-days 60 60 24)))
      (effect
```

```
            (unassert-background-script heal-preclinical-gerd $var0)
            (assert-background-script preclinical-gerd $var0
                ((extra-time (+extra-time time-increment))))))
    (if (and    (>=time-since-start gerd-time)
                (<=total-time-in-acid-reflux 1.2))
        Then
        (effect
            (unassert-background-script heal-preclinical-gerd $var0))))
```

The complication with presenting this example is that this script illustrates difficult issues of dynamic (in this case, physiological) simulation, including time management, feature-value checking and updating, cause-effect relationships, and the assertion and unassertion of interrelated scripts. Rather than simplifying the format, as we do with other structures we illustrate, it was easier to present it in its internal Lisp form and accept that it reveals as much about engineering as it does about the underlying ontological knowledge.

Not all scripts must support dynamic simulations. There are also more familiar workflow scripts that describe, for example, how a doctor should go about diagnosing a patient. These were used by the mentoring agent in the MVP application, whose task was to watch the actions of the user (who played the role of attending physician) and determine whether they conformed to good clinical practice. The same formalism is used for representing both of the above kinds of scripts.[10]

Scripts are used to support agent reasoning, including reasoning about language. Since this book is centrally about language, it is this kind of reasoning support that we are interested in. The following example illustrates the use of scripts in language-oriented reasoning. Consider the interaction: *"How was your doctor's appointment?" "Great! The scale was broken!"* Why does the second speaker say *the scale*? What licenses the use of *the*, considering that this object was not previously introduced into the discourse? The mention of a doctor's appointment prepares the listener to mentally access objects (like *scale*) and events that are typically associated with a doctor's appointment, making those objects and events primed for inclusion in the situation model (discourse context). In fact, the linguistic licensing of *the* with *scale* is evidence that such script activation actually takes place. Of course, it is our script-based knowledge that also explains why the person is happy, and it further allows us to infer the body type of the speaker. If we want LEIAs to be able to reason at this level as well, then scripts are the place to store the associated knowledge.

For further discussion of issues related to the content and acquisition of ontology, see the deep dive in section 2.8.2.

2.3.2 The Lexicon

A LEIA's lexicon for any language maps the words, phrases, and constructions of the language to the concepts in the ontology. The defining features of the current English lexicon are as follows:

- It contains both syntactic and semantic descriptions, linked by variables.
- It contains around 30,000 word senses.
- Open-domain vocabulary is covered, with some areas of specialization reflecting past application areas.
- Most semantic descriptions express meaning using ontological concepts, either by directly mapping to a concept or by mapping to a concept and then modifying it using property-based constraints. However, the meaning of some words, like the adverb *respectively,* must be dynamically computed in each context. In such cases, the lexical description includes a call to a procedure to carry out the necessary computation.
- During lexical acquisition, acquirers attempt to include sufficient constraints to enable the system to disambiguate the words of input at runtime. It would make no sense for a computational lexicon to split senses as finely as many human-oriented lexicons do if the agent has no way of choosing between them.
- To date, we have made it a priority to acquire frequent and semantically complex argument-taking words, such as *have* and *make*, because preparing agents to treat those hard cases is at the core of the scientific work. Acquiring a large number of nouns, such as kinds of birds or trees, is much simpler and could be done by less-trained individuals (as resources permit)—and even automatically by the agent itself through learning by reading.
- The lexicon covers all parts of speech: noun, verb, adjective, adverb, conjunction, article, quantifier, relative pronoun, number, pronoun, reflexive pronoun, auxiliary.
- It accommodates multiword expressions and constructions of any structure and complexity, as described in section 4.3.

Although we already briefly introduced the lexicon, its role in NLU is so important that we will present a few more example entries to reinforce the main points. Note that in these lexicon examples, we make explicit the ontological constraints on case roles that are drawn from the ontology, as they are accessible to the system when it processes inputs, and these are key to understanding how automatic disambiguation works. The first example juxtaposes two verbal senses of *address.*

```
address-v1
   def.   to talk to—usually, formally
   ex.    He addressed the audience.
   syn-struc
      subject        $var1
      v              $var0
      directobject   $var2
   sem-struc
      SPEECH-ACT
         AGENT          ^$var1 (sem HUMAN)
         BENEFICIARY    ^$var2 (sem HUMAN) (relaxable-to ANIMAL)
```

address-v3
  def. to consider, think about
  ex. He addressed the problem.
  syn-struc
    subject $var1
    v $var0
    directobject $var2
  sem-struc
    CONSIDER
      AGENT ^$var1 (sem HUMAN)
      THEME ^$var2 (sem ABSTRACT-OBJECT)

Syntactically (as shown in the syn-struc zones), both senses expect a subject and a direct object in the active voice, filled by the variables $var1 and $var2, respectively. However, the meanings of the direct objects are constrained differently, as shown in the respective sem-strucs. In address-v1 the meaning of the direct object (^$var2) is constrained to a HUMAN or, less commonly, an ANIMAL, whereas in address-v3 the meaning of the direct object is constrained to an ABSTRACT-OBJECT. This difference in constraints permits the analyzer to disambiguate. If the direct object in an input sentence is abstract, as in *He addressed the problem*, then *address* will be analyzed as an instance of the concept CONSIDER using address-v3. By contrast, if the direct object is human, as in *He addressed the audience*, then *address* will be analyzed as SPEECH-ACT using address-v1. The semantic roles that each variable fills are explicitly indicated in the sem-struc zone as well. In both of the senses presented here, the meaning of $var1 (^$var1) fills the AGENT role. In address-v1, the meaning of $var2 (^$var2) fills the BENEFICIARY role, whereas in address-v3, the meaning of $var2 (^$var2) fills the THEME role.

The examples above illustrate how lexically recorded *semantic* constraints support disambiguation, given the same syntactic structure. However, *syntactic* constraints can also support disambiguation. Consider the four senses of *see* shown below. The latter two require, respectively, an imperative construction (see-v3) and a transitive construction that includes a PP headed by *to* (see-v4). These syntactic constraints, along with the associated semantic constraints, provide strong heuristic guidance for automatic disambiguation.

see-v1
  def. to perceive visually
  ex. He saw her new car.
  syn-struc
    subject $var1
    v $var0
    directobject $var2
  sem-struc
    INVOLUNTARY-VISUAL-EVENT
      AGENT ^$var1 (sem ANIMAL)
      THEME ^$var2 (sem PHYSICAL-OBJECT)

```
see-v2
    def.   to consult with for advice
    ex.    Grandma saw her doctor.
    syn-struc
        subject        $var1
        v              $var0
        directobject   $var2
    sem-struc
        PROFESSIONAL-CONSULTATION
            AGENT          ^$var2 (sem MEDICAL-ROLE, LEGAL-ROLE)
            BENEFICIARY    ^$var1 (sem HUMAN)

see-v3
    def.   to refer to a portion of text
    ex.    For details, see chapter 3.
    syn-struc
        v              $var0 (form imperative)
        directobject   $var1
    sem-struc
        READ
            THEME      ^$var1 (sem TEXT-UNIT)

see-v4
    def.   to accompany someone somewhere
    ex.    He saw me to my car.
    syn-struc
        subject        $var1
        v              $var0
        directobject   $var2
        pp
            prep       $var3 (root 'to')
            obj        $var4
    sem-struc
        ESCORT
            AGENT          ^$var1 (sem HUMAN)
            BENEFICIARY    ^$var2 (sem HUMAN)
            DESTINATION    ^$var4 (sem PLACE) (relaxable-to PHYSICAL-OBJECT)
        ^$var3    null-sem+
```

A global rule used for disambiguation is to prefer analyses that fulfill more specific (i.e., narrower) constraints. In most cases, this rule works well: after all, when one says *I saw my doctor yesterday*, it typically refers to PROFESSIONAL-CONSULTATION—unless, of course, one adds the adjunct *at a basketball game*, in which case INVOLUNTARY-VISUAL-EVENT is the best choice. As people, we make the latter adjustment based on the knowledge that one consults with physicians in medical buildings, not at basketball games. While such knowledge about where events typically occur is recorded in the LEIA's ontology, we are still working toward compiling a sufficient inventory of reasoning rules to exploit it.

2.3.3 Episodic Memory

Episodic memory records the agent's knowledge of instances of objects and events. This knowledge can result from language understanding, vision processing, the interpretation of stimuli generated by computer simulations, the agent's recording of its own actions, and its memories of its own reasoning and decision-making. Entries in episodic memory are essentially TMRs with some additional metadata. In this book we will not discuss episodic memory in detail, but it is worth noting that managing it involves many important issues of cognitive modeling, such as consolidating information about objects and events presented at different times or by different sources, creating generalizations from repeating events, and modeling forgetting. In NLU, episodic memory is needed to support reference resolution (section 7.7), analogical reasoning (section 6.1.6), and learning (section 8.3). In discussing these capabilities, we will assume that the agent's episodic memory is, essentially, a list of remembered TMRs.

2.4 Incrementality

The original formulation of Ontological Semantics (Nirenburg & Raskin, 2004) was oriented around processing inputs as full sentences.[11] That was natural at the time: the text genre our team concentrated on was formal (typically journalistic) prose; syntactic parsers worked at the level of full sentences; and the applications being served were not time-sensitive. However, over the past fifteen years we realized that the best way for computationally oriented linguists to contribute to AI is in the area of human-AI collaboration. As a result, our research interests have shifted to agent applications, the genre of dialog, and modeling strategies that will allow agents to use language in maximally humanlike ways.

Human language understanding is incremental, as evidenced by behaviors such as finishing other people's sentences, interrupting midsentence to ask for clarification, and undertaking action before an utterance is complete (*Pass me that spatula and ...* [the interlocutor should already be in the process of spatula passing]).[12] Accordingly, LEIAs need to be able to process language incrementally *if this will best serve their goals*. This condition is important: the additional processing demands imposed by incrementality are not always needed, and it would be unwise to make LEIAs always process language incrementally simply because they can. Incrementality is just one of the many tools in a LEIA's NLU toolbox, to be used, as warranted, to optimize its functioning.

Consider the incremental analysis of the input *Audrey killed the motor*, presented, for reasons of clarity, with only a subset of details. The first word of input is *Audrey*. The system's onomasticon contains only one sense of this string, so the nascent TMR is

```
HUMAN-1
    GENDER                 female
    HAS-PERSONAL-NAME      'Audrey'
```

The next word is *killed*, so the combination *Audrey killed* is analyzed. The lexicon currently has five senses of *kill*, but only three of them permit a HUMAN to fill the subject slot:

1. cause to die: *Who do you think killed the guard?* (COCA)
2. cause to cease operating: *She slowed the ATV to a halt and killed the engine.* (COCA)
3. thwart passage of, veto: *You've killed the legislation on tobacco.* (COCA)

The other two senses can be excluded outright since one requires the subject to be an event (*When I was 10 my father died—he was a miner and lung disease killed him.* (COCA)) and the other requires it to be a nonhuman object that serves as an instrument (*The bomb killed the guy next to him …* (COCA)). The fragment *Audrey killed* offers the three equally acceptable TMR candidates, as follows:

TMR candidate 1 for *Audrey killed*

KILL-1
    AGENT   HUMAN-1
    TIME   <*find-anchor-time*

HUMAN-1
    GENDER   female
    HAS-PERSONAL-NAME   'Audrey'

TMR candidate 2 for *Audrey killed*

VETO-1
    AGENT   HUMAN-1
    TIME   <*find-anchor-time*

HUMAN-1
    GENDER   female
    HAS-PERSONAL-NAME   'Audrey'

TMR candidate 3 for *Audrey killed*

ASPECT-1
    PHASE   end
    SCOPE   OPERATE-DEVICE-1

OPERATE-DEVICE-1
    AGENT   HUMAN-1
    TIME   <*find-anchor-time*

HUMAN-1
    GENDER   female
    HAS-PERSONAL-NAME   'Audrey'

The next word of input is *the*. The LEIA does not launch a new round of semantic analysis for the fragment *Audrey killed the* because no useful information can be gleaned from function words without their heads.

The next and final stage of analysis is launched on the entire sentence *Audrey killed the motor*. Each of the three still-viable lexical senses of *kill* includes semantic constraints on the direct object: for sense 1 it must be an ANIMAL; for sense 2, an ENGINE; and for sense 3, a BILL-LEGISLATIVE. Since *motor* maps to the concept ENGINE, sense 2—'cause to cease operating'—is selected, and the final TMR for *Audrey killed the motor* is

```
ASPECT-1
    PHASE        end
    SCOPE        OPERATE-DEVICE-1

OPERATE-DEVICE-1
    AGENT        HUMAN-1
    THEME        MOTOR-1
    TIME         <find-anchor-time

HUMAN-1
    GENDER                  female
    HAS-PERSONAL-NAME       'Audrey'
```

For the above illustration we chose a simple example. In reality, most sentences involve much more midstream ambiguity, resulting in many more candidate analyses. Moreover, it is not unusual for the LEIA to be unable to fully resolve all the ambiguities, even given the full sentence, when using ontological and lexical constraints alone—other contextual information can be needed. The point here is that LEIAs can narrow down analyses midstream, just like people can, as more elements of input become available.

2.5 The Stages of NLU and Associated Decision-Making

The incrementality just described can more precisely be called *horizontal* incrementality, since it involves processing words of input as they appear in the transcribed language stream. It juxtaposes with another important manifestation of incrementality: *vertical* incrementality. This refers to the depth of analysis applied to any input fragment. When a LEIA leverages more context, this can mean either processing more elements of input (horizontal incrementality) or leveraging more knowledge resources and reasoning algorithms to analyze the given elements of input (vertical incrementality). The availability of horizontal and vertical incrementality during NLU is graphically represented in figure 2.1.

As an illustration of these notions of incrementality, consider the following examples, in which underlining separates text chunks that will be consumed sequentially during incremental semantic analysis.

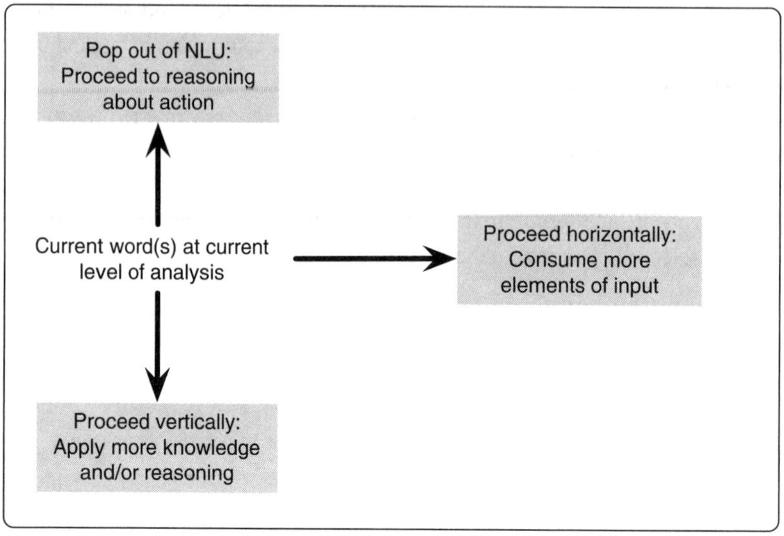

Figure 2.1
Horizontal and vertical incrementality.

(2.1) A black bear ___ is eating ___ a fish.

(2.2) My monkey ___ promised ___ he ___ wouldn't do ___ that ___ anymore!

(2.3) I ___ said, ___ "The mail ___ just came."

If you heard or read sentence (2.1) without being able to look ahead, you would probably have a single interpretation at each stage of input, something like, *A large mammal with black fur // A large mammal with black fur is ingesting // A large mammal with black fur is ingesting an aquatic animal.* You would probably not consider the possibility of nonliteral word senses, implicatures, sarcasm, humor, and the like. There is no need to invoke such extra reasoning since your basic knowledge of language and the world led to an interpretation that worked just fine.

By contrast, (2.2) requires more effort, and more context, to interpret. Since the small simians we call monkeys cannot make promises, either *monkey* or *promise* must be nonliteral. *Monkey* is a more obvious choice since people are often referred to by the name of an animal featuring a contextually relevant characteristic. Our sentence could, for example, be said jokingly about a child who is swinging dangerously from a jungle gym. In addition, although a basic interpretation can be gleaned from the sentence without contextual grounding, its full interpretation requires determining the referents for *my*, *he*, and *that.*

Example (2.3), when taken out of context, has two instances of residual ambiguity: the identity of *I* and the force of *I said.* The latter can be used

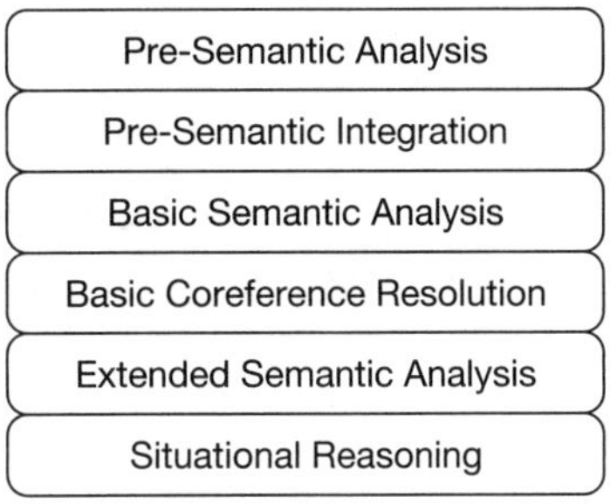

Figure 2.2
Stages of vertical context available during NLU by LEIAs.

- when the interlocutor fails to hear the original utterance, as in a noisy room or over a bad phone connection;
- as part of a story: "I said, 'The mail just came.' And he suddenly leaps out of his chair and barrels out the door!"; and
- to emphasize an indirect speech act that was not acted on originally: "The mail just came. [No reaction] I *said*, the mail just came." (The implication is that the interlocutor is supposed to go fetch it.)

All these interpretations can be computed from general lexical, semantic, and pragmatic knowledge, but choosing among them requires additional features from the speech context.

The point of these examples is that it would make no more sense to have LEIAs invoke deep reasoning to analyze (2.1) than it would to expect them to understand the pragmatic force of *I said* in (2.3) without access to contextual features. Accordingly, a key aspect of the intelligence of intelligent agents is their ability to independently determine which resources to leverage, when, and why, as well as what constitutes a sufficient analysis.

We have found it useful to organize vertical context into the six processing stages shown in figure 2.2.

These stages are detailed in chapters 3–7 but we sketch them below to serve the introductory goals of this chapter.

1. Pre-Semantic Analysis covers preprocessing and syntactic parsing. It is carried out by an externally developed tool set and includes part-of-speech tagging, morphological analysis, constituent and dependency parsing, and named-entity recognition.
2. Pre-Semantic Integration adapts the abovementioned heuristic evidence so that it can best serve semantic/pragmatic analysis. Component functions establish linkings between input strings and lexical senses; reambiguate certain decisions that are inherently semantic and, therefore, cannot be made confidently during syntactic analysis (prepositional phrase attachments, nominal-compound bracketing, and preposition/

particle tagging); attempt to recover from noncanonical syntactic parses; and carry out the first stage of new-word learning.

3. Basic Semantic Analysis carries out lexical disambiguation and semantic dependency determination for sentences taken individually. It computes what some call "sentence semantics" by using the static knowledge recorded in the lexicon and ontology. It often results in residual ambiguity (multiple candidate analyses) and often contains underspecifications (e.g., *he* is some male animal whose identity must be contextually determined).
4. Basic Coreference Resolution includes a large number of functions aimed at identifying textual coreferents for overt and elided referring expressions. During this stage, the agent also reconsiders lexical disambiguation decisions in conjunction with what it has learned about coreference relations.
5. Extended Semantic Analysis invokes more knowledge bases and more reasoners to improve the semantic/pragmatic analysis of not only individual sentences but also multisentence discourses.
6. Situational Reasoning attempts to compute all outstanding aspects of contextual meaning using the agent's situational awareness—that is, its interpretation of nonlinguistic percepts, its knowledge about its own and its interlocutor's plans and goals, its mindreading of the interlocutor, and more.

For purposes of this introduction, two aspects of these stages deserve further comment: (a) the first five stages form a functional unit, for both theoretical and practical reasons, and (b) the conclusion of each stage represents a decision point for the agent. We consider these issues in turn.

a. *The first five stages form a functional unit* since they represent all of the language analysis that the agent can bring to bear without having specialized knowledge of the domain or carrying out situational reasoning, which can go far beyond language per se (involving plans, goals, mindreading, and so on). We find Harris's (in press) notion of "semantic value" a useful description of this level of processing—although, his definition assumes that sentences are considered individually, whereas LEIAs can analyze multisentence inputs together using these first five stages. Harris defines a sentence's semantic value as "not its content but a partial and defeasible constraint on what it can be used to say."

For example, *Give him a shot* can refer to a medical injection, a scoring opportunity, the imbibing of a portion of alcohol, or the opportunity to carry out some unnamed action that the person in question has not yet had the opportunity to try. (The fact that, given this input in isolation, individual people might not recognize the ambiguity and might default to a particular analysis reflects aspects of their individual minds, experiences, and interests.) If the agent has access to the preceding context, and if that context happens to mention something from the medical realm (e.g., a patient, a hospital), then the agent will prefer the *injection* interpretation. But if this evidence is not available, then residual ambiguity is entirely correct as the result of the fifth stage of processing.

Apart from being theoretically justified, there is a practical reason to bunch stages 1–5: these, but not stage 6, can be profitably applied to texts in the open domain as a means of validating and enhancing microtheories (see chapter 9).

b. *The conclusion of each stage represents a decision point for the agent*, represented by the control flow in figure 2.3. Decisions about actionability rely on the particular plans and goals of a particular agent at a particular time—a topic discussed in chapter 8.

Although it is, in some respects, premature to describe the agent's post-stage decision-making before detailing what happens in each stage, we have found in teaching this material that it is important to motivate, right from the outset, why the stages are formally delineated rather than merged into one all-encompassing program. We will provide this motivation using examples, with the caveat that readers are not expected to understand all the details. Instead, they should aim to understand the gist—and then plan to return to this section later on, having absorbed the material in chapters 3–7.

Specifically, each subsection below provides multiple examples of the decision points labeled 1–6 in figure 2.4.

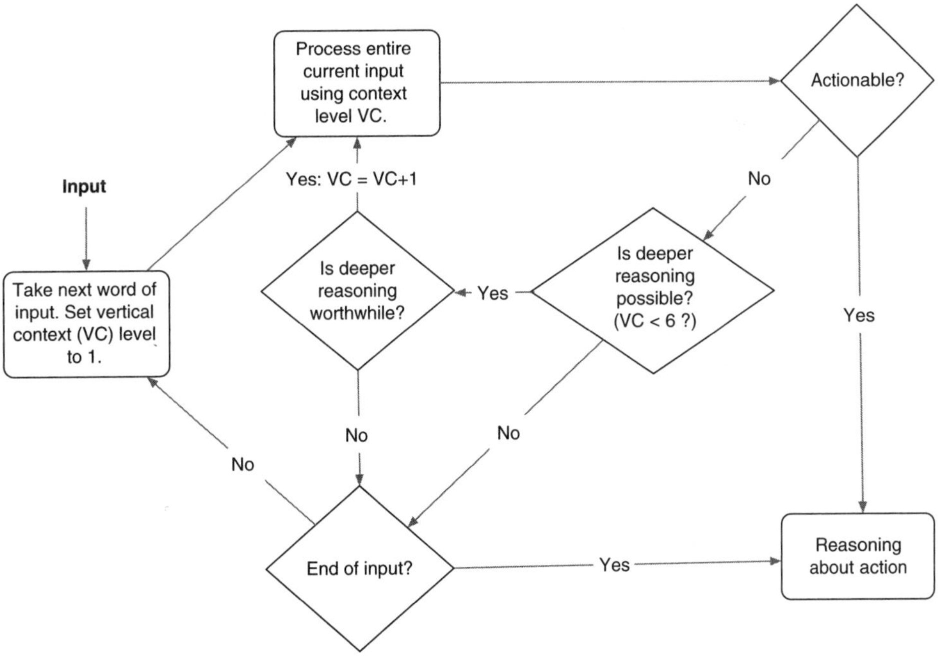

Figure 2.3
The control flow of decision-making during semantic analysis.

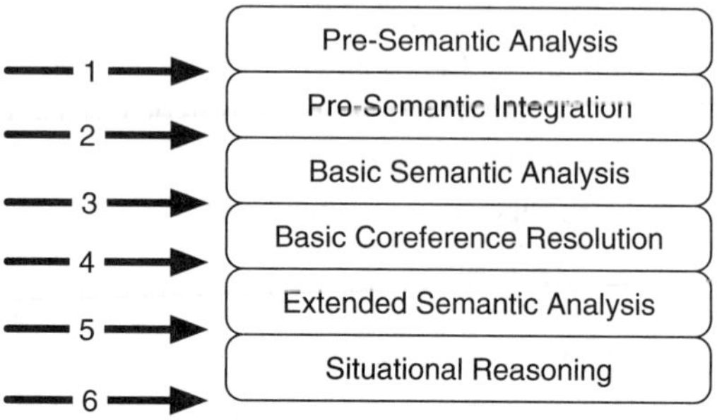

Figure 2.4
Decision points during vertical-incremental processing.

Some of the examples are relevant for most applications, whereas others envision applications of a particular profile. For example, high-risk applications require the agent to be more confident in its analyses than do low-risk applications. Similarly, applications that are expected to involve extensive off-topic exchanges impose different challenges than those in which participants are expected to reliably stay on topic.

2.5.1 Decision-Making after Pre-Semantic Analysis

Example 1. The agent is a furniture-assembly robot. It has multiple human collaborators who chat about this and that to pass the time. The agent is modeled to "skim" what it hears and only semantically analyze inputs that might be task relevant. To operationalize this skimming, the agent has a precompiled list of words and phrases of interest. The list was generated by searching the lexicon for all words and/or phrases that map to concepts of interest. Since the task is not urgent, the agent does not use horizontal incrementality—it waits until the end of each sentence to do any processing at all. At that point, it runs Pre-Semantic Analysis and, thanks to the morphological analyzer, it obtains base (dictionary) forms for all input words. It compares those with its list of words and phrases of interest. If there is little to no overlap, it treats the input as actionable with the action being *Ignore this utterance; it is outside of purview.* This will be the agent's decision when, for example, it hears, "Did you see the Pens last night? The third period was a heartbreaker!"[13]

Example 2. The agent is a military robot engaged in combat. It works with only one human teammate at a time, so there is no chance of off-topic conversation. The task is maximally high-risk and time-sensitive. The agent must understand absolutely everything its human teammate says. The agent analyzes each word immediately (horizontal incrementality) and as deeply as possible (vertical incrementality). At the stage of Pre-Semantic Analysis, if any word of input cannot be clearly recognized (e.g., due to background noise), the agent interrupts immediately for clarification.

Example 3. The agent is tasked with expanding its lexicon off-line. In order to focus on specific words, it must skim a lot of text. The morphological analyzer and part-of-speech tagger, which count among the preprocessing tools, can identify sentences that contain word forms of interest. Having compiled a set of potentially interesting contexts using these shallow analysis methods, the agent can then apply deeper analysis to them in service of learning.

2.5.2 Decision-Making after Pre-Semantic Integration

Example 1. Returning to our furniture-building robot, we said that one way to implement skimming was to precompile a list of words and phrases of interest and check new inputs against it. Another way to do this (without precompiling a list) is to carry out Pre-Semantic Integration, during which the agent looks up the words of input in its lexicon and can determine if they map to concepts of interest.

Example 2. Returning to our military robot who must fully understand everything, if it realizes, as a result of lexical lookup, that a word is missing from its lexicon, it can interrupt its human partner and immediately initiate a new-word-learning dialog rather than engage in the normal, multistage process of new-word learning. The latter process would require that the agent wait until the end of the utterance to attempt the learning—which might be too long, depending on the urgency of the application.

Example 3. Three of the procedures launched during Pre-Semantic Integration attempt to recover from bad (irregular, incomplete) parses. If those procedures (see sections 3.2.2, 3.2.4, and 3.2.6) work reasonably well, then processing can proceed as usual. However, if they don't, then the typical syntax-informs-semantics approach to NLU will not work. In such cases, the agent will skip stages 3–5 and proceed directly to stage 6, Situational Reasoning, where it will attempt to cobble together the intended meaning using minimal syntactic evidence (essentially, noun-phrase boundaries and word order) supported by extensive semantic and pragmatic analysis. This decision is used by all agents in all applications since there is no way to strong-arm stages 3–5 given an irreparably bad syntactic parse.

Example 4. Many agents will be tasked with learning new words on the fly. If there is a maximum of one unknown word per clause, this learning process has a chance of working well. If, by contrast, there are multiple unknown words in a given clause, it is unlikely that new-word learning will produce useful results. So, upon detecting the multiple-unknown-words case, a learning agent can choose to either ignore the sentence (under the assumption that so many unknown words would make it out of purview) or tell its human teammate that the input contains too many unknown words to be treated successfully.

2.5.3 Decision-Making after Basic Semantic Analysis

Example 1. Back to our chair-building robot, one of its tasks is to determine which inputs can be ignored because they are out of purview. We already saw two methods of making this determination. This stage presents us with a third, which offers even higher confidence. The agent can go ahead and carry out Basic Semantic Analysis—which usually results in multiple candidate analyses—and check whether any of them aligns with its in-purview ontological scripts. This method is more reliable than previous ones because it looks not only for concepts of interest but for combinations of concepts of interest. For example, the humans in the scene might be having off-topic conversations about chairs—for example, they might be discussing the selection of chairs for someone's new home. So the mention of chairs does not guarantee that the input is relevant to the agent.

Example 2. For some inputs, Basic Semantic Analysis is sufficient to generate a full and confident interpretation—as for "Chairs are so easy to build!" This sentence matches a construction (X is (ADV) easy to Y) that guides semantic analysis. The words *chair* and *build* are ambiguous, but the highest-scoring semantic analysis combines the physical meaning of *build* with the furniture meaning of *chair*. None of the entities requires coreference, and there are no indicators that a nonliteral interpretation should be sought, so the analysis is finished and the agent will delve no deeper.

Example 3. Consider, again, the lexicon-expansion agent that has identified sentences of interest (thanks to decision-making after Pre-Semantic Analysis) and must now analyze them more deeply. In most cases, Basic Semantic Analysis is sufficient to support new-word learning. The reason is that, given a large corpus, the agent can choose to avoid difficult examples, such as those that require coreference resolution. For example, it can learn the meaning of *cupuacu* from the simple and direct "Some people think that *cupuacu* tastes good" rather than attempt to figure it out from the trickier "There are many exotic fruits that most Americans have never tried. Take, for example, *cupuacu*."

2.5.4 Decision-Making after Basic Coreference Resolution

Example 1. Basic Semantic Analysis was sufficient to generate a full and confident interpretation for some inputs (e.g., "Chairs are so easy to build!"). Basic Coreference Resolution expands the inventory of inputs that are fully and confidently interpreted. For example, if we modify our example to "Chairs are great; they are so easy to build!", establishing the coreference between *chairs* and *they* is all that is needed on top of Basic Semantic Analysis to generate a full and confident interpretation. Similarly, verb phrase ellipsis in the following example can be resolved during Basic Coreference Resolution, resulting in a full interpretation of "Disassembling chairs is frustrating but it's sometimes necessary."

Example 2. We said earlier that our lexicon-expansion agent will generally avoid inputs that contain underspecified referring expressions, such as pronouns. However, it can choose to include such inputs in its corpus for learning if (a) the coreference relations can be easily and confidently established or (b) the word is so rare that the only available examples include pronouns. Returning to our example with *cupuacu*, a sentence like "Some people like *cupuacu* and eat it regularly," the coreference between *it* and *cupuacu* can be reliably established, so that the agent can use the understood proposition *they eat cupuacu* as its example for learning.

2.5.5 Decision-Making after Extended Semantic Analysis

Example 1. Continuing with the lexicon-expansion agent, after Basic Semantic Analysis, even supplemented by Basic Coreference Resolution, the agent will often be dealing with as-yet incomplete analyses involving residual ambiguity (i.e., multiple viable candidates), incongruity (no high-scoring candidates), or underspecification. This stage offers many functions to improve such analyses. For example, the agent has methods to select the correct meaning of *hand* in "If you see a mechanical clock on the wall and its *hands* are not moving, try winding it and resetting it." It will also be able to fully analyze the nominal compound *physician neighbor* found in an example like "My physician neighbor came straight over when I called saying that my dog was bleeding" (the compound means HUMAN (HAS-SOCIAL-ROLE PHYSICIAN, NEIGHBOR)). More accurate analyses support more correct and more specific new-word learning.

Example 2. These first five stages of analysis can be launched on texts in any domain, outside of comprehensive agent applications. This is because a considerable amount of reasoning related to language understanding relies on basic knowledge of lexicon and ontology. Therefore, whenever language understanding is undertaken outside of a full-fledged agent application, if the application is not time-sensitive, the most natural decision is to run the inputs through Extended Semantic Analysis to achieve the best possible outcome.

2.5.6 Decision-Making after Situational Reasoning

Example 1. The chair-building robot will need to ground many referring expressions in the inputs that are relevant to it: for example, "Grab that hammer and give it to me." It will also need to carry out such grounding as it continues the task of determining which inputs are off-topic. For example, someone's dialog turn might be, "I think we should move all the chairs there first, since they are light," which may or may not be relevant to the agent depending on the referents for *we*, *the chairs*, and *there*. Once these referents have been established, the agent can determine that, in the following context, this utterance is outside of purview: "My wife and I decided to remodel the first floor of our house. When we refinish the floors, we are thinking of putting all the furniture in the basement to get it out of the way. I think we should move all the chairs there first, since they are light." This is part of Situational Reasoning, not Basic Coreference Resolution, because of the grounding needs.

Example 2. Our envisioned military robot is likely to process many inputs that require full reference resolution, defined as linking mentioned entities to both the real-world environment and to its short-term memory. These are both part of Situational Reasoning. For example, if the robot is told, "Pick up that grenade and throw it into the building," the robot will need to ground the referents of *that grenade* and *the building* (it will have coreferred *it* with *the grenade* during Basic Coreference Resolution). If the agent has any doubt about its interpretation, it will need to clarify with its human teammate.

Example 3. Human dialog is often syntactically irregular and fragmentary. As mentioned with respect to the decision-making after stage 2, syntactically irregular inputs that cannot be normalized or worked around to a reasonable degree will sidestep stages 3–5. At this stage, the agent will try to make sense of them by cobbling together candidate semantic analyses and comparing those candidates against the expectations of task-relevant scripts. For example, upon hearing the input, "That screwdriver … no, wait, maybe a hammer is better? Yeah, give me one," the agent will need to understand that the intended command is "Give me a hammer." If the agent is capable of doing only a limited number of things, among which is giving a human a hammer, it can proceed to act on this interpretation.

This concludes our brief and, in some sense, premature overview of agent decision-making during language analysis. The goals were modest: to provide an initial motivation for dividing the process of NLU into distinct stages and to highlight the kinds of decisions the agent must make to proceed after each one.

2.6 Microtheories

We use the term *microtheories* to refer to computational linguistic models that address individual issues of language processing, such as word sense disambiguation, coreference resolution, and indirect speech act interpretation—to name just a few. Microtheory development involves an initial top-down analysis of the whole problem space (or a realistic approximation of it) followed by deeper, algorithm-supported treatments that are implemented and iteratively enhanced. Among the methodological preferences in building microtheories are striving for a natural, waste-free progression over time (rather than hacking partial solutions that will necessarily be thrown away) and prioritizing phenomena that are most readily treatable and/or most urgent for a particular application or research goal.

It would be impossible to overstate how great a distance there can be between simple and difficult manifestations of a linguistic phenomenon. Consider nominal compounding. Analyzing a compound can be as simple as looking up a sense stored in the lexicon (*coffee cup*) or as difficult as attempting to learn the meanings of two unknown words on the fly before semantically combining them in a context-appropriate way. The difference in timelines for achieving confident results across so broad a spectrum underscores why it is so important to at least sketch out the entire problem space when initially developing a microtheory.

Saying that linguistic phenomena will be treated over time does not imply that we expect each one to submit, quickly and gracefully. Quite the opposite. The problems will only get more difficult as we approach the hardest 5% of instances. Our objective is to make LEIAs well-rounded enough in their situational understanding to be able to adequately deal with the hardest cases in the same ways that a person does when faced with uninterpretable content (even though what is uninterpretable for LEIAs will, in many cases, be different from what is uninterpretable for people).

Computational linguistics, like its parent discipline, computational cognitive science, is an applied science. This means that it leverages scientific understanding in service of practical applications. However, computational linguistics presents not a clear dichotomy between *theory* and *system*[14] but more of a trichotomy between *theory*, *model*, and *system*. As a first approximation, theories in cognitive science are abstract and formal statements about how human cognition works; models account for real data in computable ways and are influenced as much by practical considerations as by theoretical insights; systems, for their part, implement models within the real-world constraints of existing technologies. Since this trichotomy is at the heart of our story, let us flesh it out just a bit further.[15]

Theories. The most general statement of our view is that theories attempt to explain and reflect reality as it is, albeit with great latitude for underspecification. Another important property of theories for us is that they are not bound by practical concerns such as computability or the attainability of prerequisites. We share the position, formulated in Winther (2016), that "laws of nature are rarely true and epistemically weak. Theory as a collection of laws cannot, therefore, support the many kinds of inferences and explanations that we have come to expect it to license." This position ascends to Cartwright's (1983) view that "to explain a phenomenon is to find a model that fits it into the basic framework of the theory and that thus allows us to derive analogues for the messy and complicated phenomenological laws which are true of it" (p. 152).

Theories guide developers' thinking in developing models and interpreting their nature, output, and expectations. In our work we are guided by the theory of Ontological Semantics (Nirenburg & Raskin, 2004) that proposes the major knowledge and processing components of the phenomena in its purview, which is language understanding. However, the lion's share of our work is on developing models (microtheories) and systems.

Models. Computational cognitive models of language processing formally describe specific linguistic phenomena as well as methods for LEIAs to process occurrences of these phenomena in texts.[16] The most important property of such models is that they must be computable. This means that they must rely exclusively on types of input (e.g., property values) that can actually be computed using technologies available at the time of model construction. If some feature that plays a key role in a theory cannot be computed, then it either must be replaced by a computable proxy, if such exists, or it must be excluded from the model. In other words, models, unlike theories, must include concrete decision algorithms and computable heuristics. To take just one example from the realm of pronominal coreference, although the notions *topic* and *comment* figure prominently in theoretical-linguistic descriptions of coreference, they do not serve the modeling enterprise since their values cannot be reliably computed in the general case. (To date, no adequate model has been proposed for deriving such values.)

Models must account for the widest possible swath of data involving a particular linguistic phenomenon—which is a far cry from the neat and orderly examples found in dictionaries, grammars, and textbooks.

Models should embrace well-selected simplifications, drawing from the collective experience in human-inspired machine reasoning, which has shown that it is counterproductive to populate decision functions with innumerable parameters whose myriad interactions cannot be adequately accounted for (Kahneman, 2011).

Models of natural language should reflect the fact that people are far from perfect both in generating and in understanding utterances, and yet successful communication is the norm rather than the exception. So, models of language processing need to account for both the widespread imperfection and the overwhelming success of language use. The notions of *cognitive load* and *actionability* are instrumental in capturing this aspect of modeling. Cognitive load describes how much effort humans have to expend to carry out a mental task. As a first approximation, a low cognitive load for people should translate into a simpler processing challenge for machines and, accordingly, a higher confidence in the outcome. Of course, this is an idealization since certain analysis tasks that are simple for people (such as reasoning by analogy) are quite difficult for machines; however, the basic insight remains valid. Actionability captures the idea that people can often get by with an imperfect and incomplete understanding of both language and situations. So, the approximations of cognitive load, and the associated confidence metrics, do not carry an absolute interpretation. For example, in the context of off-task chitchat, an agent might decide to simply keep listening if it doesn't understand exactly what its human partner is saying since the risk of incomplete understanding is little to none. By contrast, in the context of military combat, anything less than full confidence in the interpretation of an order to aggress will necessarily lead to a clarification subdialog to avoid a potentially catastrophic error.

Finally, models must operationalize the factors identified as most important by the theory. Cognitive load and actionability provide useful illustrations of this requirement. The cognitive load of interpreting a given input can be estimated using a function that considers the number and complexity of each contributing language-analysis task. Let us consider one example from each end of the complexity spectrum. The sentence *John ate an apple* will result in a low-complexity, high-confidence analysis if the given language understanding system generates a single, canonical syntactic parse, finds only one sense of *John* and one sense of *apple* in its lexicon, and can readily disambiguate between multiple senses of *eat* given the fact that only one of them aligns with a human agent and an ingestible theme. At the other end of the complexity (and confidence) spectrum is the analysis of a long sentence that contains multiple unknown words, does not yield a canonical syntactic parse, and offers multiple similarly scoring semantic analyses of separate chunks of input.

The above demonstrates the importance of gauging the simplicity of language material. Whether we are building a model of verb phrase ellipsis resolution, nominal compound interpretation, lexical disambiguation, or new-word learning, we can start by asking, Which kinds of attested occurrences of these phenomena are *simple*, and which feature values manifest that simplicity? Then we can start our model development with the simpler phenomena and proceed to the more complicated ones once the basics of the nascent microtheory have been sketched.

"Simpler-first" modeling can be guided by various linguistic principles, such as parallelism and prefabrication (e.g., remembered expressions and constructions). Consider, in this regard, the pair of examples (2.4) and (2.5), which illustrate the type of verbal ellipsis called *gapping*.

(2.4) Delilah is studying Spanish and Dana __, French.

(2.5) ? Delilah is studying Spanish and my car mechanic, who I've been going to for years, __, fuel-injection systems.

Gapping is best treated as a *construction* (a prefabricated unit) that requires the overt elements in each conjunct (i.e., the arguments and adjuncts) to be syntactically and semantically *parallel*. It also requires that the sentence be relatively *simple*. The infelicity of (2.5), indicated by the question mark, results from the lack of simplicity, the lack of syntactic parallelism (the second clause includes a relative clause not present in the first), and the lack of semantic parallelism (languages and fuel-injection systems are hardly comparable).

Simpler-first modeling carries one absolute requirement: that the models enable agents to independently determine which examples are covered by the model and with what confidence value. There is no oracle to tell agents that they *can* understand *this* example, but *that* example is too hard. Every model described in this book includes methods for automatically detecting which examples are covered and with what confidence. (This topic is explored in particular depth in section 9.3.) Of course, agents need to treat every input, by hook or by crook, but the treatment can involve "consciously" generating low-confidence analyses. How agents will act on such analyses will be decided in non-NLU modules of the cognitive architecture.

Computing confidence in overall language analysis is not a simple matter. It requires establishing the relative importance of all subtasks that contribute to overall processing and accounting for their interactions. For example, even if an input's syntactic parse is suboptimal, an agent can be confident in a candidate interpretation if the latter works out semantically and is situationally appropriate—meaning that it aligns with the agent's expectations about what should happen next in a workflow script.

As concerns actionability, it can only be judged on the basis of an agent's assessment of its current plans and goals, its assessment of the risk of a mistake, and so on—which means that the modeling of language must necessarily be integrated with the modeling of all other cognitive capabilities (see chapters 7 and 8).

Systems. The transition from models to systems moves us yet another step away from the neat and abstract world of theory. Models are dedicated to particular phenomena, while the overall task of natural language understanding involves the treatment of many phenomena in a single process. Thus, the first challenge of building comprehensive NLU systems is integrating the computational realizations of the models of individual phenomena. This, in turn, requires managing inevitable cross-model incompatibilities. Since the idea of cross-model incompatibilities might not be self-evident, we will unpack it.

Any program of R&D must take into account economy of effort. In the realm of knowledge-based NLU, if components and tools for computing certain heuristic values exist, then developers should at least consider using them. However, importation comes at a cost. Externally developed components and tools are likely to implement different explicit or implicit linguistic models, thus requiring an added integration effort. For example, different systems rely on different inventories of parts of speech, syntactic constituents, and semantic dependencies. So, if an off-the-shelf preprocessor and syntactic parser are to be imported into an NLU system, the form and substance of the primitives in the source and target models must be aligned—which not only requires a significant effort but also often forces modifications to the target model, not necessarily improving it.

There is no generalized solution to the problem of cross-model incompatibility since there is no single correct answer to many problems of language analysis—and humans are quite naturally predisposed to hold fast to their individual preferences. So, dynamic model alignment is an imperative of developing computational-linguistic systems that must be proactively managed. However, its cost cannot be underestimated. We are talking here not only about the initial integration effort. The output of imported modules often requires modification, which requires additional processing. (See section 3.2.5 for examples.) In fact, the cost of importing processors has strongly influenced our decision to develop most of our models and systems in-house.

Another challenge of system building is that all language processing subsystems—be they imported or developed in-house—are error-prone. Even the simplest of capabilities, such as part-of-speech tagging, are far from perfect at the current state of the art. This means that downstream components must anticipate the possibility of upstream errors and prepare to manage the overall cascading of errors—all of which represents a conceptual distancing from the model that is being implemented.

Because of the abovementioned and other practical considerations, implemented systems are unlikely to precisely mirror the models they implement. This complicates the task of assessing the quality of models. If one were to seek a "pure" evaluation of a model, the model would have to be tested under the unrealistic precondition that all upstream results used by the system that implemented the model were correct. In that case, any errors would be confidently attributed to the model itself. However, meeting this precondition typically requires human intervention, and introducing such intervention into the process would render the system not truly computational in any interesting or useful sense of the word. The system would amount to a theory-model hybrid rather than a computational linguistic system. So, as long as one insists that systems be fully automatic (we do), any evaluation will be namely an evaluation of a system, and, in the best case, it will provide useful insights into the quality of the underlying model.

To summarize this section on microtheories: Our microtheories are explanatory, broad-coverage, heuristic-supported treatments of language phenomena that are intended to be implemented and enhanced over time.

An obvious question is, *Why not import microtheories?* It would be a boon if we could, but, unfortunately, the majority of published linguistic descriptions either are not precise enough to be implemented or rely on preconditions that cannot be automatically fulfilled—as discussed in chapter 1. If our vision of Linguistics for the Age of AI takes hold, linguists *will* take up the challenge of developing these kinds of microtheories, which we will be only too happy to import.

2.7 "Golden" Text Meaning Representations

The idea of "golden" (also known as gold and gold-standard) TMRs comes from the domain of corpus annotation. Golden annotations are those whose correctness has been attested by people, either because people manually created the annotations to begin with or because they checked and corrected the results of automatic annotation. Golden annotations are a foundation of supervised machine learning.

Analogously, we call a TMR golden if, in generating it, (a) the LEIA has leveraged the current state of the knowledge bases correctly and (b) those knowledge bases, along with supporting NLU rule sets, were sufficient to generate a high-quality analysis (McShane et al., 2005c).

However, it is not the case that the LEIA will always, over time, generate the same golden TMR for a given input. The reason is that knowledge engineering is an ongoing process, and knowledge can be recorded in different ways and at different grain sizes. This means that the golden TMR generated for a given input in 2020 might not be exactly identical to the one generated in 2025.

For example, at the time of writing, LEIAs have not been used in applications specifically involving hats, so the English words for all kinds of hats (*beret*, *fedora*, *baseball cap*, and so on) are mapped to the concept HAT with no distinguishing property values listed. After all, it takes time and energy to record all those property values. The current analyzer, therefore, generates the same TMR for the sentence variants "The hat <beret, fedora, baseball cap> [i.e., HAT] blew off his head." However, if a hat manufacturer came along in 2024 with a request for a hat-sales app, then the ontological tree for HAT might be expanded into all subtypes with their relevant property values. The 2025 analyzer would then generate different analyses for the variants of our hat example.

All of this might not seem important until one considers the potential utility of TMRs for the long-term support of both LEIAs themselves and the NLP community at large. As we describe in section 6.1.6, LEIAs can use their repository of past TMRs to help with certain aspects of analyzing new inputs. However, the sliding scale of correctness must factor into this analogical reasoning. Similarly, a sufficiently large repository of TMRs, created over time, could seed machine learning in service of subsequent LEIA-style NLU, in the way that manual annotation efforts currently do—but again, with the caveat that levels of precision might be different. In short, golden TMRs could serve as excellent, semantically oriented corpus annotations, but they would have to be linked to the specific version of each knowledge base used to generate them, since *correct* with one knowledge base might not be with another.

An interesting issue with respect to golden TMRs involves paraphrase, both in language and in the ontological metalanguage. We address this topic in the deep dive in section 2.8.4.

2.8 Deep Dives

2.8.1 The LEIA Knowledge Representation Language versus Other Options

Over the history of AI, many opinions about the optimal relationship between natural language (NL) and knowledge representation languages (KRLs) have been expressed, and many approaches to KRLs have been tried. There are at least three substantially different ones, along with many variations on the themes. We briefly review them below.

1. *NL →KRL →NL translation*. This is the approach advocated in this book and, for purposes of this survey, it is sufficient to list its key advantages and challenges. *Advantages*: The NL input and output are readable by people. The KRL is readable by trained people. NL knowledge bases are available as the source of translations into the KRL. The KRL is suitable for reasoners. The KRL representations are language independent, thus fostering multilingual applications. *Challenges*: The KRL must be expressive and influenced by NL. The translation requires large, high-quality knowledge bases, high-quality analyzers and generators, and extensive reasoning about language, the world, and the situation.

2. *NL is KRL (or KRL is NL)*. This position ascends to Wittgenstein (1953) and essentially declares that the symbols used in KRLs are ultimately taken from NLs and cannot be taken from anywhere else. As a result, KRLs retain NL-like features, such as ambiguity and vagueness, no matter how carefully these languages are designed. The NL-is-KRL movement among computational logicians and proponents of controlled languages is relatively recent and has been formulated as a research program: How to make KRLs more NL-like? Yorick Wilks believes that KRLs are already, in the final analysis, NLs. He has consistently argued (e.g., Wilks 1975, 2009) that knowledge representations are, by nature, ambiguous and vague and that it is, in principle, impossible to eliminate such "language-like features" as ambiguity from ontologies and conceptual structures. This position sounds the death knell for standard automatic reasoning techniques because it essentially states that any automatic reasoning will be indeterminate.

At the same time, Wilks claims, and rightly so, that ambiguous, incomplete, and inconsistent knowledge resources can be, and still are, useful for NLP. As his main example, he cites WordNet (Miller, 1995), as it has been used widely by corpus-based NLP practitioners even though it is demonstrably challenging when used as a knowledge base for NLP (Nirenburg et al., 2004). The catch is that the types of applications Wilks has in mind rely only partially on semantics. Take the example of a personal conversational assistant (Wilks, 2004; Wilks et al., 2011), which can, to a degree, fake understanding and, when needed, change the topic of conversation to something that it can better handle. This is an admirable sleight-of-hand strategy that works in applications whose purpose is largely phatic communication. Strategically similar approaches can pass the responsibility for understanding to a human in the loop: for example, people can often make sense of noisy output from machine translation and summarization systems. However, such detour strategies will not work when understanding by the intelligent agent is crucial. We have commented in detail on the NL-is-KRL opinion in Nirenburg and Wilks (2001) and Nirenburg (2010); here we will limit our remarks to a few relevant methodological points.

While Wilks cited the extraction and manipulation of text meaning as the major scientific objective of his work, ambiguity of representation was not a central issue for him. The desire to nudge the evaluation results of systems like word sense disambiguation engines into the 90% range (cf. Ide & Wilks, 2006) led him to claim that, for NLP, only word sense distinctions at the coarse-grained level of homographs are important.[17] Such a claim may work in the world of Semeval (https://www.wikiwand.com/en/SemEval) and similar competitions but, in reality, the situation with sense delimitation is much murkier.

For example, the English word *operation* has eleven senses in the *American Heritage Dictionary* and (so far) three senses in the LEIA's lexicon—roughly, *military operation*, *surgery*, and *general state of functioning*. In the LEIA's ontological metalanguage, different concepts correspond to these meanings, each with its own set of properties and value sets. If, by contrast, this three-way ambiguity were retained in the representation, then, to gain more information about *the operation*, the reasoner would not know whether to ask, "Was general anesthesia administered?" or "Was a general in command?" Of course, it is entirely possible that, at any given time, some of the property-based distinctions needed to avoid confusing the reasoning engine have not yet been introduced. It follows that, if certain distinctions are not required for any reasoning purposes, such benign ambiguities may be retained in the representation. This is clearly an operational, application-oriented approach, but we have to live with it because the field has not yet come up with a universal theoretical criterion for sense delimitation.

It is reasonable to hope that the balance between short-term and long-term research in NLP and reasoning is on the road to being restored. Even in the currently dominant empiricist research paradigm, researchers recognize that the core prerequisite for the improvement of their application systems (which today achieve only modest results) is not developing better machine learning algorithms that operate on larger sets of training data but, rather, enhancing the types of knowledge used in the processing. The terminology they prefer is judicious selection of distinguishing features on which to base the comparisons and classifications of texts. As Manning (2004) notes, "In the context of language, doing 'feature engineering' is otherwise known as doing linguistics. A distinctive aspect of language processing problems is that the space of interesting and useful features that one could extract is usually effectively unbounded. All one needs is enough linguistic insight and time to build those features (and enough data to estimate them effectively)." These features are, in practice, the major building blocks of the metalanguage of representation of text meaning and, therefore, of KRL.

To summarize, the advantages of *NL is KRL* are that people find it easy to use and NL knowledge bases are available. The main problem is that reasoning directly in NL amounts to either matching uninterpreted strings or operating in an artificially constrained space of "allowed" lexical senses and syntactic constructions (see next section). This is why the machine reasoning community has for decades been writing inputs to its systems by hand.

3. *Controlled NL as KRL*. The notion of controlled languages ascends at least to the "basic English" of Ogden (1934). Controlled languages have a restricted lexicon, syntax, and semantics. Dozens of controlled languages, deriving from many natural languages, have been developed over the years to serve computer applications. A typical application involves using a controlled language to write a document (e.g., a product manual) in order to facilitate its automatic translation into other languages. Presumably, the controlled-language text will contain fewer lexical, grammatical, semantic, and other ambiguities capable of causing translation errors. Controlled languages are usually discussed together with authoring tools of various kinds—spell checkers, grammar checkers, terminology checkers, style checkers—that alleviate the difficulties that authors face when trying to write in the controlled language.

Controlled languages can also be used as programming languages. According to Sowa (2004), the programming language COBOL is a controlled English. The controlled languages especially relevant to our discussion are computer processable controlled languages (Sukkarieh, 2003), modeled after Pulman's (1996) Computer Processable English. The defining constraint of a computer processable controlled language is "to be capable of being completely syntactically and semantically analysed by a language processing system" (Pulman, 1996). Work in this area involves building tools to facilitate two aspects of the process: authoring texts in the controlled language and carrying out specialized types of knowledge acquisition—for example, compiling NL-NL_c dictionaries that specify which senses of NL words are included in the controlled language (NL_c). One benefit of this approach is that such dictionaries can, if desired, be completely user dependent, which means that different kinds of reasoning will be supported by the same general apparatus using these idiolects of NL_c.

A large number of computer processable controlled languages have been proposed, among them the language used in the KANT/KANTOO MT project (Nyberg & Mitamura, 1996), Boeing's Computer Processable Language (P. Clark et al., 2009), PENG Light Processable ENGlish (http://web.science.mq.edu.au/~rolfs/PENG-Light.html), the Controlled English to

Logic Translation (CELT) language (Pease & Murray, 2003), Common Logic Controlled English (Sowa, 2004), and Attempto (Fuchs et al., 2006). While there are differences between them, strategically all of them conform to the methodology of relying on people, not machines, to disambiguate text. The disambiguated text can be represented in a variant of first-order logic—possibly with some extensions—and used as input to reasoning engines. Having no ontological commitment broadens the opportunity for user-defined applications that can bypass the automatic analysis of open text. However, the research and development devoted to the use of controlled languages is, in our opinion, primarily technology-oriented and contributes little to the long-term goal of creating truly automatic intelligent agents, which is predicated on the capability of understanding unconstrained language.

A variation on the theme is a multistep translation method favored by some logicians. The goal is to constrain NL inputs to reasoners to whatever can be automatically translated into first-order logic. So, the process involves human translation from NL to a controlled NL, for which a parser into KRL is available. As such, having a controlled NL (NL_c) and its associated parser becomes equivalent to having a KRL (e.g., Fuchs et al., 2006; Pease & Murray, 2003; Sukkarieh, 2003; McAllester & Givan, 1992; Pulman, 1996). It is assumed that the overall system output is formulated in NL_c and that this will pose no problems for people. The problem, of course, is that a human must remain in the loop indefinitely.

In sum, the advantage of *controlled NL as KRL* is that texts in a controlled NL can be automatically translated, without loss, into the KRL—and this can be practical in some applications. The problem is that orienting around a controlled language is impractical for the general case of deriving knowledge from text, since the vast majority of texts will never be written in the given controlled language. In addition, the quality of the translation depends on the ambiguity of the controlled NL text, since writing with a complete absence of ambiguity is not always achievable. Finally, writing texts in controlled languages is notoriously difficult for people, even with training, and people must always remain in the loop, being responsible for the very first step: NL-to-controlled-NL translation.

Acknowledging the partial utility of the above options, we believe that fully automatic NL → KRL → NL translation is indispensable for sophisticated, reasoning-oriented applications. Moreover, it is the most scientifically compelling way of approaching the correlation between language and its meaning.

2.8.2 Issues of Ontology

Many issues involving the content and form of ontology are worth discussing, as evidenced by the extensive philosophical and NLP-oriented literature on ontology (see Nirenburg & Raskin, 2004, Chapter 5, for broad discussion; Kendall & McGuinness, 2019, for an engineering-oriented contribution). We constrain the current section to questions that regularly arise with respect to the ontology used by LEIAs.

Issue 1. *How do concepts differ from words?* Ontological concepts, unlike words of a language, are unambiguous: they have exactly one meaning, defined as the combination of their property-facet-value descriptions. For example, the concept SHIP in the LEIA's ontology refers only and exactly to a large sailing vessel, not a spaceship, or the act of transporting something, or any other meaning that can be conveyed by the English word *ship*. Contrast this lack of ambiguity in the LEIA's ontology with the extensive ambiguity found in machine-readable dictionaries (human-oriented dictionaries that are computer accessible) and wordnets (hierarchical inventories of words, e.g., WordNet; Miller, 1995). In all of these resources, a given string can refer to different parts of speech and/or meanings. The problems with using multiply ambiguous lexical resources as a substrate for intelligent agent reasoning are well-documented (see, among many, Bar Hillel, 1970; Ide & Véronis, 1993; Nirenburg et al., 2004).

Issue 2. *How do concepts, stored in the ontology, differ from concept instances, stored in the LEIA's episodic memory?* Concepts represent *types* of objects and events, whereas concept instances represent real-world (or imaginary-world) *examples* of them. In most cases, the distinction between concepts and instances is clear: CAT is a concept, whereas our fourteen-year old multicolor domestic shorthair named Pumpkin is an instance—perhaps recorded as CAT-8 in some particular agent's memory. For LEIAs, this distinction is formally enforced by storing information about concepts in the ontology and information about instances in the episodic memory. Note that some ontologies do not make this distinction: for example, Cyc (Panton et al., 2006) contains at least some instances in its ontology. Although the concept versus instance distribution is usually clear, there are difficult cases. For example, are specific religions, or specific makes of cars, or specific sports teams (whose players and coaches can change over time) concepts or instances? There are arguments in favor of each analysis, and we are making the associated decisions on a case-by-case basis, to the degree needed for our practical work.

Issue 3. *How was the ontology acquired, and how is it improved?* Most of the current ontology was acquired manually in the 1990s. We are not an ontology development shop and, with the exception of project-specific additions—such as extensive scripts for the Maryland Virtual Patient application—we do not pursue general ontology acquisition (though we could mount a large-scale acquisition project if support for that became available). Instead, we focus resources on advancing the science of cognitive modeling and NLU. That being said, we are working toward enabling LEIAs to learn ontological knowledge during their operation (see chapter 8). We agree with the developers of Cyc that "an intelligent system (like a person) learns at the fringes of what it already knows. It follows, therefore, that the more a system knows, the more (and more easily) it can learn new facts related to its existing knowledge" (Panton et al., 2006).

Issue 4. *Why is there no concept for "poodle" in the current ontology?* Since we do not pursue general ontology acquisition, not all concepts that would ideally be described in the LEIA's world model have yet been acquired. In some such cases, the associated English words are treated by listing them as hyponyms of other words in the lexicon. For example, most kinds of dogs are listed as hyponyms of the word *dog* (which is mapped to the concept DOG) in the lexicon, which means that LEIAs will understand that *poodle* and *Basset hound* belong to the class DOG, but they will not know what distinguishes a poodle from a Bassett hound. In other cases, given entities—particularly those belonging to specialized domains, such as airplane mechanics—are entirely missing from our resources, which means that LEIAs will need to treat them as unknown words.

A methodological tenet is that concepts should not be acquired if they do not differ from their parents or siblings with respect to substantive property values—that is, property values apart from those indicating the place of the concept in the ontological hierarchy (IS-A and SUBCLASSES). For example, the concept POODLE should not be acquired unless the acquirer has the time to list at least some of the feature values that differentiate it from other breeds of dog, such those shown below.

POODLE

COLOR	default	black, white, brown, tan
	sem	silver, orange
INTELLIGENCE	default	.9[18]
FRIENDLINESS	default	.9
FUR-TYPE	value	curly-hair
EASE-OF-TRAINING	default	.9

The reason for this tenet is that a concept means only and exactly what its set of property-facet-value triples says it means. As such, if two concepts have the same inventory of property values, then they are, for purposes of agent reasoning, identical.

Issue 5. *Should something as specific as "eating hot liquids with a spoon" be a concept?* If this action needs to be described in great detail, as to support the simulation of a character who eats hot liquids with a spoon, then this has to be a concept whose description will be recorded using a script. In contrast to eating with a fork, eating with a spoon does not permit lifting the object by poking it; and in contrast to eating cold liquids, eating hot liquids often involves cooling the liquid, either by holding a spoonful in the air for a while or by blowing on it. Of course, an EATING-HOT-LIQUIDS-WITH-A-SPOON concept/script could inherit much from its ancestors, but certain things would be locally specified, thus justifying its existence as a separate concept.

Issue 6. *Are all values of all properties locally defined in all frames?* No, for two reasons—one theoretical and the other practical. The theoretical reason is that, in many cases, a particular property cannot be constrained any more narrowly for a child than for its parent. For example, the concept NEUROSURGERY inherits the property-facet-value triple "LOCATION default OPERATING-ROOM" from its parent, SURGERY, because neurosurgery typically happens in the same place as any other surgery. ORAL-SURGERY, by contrast, overrides that default value for LOCATION because it is typically carried out in a DENTIST-OFFICE. On the practical level, we have not had the resources to fully specify every local property value of every concept. In part, this can be automated, but it would also benefit from an industrial-strength manual or semi-automatic acquisition effort.

Issue 7. *Is the LEIA's ontology divided into upper and lower portions?* When developers divide ontologies into upper (top-level, domain-independent) and lower (domain-specific) portions, the goal is to have a single, widely agreed on upper ontology that can serve as the core to which domain-specific ontologies can link.[19] We have not entered into the arena of ontology merging and have made no formal division between upper and lower portions of the LEIA's ontology. In fact, we would argue that the large expenditure of resources on ontology merging is misplaced, since most efforts in that direction do not adequately address core semantic issues that currently cannot be resolved by automatic methods.

Issue 8. *Are other ontologies used when building the LEIA's ontology?* Over the years we have tried to use external resources when building both the lexicon and the ontology. In most cases, we have found that the overhead—learning about the resource, converting formats, and, especially, carrying out quality control—was not justified by the gains.[20] Our use of external resources resonates with the findings of P. Cohen et al. (1999), who attempted to determine experimentally how and to what extent the use of existing ontologies can foster the development of other ontologies. They found that (a) the most help was provided when acquirers were building domain-specific ontologies and (b) many questions still remain about how best to use existing resources for the building of new resources.

Issue 9. *Why is the LEIA's ontology not available as freeware?* As mentioned earlier, we use the ontology as a substrate for research; we are not developing it as a resource. It has idiosyncrasies that are not documented, and, most importantly, we are not in a position to provide user support.

Issue 10. *Doesn't every culture, and even every individual person, have a different ontology?* Perhaps, but the importance of this consideration for LEIA development is quite limited and specific. That is, we are not especially interested in how many colors are recognized as basic by speakers of different languages or whether the nuances of different verbs of motion should be recorded as different concepts or be described in the lexical senses of the associated words. In this, we agree with McWhorter's (2016) critique of Neo-Whorfianism to the effect that minor crosslinguistic differences should not be blown out of proportion. What *is* important for LEIAs is that they can learn new knowledge, and even learn incorrect knowledge (like people do) during their operation—all of which can have interesting consequences.

Issue 11. *Are there other examples of ontologically grounded deep NLU systems?* Yes, but they better serve as points of juxtaposition than comparison. For example, Cimiano, Unger, and McCrae (2014; hereafter, CU&M) describe an approach to NLU with a strong reliance on handcrafted ontological and lexical knowledge bases. They use a domain-specific ontological model to drive lexical acquisition, and they constrain the interpretations of inputs to those applicable to the domain.

> The interpretation process generates only interpretations that are compatible with the ontology (in terms of corresponding to some element in the ontology) and that are aligned to the ontology (in the sense of sharing the semantic vocabulary). … From the point of view of an application that builds on a given ontology, any semantic representation that is not aligned to the ontology is useless as it could not be processed further by the backend system or application in a meaningful way. (p. 141).

This approach has noteworthy merits: it takes seriously the interrelated needs of ontological modeling and lexical acquisition in service of language understanding; it attempts to foster field-wide acceptance and collaboration by using standard ontologies (e.g., SUMO; Niles & Pease, 2001) and representation formalisms (e.g., OWL, RDF); it uses a modular architecture; and it is the seed of what CU&M envision as "an ecosystem of ontologies and connected lexica, that become available and accessible over the web based on open standards and protocols" (p. 143). CU&M's narrative is grounded in examples from the domain of soccer and offers an accessible developer's view of how to build a knowledge-based system.

On the flip side, as the authors themselves acknowledge, "the instantiation of our approach that we have presented in this book lacks robustness and coverage of linguistic phenomena," which they analyze "not as a deficiency of our approach itself, but of the particular implementation which relies on deterministic parsers, perfect reasoning, and so on" (CU&M, p. 142). They suggest that the answer lies in machine learning, which will be responsible for computing "the most likely interpretation of a given natural language sentence," being trained on ontology-aligned semantic representations (pp. 142–143).

For reasons that will become clear in this book, we find this last bit problematic. That is, CU&M expect machine learning to *somehow* solve what is arguably the biggest problem in all of natural language understanding—lexical disambiguation. Note also that since this approach accommodates only domain-specific inputs and their domain-specific interpretations, it appears to block the possibility of systems learning new things over time, thus forever making human acquisition the only road to system enhancement. In short, although we applaud the direction of this research and its goals, we insert a cautionary statement regarding inflated expectations of machine learning.

To conclude this discussion of ontology, the main thing to remember about intelligent agents is that they cannot be expected to function at a human level without significant domain knowledge. The necessary depth of knowledge cannot be expected to be available for all domains at the same time, any more than we can expect physical robots to be configured to carry out 10,000 useful physical maneuvers at the blink of an eye. This motivates our decision to relegate ontology development to a needs-based enterprise and to work toward enabling agents to acquire this knowledge through lifelong learning by reading and interacting with people. Learning is a multistage process that is addressed in practically all upcoming chapters of the book.

2.8.3 Issues of Lexicon

As with the discussion of ontology, we constrain the discussion of lexicon to issues that are particularly important for agent-building work. For a more comprehensive introduction to lexicon-oriented scholarship, see Pustejovsky and Batiukova (2019).

Issue 1. *Enumerative versus generative lexicons.* An enumerative lexicon explicitly lists all word senses that are to be understood by a language processing system. A generative lexicon (as described, e.g., in Pustejovsky, 1995) encodes fewer senses but associates compositional functions with them, such that certain word meanings can be computed on the fly. In short, a generative lexicon has rules attached to its entries, whereas an enumerative lexicon does not. A big difference? In practical terms, not really.

Anyone who has acquired a lexicon for use in NLU knows that rules play a part. It is highly unlikely that the acquirer of an English lexicon will explicitly list all regular deverbal nouns alongside their verbs (e.g., *hiking* < *hike*) or all regular agentive nouns alongside their verbs (e.g., *eater* < *eat*). At some point during knowledge acquisition or running the language processing engine, lexical and morphological rules expand the word stock to cover highly predictable forms. This process generally does not yield 100% accuracy, and, depending on the application, errors might need to be weeded out by acquirers (for a discussion of the practical implications of using lexical rules at various stages of acquisition and processing, see Onyshkevych, 1997). In short, it would be a rare enumerative lexicon that would not exploit lexical rules at all.

What *is* important about lexical rules, whether they are embedded in an enumerative lexicon or put center stage in a generative lexicon, is that they have to be known in advance. This means that the supposedly novel meanings that generative lexicons seek to cover are actually not novel at all. If they were novel, they would not be covered by generative lexicons either. As Nirenburg and Raskin (2004) suggest,

> a truly novel and creative usage will not have a ready-made generative device for which it is a possible output, and this is precisely what will make this sense novel and creative. Such a usage will present a problem for a generative lexicon, just as it will for an enumerative one or, as a matter of fact, for a human trying to treat creative usage as metaphorical, allusive, ironic, or humorous at text-processing time. The crucial issue here is understanding that no lexicon will cover all the possible senses that words can assume in real usage. (pp. 119–120)

The primary difference between the ultimate (listed or inferred) lexical stocks of, say, Pustejovsky's generative lexicon and the LEIA's lexicon lies in sense "permeation," to use Pustejovsky's term. Pustejovsky argues that a verb like *bake* actually has two meanings—*bring into existence*, as for *cake*, and *heat up*, as for *potato*—and that these senses permeate each other, no matter which one is dominant in a given context. So, for Pustejovsky, *bake a potato* primarily means *heat up* but with a lingering subsense of *bringing into existence*. Ontological Semantics, by contrast, rejects this notion of sense permeation on the following grounds: (a) deriving one meaning from the other dynamically is too costly to be worth the effort; it is preferable to list multiple senses and the semantic constraints that support their automatic disambiguation; (b) real language use tends to avoid, not introduce, ambiguity; in fact, speakers generally have a hard time detecting ambiguity even when asked to do so; and (c) we do not see any practical, agent-oriented motivation for introducing a sense-and-a-half situation (Nirenburg & Raskin, 2004, p. 120). The lexicon used by LEIAs reflects a combination of enumerated senses and senses that are dynamically generated as a runtime supplement.

Issue 2. *Manual versus automatic acquisition of lexicon*. Ideally, all static knowledge resources would be acquired either fully automatically or primarily automatically with modest human post-editing. This is, one might say, the holy grail of knowledge-rich NLP. In the late 1980s to early 1990s, NLP centrally concentrated on trying to maximally exploit machine-readable dictionaries that were oriented toward people; however, the results were unsatisfying and the direction of work was ultimately abandoned.[21] As reported by Ide and Véronis (1993) in their survey of research involving machine-readable dictionaries (which bears the suggestive title "Extracting knowledge bases from machine-readable dictionaries: Have we wasted our time?"), "The previous ten or fifteen years of work in the field has produced little more than a handful of limited and imperfect taxonomies." In fact, Ide and Véronis share our group's long-standing belief that it is not impossible to manually build large knowledge bases for NLP, as lexicographers can be trained to do so efficiently. And, ultimately, one needs lexicographers, even in the automatic-acquisition loop, for quality control. The crucial prerequisite for the success of any human-aided machine acquisition is that the output not be too errorful, since that can make the task faced by the human inspector overwhelming. A comparison that will be appreciated by computer programmers is the challenge of debugging someone else's insufficiently commented code.

The LEIA's core lexicon was acquired manually, but, to speed up the process, the acquirers selectively consulted available resources such as human-oriented dictionaries, thesauri, and WordNet. For example, in domains that are of relatively less interest to our current applications, such as types of hats or dogs, we list hyponyms found in WordNet in the lexical senses for *hat* and *dog* without attempting to encode their distinguishing features. Of course, the decision regarding grain size can be changed as applications require, and, ideally, all hyponyms and near synonyms ("plesionyms," to use the term coined by Hirst, 1995) will be distinguished by property values. In addition, we sometimes use the meronymic information in WordNet to jog our memories about words relevant to a domain, and we consult thesauri in order to treat semantic clusters of words at a single go.

The lexicon is acquired in conjunction with the ontology, and decisions about where and how to encode certain types of information are taken with the overall role of both resources in mind. Much of lexical acquisition is carried out in response to lacunae during text processing by the language analyzer. Often, a given text-driven need leads to the simultaneous coverage of a whole nest of words.

Two widespread but ungrounded assumptions (or, perhaps, traces of wishful thinking) about lexicon acquisition are worth mentioning at this point: (a) lexicons will automatically materialize from theories (they won't; lexicons are part of models, not theories)[22] and (b) once the issues discussed in the literature of formal semantics have been handled, the rest of the NLU work will be trivial. In fact, the paucity of discussion of a host of difficult lexical items on scholarly pages is as unfortunate as the overabundance of attention afforded to, say, the grinding rule (the correlation between the words indicating animals and the words indicating their meat) or the interpretation of the English lexeme *kill*. We agree with Wilks et al.'s (1996) critique of the debate over whether *kill* should be represented as CAUSE-TO-DIE or CAUSE-TO-BECOME-NOT-ALIVE:

> The continuing appeal to the above pairs not being fully equivalent (Pulman, 1983a) in the sense of biconditional entailments (true in all conceivable circumstances) has led to endless silliness, from Sampson's (1975) claim that words are "indivisible," so that no explanations of meaning can be given, let alone analytic definitions, and even to Fodor's (1975) use of nonequivalence to found a theory of mental language rich with innate but indefinable concepts like "telephone"! (p. 58).

We make decisions about what to acquire and how deeply to describe given entities based on the usual pragmatic considerations: cost and the needs of applications. Hyponyms of *frog* will remain undifferentiated until we either must differentiate them for a herpetology-related application or we have a cheap way of carrying out the work—as by using semiautomated methods of extracting the properties of different species of frogs from texts (see chapter 7 for a discussion of learning by reading). We have dealt with many difficult lexical issues using the microtheories described in this book, but an important methodological choice in pursuing them is to achieve a sufficient descriptive depth and breadth of coverage while tempering our sometimes overambitious academic enthusiasms.

Issue 3. *General-purpose versus application-specific lexicons.* It is difficult to build useful NLP lexicons without knowing ahead of time what processors or applications they will serve. In fact, Allan Ramsay (cited in Wilks et al., 1996, p. 135) has called this impossible due to the extremely narrow "least common denominator" linking theoretical approaches. The difficulty in finding least common denominators has been met repeatedly in the sister domain of corpus annotation: due to the expense of corpus annotation, effort has been expended to make the results useful for as many practitioners as possible. However, by the time the inventory of markers is limited, on the one hand, by theoretically agreed-upon constructs and labels, and, on the other hand, by the ability of annotators to achieve consistency and consensus, the actual markup is less deep or robust than any individual group or theory would prefer. By committing to a known processing environment, and developing that environment along with the knowledge bases it uses, we have made the endeavor of NLU more feasible.

One way of fully appreciating the advantages of environment specificity is to look at lexicons that were bound by environment-neutral ground rules—as was the case with the SIMPLE project (Lenci et al., 2000). SIMPLE developers were tasked with building 10,000-sense "harmonised" semantic lexicons for twelve European languages with no knowledge of the processors or theoretical frameworks they might need to ultimately serve.[23] For commentary on this difficult task, see McShane et al. (2004).

Issue 4. *The largely language independent lexicon.*[24] Saying that a lexicon is *largely language independent* should raise eyebrows: after all, conventional wisdom has it that whereas ontologies are language independent, lexicons are language dependent. However, it turns out that both parts of this statement require qualification: many if not most resources that are called ontologies these days are language dependent; and at least some computational lexicons have substantial language independent aspects. In the domain of knowledge-rich NLU, the importance of language independent (i.e., crosslinguistically reusable) resources cannot be overstated. Building high-quality resources requires large outlays of human resources, which can be justified if one can significantly reduce the effort of producing equivalent resources in languages beyond the first (for examples, see Nirenburg & McShane, 2009).

In our approach, the most difficult aspect of lexical acquisition—describing the meaning of words and phrases using a large inventory of expressive means—needs to be done only once. The linking of these semantic structures to words and phrases of a given language is far simpler, even though it sometimes requires tweaking property values to convey special semantic nuances. Below we describe some of the expressive means available in a LEIA's lexicon and their implications for a largely language independent lexicon.

Property-modified sem-strucs serve as virtual ontological concepts. The most obvious way to represent lexical meaning in an ontological-semantic environment is to directly map a lexeme to an ontological concept: for example, the canine meaning of the word *dog* maps to the concept DOG. The ontological description of DOG includes the fact that it has all the typical body parts of its parent, CANINE; that it is the AGENT-OF BARK and WAG-TAIL; that it is the THEME-OF both CYNOMANIA (intense enthusiasm for dogs) and CYNOPHOBIA (fear of dogs); and so on. In short, direct ontological mapping does not constitute upper-case semantics in the sense used by logicians because the concept DOG is backed up by a richly informative knowledge structure. In the case of argument-taking lexemes, the syntactic arguments and semantic roles need to be appropriately associated using variables, as shown in our examples of *address* and *see* presented in section 2.3.2.

A variation on the theme of direct concept mapping is to map a lexeme to a concept but, in the lexicon, further specify some property value(s). For example:

- *Zionist* is described as a POLITICAL-ROLE that is further specified as the AGENT-OF a SUPPORT event whose THEME is the NATION that HAS-NAME 'Israel'.
- The verbal sense of *asphalt* (as in *They asphalted our road*) is described as a COVER event whose INSTRUMENT is ASPHALT and whose THEME is ROADWAY-ARTIFACT. The ontological

description of COVER, by contrast, indicates that this concept has much broader applicability, permitting its INSTRUMENT and THEME to be any PHYSICAL-OBJECT.

- The verb *recall*, as used in *They recalled the high chairs*, is described as a RETURN-OBJECT event that is CAUSED-BY a REQUEST-ACTION event whose AGENT is a FOR-PROFIT-CORPORATION and whose THEME is ARTIFACT, INGESTIBLE or MATERIAL. Here, too, the constraints on CAUSED-BY and THEME are narrower than are required by the concept RETURN-OBJECT.

The lexical constraining of ontological property values can be viewed as creating virtual, unlabeled concepts.

The question, then, is how does one determine whether a given meaning should be recorded as a new ontological concept (say, ASPHALT-EVENT) or be expressed lexically by tightening constraints on an existing concept (as in the description of *asphalt* above)? There are two nonconflicting answers to this question. On the one hand, the ontology, as a language-neutral resource, should contain only those meanings that are found in a large number of languages; so whereas DOG is a good candidate for an ontological concept, ASPHALT-EVENT is not. Naturally, this is not a precise criterion, but we do not know how it could be otherwise without embarking on an entire program of study devoted to this decision-making process. On the other hand, for LEIAs, it does not matter whether meanings are described in the ontology or the lexicon since the resources are always leveraged together. So, we could alternatively treat the verbal sense of *asphalt* by creating the ontological concept ASPHALT-EVENT as a child of COVER whose INSTRUMENT is ASPHALT and whose THEME is ROADWAY-ARTIFACT. Then the lexical sense for the verb *asphalt* would include a direct mapping to this concept.

The possibility of creating virtual ontological concepts within the sem-struc zone of lexicon entries is the first bit of evidence that the LEIA's lexicon is not entirely language specific: after all, there are other languages that have lexemes meaning *Zionist*, *to asphalt (some roadway)*, and *to recall (some product)*. Once a sem-struc expressing such a meaning is developed for one language, it can be reused in others.[25]

Alternative semantic representations. Knowledge acquisition requires decision-making at every turn. Consider the multiword expression *weapons of mass destruction*, for which two lexicon acquirers on our team, working separately, came up with different but equally valid descriptions: one was a set containing CHEMICAL-WEAPON and BIOLOGICAL-WEAPON, and the other was WEAPON with the potential (a high value of potential modality) to be the INSTRUMENT of KILLing more than 10,000 HUMANS.[26] These descriptions would both support roughly the same types of reasoning. Since we were not pursuing the domain of warfare in any great detail at the time, these descriptions were created without much ado and either one of them—or a combination of both—aligns with the overall grain size of description of the resources. By contrast, if we had been deeply invested in a warfare-oriented application, we would have sought counsel from a domain expert regarding the best representation of this, and all other relevant, concepts.

The goal for multilingual lexical acquisition is to avoid (a) duplicating the work of analyzing such entities during the acquisition of each new language or, worse yet, (b) endlessly quibbling over which of competing analyses is best. For example, if we had gone with the "KILL > 10,000 HUMANS" approach to describing *weapons of mass destruction*, should the number have been 10,000 or 20,000? What about nonhuman ANIMALS? And PLANTS? Should we have used value ranges instead of single values? The strong methodological preference for reusing semantic descriptions once formulated does not imply that the first lexicon acquired is expected to reflect all the right answers or that the semantic description for that language should be set in stone. However, it does mean that acquirers should choose their semantic battles carefully in order to support practical progress in real time.

Complex semantic descriptions. The more complex a semantic description is, and the more time and thought it takes to create, the greater the benefit of reusing it crosslinguistically. Let us consider just a handful of such examples. Consider the adverb *overboard*, whose lexical sense is shown below.

```
overboard-adv1
   def.   indicates that the source of the motion is a boat and the destination is a body of water
   ex.    They threw the rotten food overboard. He jumped overboard.
   syn-struc
      root    $var1 (cat v)
      mods    $var0
   sem-struc
      ^$var1 (sem MOTION-EVENT)
         SOURCE          SURFACE-WATER-VEHICLE
         DESTINATION     BODY-OF-WATER
```

Overboard is described as modifying a MOTION-EVENT whose SOURCE is SURFACE-WATER-VEHICLE and whose DESTINATION is BODY-OF-WATER. So, if one throws rotten food overboard, THROW fulfills the requirement that $var1 represent a MOTION-EVENT, and the meaning of *overboard* adds to the TMR an indication of the SOURCE and DESTINATION.

Although this description, once formulated, might seem quite transparent, only an experienced lexicographer is likely to devise it. Moreover, it requires that the acquirer find the three relevant key concepts in the ontology. Together, this is far more work for the first lexicon than replacing the headword (*overboard*) with its equivalent in other languages—such as the prepositional phrase *za bort* in Russian.

Issue 5. *Pleonastics and light verbs.* As we have seen, not every word of a language directly maps to an ontological concept. In fact, many words used in certain constructions do not carry any individual meaning at all. Such is the case, for example, of pleonastic pronouns in examples like *It is snowing (raining)*, *It is thought (known, understood)* *that* ..., and *Someone finds it crazy (strange) that* .... In all these cases, *it* carries no meaning. The best way to record its nonsemantic function is to create multiword lexical senses for all such constructions and, in those sense descriptions, explicitly indicate that *it* is not referential. Formally, we do that using the descriptor *null-sem+*, which means *this element has null semantics*. The meaning of the rest of the construction is expressed in the normal way.

```
rain-v1
   def.   the construction for expressing rain events
   ex.    It is raining. It rains a lot here. It has just rained.
   syn-struc
      subject   $var1 (root it)
      v         $var0
   sem-struc
      RAINSTORM
         ^$var1    null-sem+
```

This construction, like all others, allows for inputs to have different values for mood, tense, and aspect, as well as free modification of all elements that have not been null-semmed.

While an acquirer of a Russian lexicon will not exploit this lexical sense (Russian conveys this meaning using an expression literally translated as *Rain goes*), an acquirer of a French lexicon can, since syntactically and semantically *Il pleut* is equivalent.

Light verbs in multiword expressions are treated similarly. Light verbs are defined as verbs that, at least in certain constructions, carry little meaning, instead relying on their complement noun to provide the meaning. Although English does not use light verbs as widely as, say, Persian, it does have a few, such as *take* in collocations like *take a bath <a shower, a break, a nap, a vacation, a breather>* and *have* in collocations such as *have a fight <an argument, a nice time, a party>*. In addition to these canonical light verbs, some verbs can function as light verbs in certain constructions. For example, there is little difference between saying *someone had a stroke <a heart attack, a seizure>* and *someone suffered a stroke <a heart attack, a seizure>*. The lexical sense for the construction *suffer* + MEDICAL-EVENT is as follows.

```
suffer-v1
   def.   the construction 'to suffer a MEDICAL-EVENT'
   ex.    He suffered a heart attack (a seizure, a stroke).
   syn-struc
      subject        $var1
      v              $var0
      directobject   $var2
   sem-struc
      ^$var2 (sem MEDICAL-EVENT)
         EXPERIENCER    ^$var1
   synonyms    have
```

As this structure shows, the meaning representation is headed by the meaning of the MEDICAL-EVENT that serves as the direct object; the EXPERIENCER of that event is the meaning of the subject of the construction. Note that the function/meaning of *suffer* is folded into the meaning representation. The reason there is no indication of null-sem+ is that it is formally impossible to null-sem the head (i.e., $var0) of a lexical sense since the entire meaning representation expresses the meaning of that string as used in the given construction.

In sum, preparing an analysis system *not* to assign semantics to certain text elements in certain constructions is as important as preparing it *to* assign semantics. The natural place to provide such construction-sensitive information is the lexicon. Since certain types of constructions are prevalent in certain families of languages, sem-strucs that predict the null semantics of given entities can be reused in related languages.

Issue 6. *Tactics for reusing lexical descriptions across languages.* We have already explained that sem-strucs—no matter what lexicon they originate from—represent the meanings of words and phrases in natural languages and that these meanings are largely applicable across languages. The latter derives from the Principle of Practical Effability (Nirenburg & Raskin, 2004), which states that what can be expressed in one language can be expressed in all other languages, using expressive means available in those languages—a word, a phrase, or a lengthy description. That is, we proceed from the assumption that word and phrase meanings across languages are *mostly similar* and that different semantic nuances across languages are an important but less frequent occurrence. Of course, one could also proceed from the assumption that the meanings of words in different languages are *mostly different* and, accordingly, require lexicon acquirers to chase down those differences. The latter approach, while no doubt an enticing opportunity for linguists to demonstrate their language analysis chops, has little place, we believe, in practically oriented tasks—from teaching humans a nonnative language to teaching computers to collaborate with us. In all cases, small nuances of difference should become the focus only long after the basics have been mastered.

All this being said, the job of writing a lexicon for language L2 based on the lexicon for language L1 should, in large part, be limited to providing an L2 translation for the headword(s); making any necessary syn-struc adjustments; and checking/modifying the linking among variables between the syn- and sem-strucs. For example:

- The first noun entry alphabetically in the English lexicon is *aardvark*, which is a simple noun mapping to the concept AARDVARK. If L2 has a word whose meaning corresponds directly to the English word *aardvark* (e.g., Russian has *aardvark*), the acquirer can simply substitute it in the header of the entry.
- The noun *table* has two entries in the English lexicon: a piece of furniture mapping to TABLE, and a structured compilation of information mapping to CHART. The corresponding entries in a Hebrew lexicon will be recorded under two different headwords: *shulhan* and *luah*, respectively.

- Lexical entries for verbs involve more work, mostly because their subcategorization properties must be described. The entry for *sleep* maps to SLEEP and indicates that the subject is realized as the filler of the EXPERIENCER case role. The corresponding entry in the French lexicon will be very similar, with *dormir* substituted for *sleep* in the header of the entry. This is because French, just like English, has intransitive verbs, and *dormir* is intransitive, just like *sleep*.
- If the lexical units realizing the same meaning in L2 and English do not share their subcategorization properties, the acquirer will have to make necessary adjustments. For example, in English the verb *live* meaning *inhabit* indicates the location using a prepositional phrase, whereas in French the location of *habiter* is expressed using a direct object. Even though this slight change to the syn-struc must be entered, this is still much faster than creating the entry from scratch.
- In some cases, a language has two words for a single concept. For example, Russian expresses *man marries woman* using one construction (X_{Male} *zhenitsja na* Y_{Female}) but *woman marries man* using another (X_{Female} *vyxodit zamuzh za* Y_{Male}). In both cases, they map to the concept MARRIAGE, so this simply requires making two lexical senses instead of one.
- The base lexicon might not include a word or phrase needed in the L2 lexicon. In this case, the task is identical to the task of acquiring a base lexicon to begin with.

To test out how realistic bootstrapping our English lexicon to another language would be—knowing from experience that devils lurk in the most unexpected of details—we carried out a small experiment with Polish. While this experiment suggested that a combination of automated and manual bootstrapping would be useful, it also revealed the need for nontrivial programmatic decisions like the following:

- Should the L1 lexicon be treated as fixed (uneditable), or should the L2 acquirer attempt to improve its quality and coverage while building the L2 version? The organizational complexity of working on two or more resources simultaneously is easy to imagine.
- Should L2 acquisition be driven by correspondences in headwords or simply by the content of sem-struc zones? For example, all English senses of *table* will be in one head entry and typically acquired at once. But should all senses of all L2 translations of *table* be handled at once during L2 acquisition, or should the L2 acquirer wait until he or she comes upon sem-strucs that represent the given other meanings of the L2 words?
- To what extent should the regular acquisition process—including ontology supplementation and free-form lexical acquisition—be carried out on L2? Ideally, it should be carried out full-scale for each language acquired with very close communication between acquirers. A high-quality integrated interface that linked all languages with all others would be desirable.

The answers to all these and related questions depend not only on available resources but also on the personal preferences of the actual acquirers of lexicons for given languages working on given projects at given times. Of course, the matter of acquisition time is most pressing for low- and mid-density languages since there tends to be little manpower available to put to the task.

2.8.4 Paraphrase in Natural Language and the Ontological Metalanguage

Our discussion of options for knowledge representation has already touched on paraphrase. Here we delve a bit deeper into that issue, which actually has two parts: linguistic paraphrase and ontological paraphrase.

Natural languages offer rich opportunities for paraphrase. For example, one can refer to an object using a canonical word/phrase (*dog*), a synonymous or nearly synonymous generic formulation (*mutt, pooch, man's best friend*), a proper name (*Fido*), an explanatory description (*a pet that barks*), or a pointer (*him*). Similarly, one can express a proposition using canonical sentences (*My dog gives me such pleasure*; *I get such pleasure from my dog*; *My dog is such a source of pleasure for me*), special constructions (*My dog, what a source of pleasure!*; *You know what you are, Fido? Pure pleasure!*), and so on. Some of these locutions—such as active and passive voice pairs (*The dog chased the cat / The cat was chased by the dog*)—will generate exactly the same meaning representation. However, even those that do not generate *exactly* the same meaning representation are, in many cases, *functionally* equivalent, meaning that they will serve the same reasoning-oriented ends for the LEIA.[27]

A standing task for LEIAs is determining how newly acquired knowledge correlates with what has previously been learned. Among the simpler eventualities are that the new meaning representation is identical to a representation stored in memory; it is identical except for some metadata value (such as the time stamp); it contains a subset or superset of known properties that unify unproblematically; or it is so completely different from everything known to date that the question of overlap does not arise. None of these eventualities directly involves paraphrase. Two eventualities that *do* involve paraphrase are these: the new meaning representation is related to a stored representation via ontological paraphrase, and the new meaning representation (or a component of it) is related to a stored representation via ontological subsumption, meronymy, or location. We will consider these in turn.

The newly acquired meaning representation is related to a stored memory via ontological paraphrase. Ontological paraphrase occurs when more than one metalanguage representation means the same thing. Ontological paraphrase is difficult to avoid because meaning representations are generated from the lexical senses for words and phrases appearing in the sentence, so the choice of a more or less specific word can lead to different meaning representations. For example, one can report about a trip to London saying *go to London by plane* or *fly to London*, whose respective meaning representations are as follows.[28]

go to London by plane

```
MOTION-EVENT-7
    DESTINATION      CITY-50
    INSTRUMENT       AIRPLANE-1
CITY-50
    HAS-NAME         'London'
```

fly to London

```
AERIAL-MOTION-EVENT-19
    DESTINATION      CITY-50
CITY-50
    HAS-NAME         'London'
```

Comparing these paraphrases, the word *go* instantiates a MOTION-EVENT, whereas the more specific word, *fly*, instantiates the more specific AERIAL-MOTION-EVENT. However, the *go* paraphrase includes the detail *by plane*, which provides the key property that distinguishes AERIAL-MOTION-EVENT from its parent, MOTION-EVENT, in the LEIA's ontology. That is, the locally specified (not inherited) properties of AERIAL-MOTION-EVENT in the ontology are

```
AERIAL-MOTION-EVENT
    IS-A          value     MOTION-EVENT
    INSTRUMENT    default   AIRPLANE
                  sem       HELICOPTER, BALLOON-TRANSPORTATION
```

This says that AERIAL-MOTION-EVENT is a child of MOTION-EVENT and that the most common instrument of flying is an AIRPLANE, but HELICOPTER and BALLOON-TRANSPORTATION are possible as well. If the agent uses this information to fill in a property value that was not attested in the input *fly to London*, the resulting representation will be almost identical to the representation for *go to London by plane.*

go to London by plane

```
MOTION-EVENT-7
    DESTINATION      CITY-50
    INSTRUMENT       AIRPLANE-1
CITY-50
    HAS-NAME         'London'
```

fly to London

```
AERIAL-MOTION-EVENT-19
    DESTINATION      CITY-50
    INSTRUMENT       default AIRPLANE
                     sem     HELICOPTER, BALLOON-TRANSPORTATION
CITY-50
    HAS-NAME         'London'
```

The main difference between these representations, which will not affect most types of reasoning, is that the ontologically extracted information is in the form of concepts, not concept instances, so the agent recognizes the information about the instrument as being generic and not part of a remembered TMR. If, in a subsequent utterance, it becomes clear which mode of transportation was used, that specific knowledge would override the generic knowledge posited in the LEIA's memory about this event.

Detecting whether pairs of meaning representations are paraphrases can be accomplished using a fairly simple heuristic: if the events in question are in an immediate (or very close) subsumption relationship, and if the specified properties of the more generic one match the listed ontological properties of the more specific one, then they are likely to be ontological paraphrases. Note that it is not enough for the properties of the events to simply unify—this would lead to too many false positives. Instead, the generic event must be supplied with the property value(s) that define the more specific one. Absent this, the most the LEIA can say is that two events might be related but they are not paraphrases.

The new meaning representation (or a component of it) is related to a stored memory via ontological subsumption, meronymy, or location. When attempting to match new textual input with a stored representation, the question is, *How close do the compared meaning representations have to be in order to be considered a match?* An important consideration when making this judgment is the application in which the system is deployed. In the Maryland Virtual Patient application (see chapter 8), in which virtual patients communicate with system users, sincerity conditions play an important role. That is, virtual patients expect users, who play the role of attending physician, to ask them questions that they can answer; therefore, they try hard to identify the closest memory that will permit them to generate a response.

One foothold for the associated analysis involves the ontological links of subsumption (IS-A), meronymy (HAS-AS-PART), and location (LOCATION). Consider the example of the user (who plays the role of attending physician) asking the virtual patient, *Do you have any discomfort in your esophagus?* The meaning representation for this question is

```
REQUEST-INFO-1
    AGENT               HUMAN-2    ("the physician")
    THEME               MODALITY-1.VALUE
    BENEFICIARY         HUMAN-1    ("the virtual patient")
    MODALITY-1
        TYPE            epistemic
        SCOPE           DISCOMFORT-1
    DISCOMFORT-1
        EXPERIENCER     HUMAN-1    ("the virtual patient")
        LOCATION        ESOPHAGUS-1
    ESOPHAGUS-1
        PART-OF-OBJECT  HUMAN-1    ("the virtual patient")
```

The interrogative mood in the input gives rise to the instance of the speech act REQUEST-INFO, whose THEME is the value of epistemic MODALITY that scopes over the proposition headed by DISCOMFORT-1. Formalism aside, this represents a yes/no question: if the event actually happened, then the value of epistemic MODALITY is 1; if it did not happen, then the value is 0. The event DISCOMFORT-1 is experienced by HUMAN-1, which will be linked to a specific HUMAN (the virtual patient) via reference resolution. The LOCATION of the DISCOMFORT is the ESOPHAGUS of that HUMAN.

If we extract the core meaning of this question, abstracting away from the interrogative elements, we have the meaning for *discomfort in your esophagus*.

DISCOMFORT-1
  EXPERIENCER HUMAN-1
  LOCATION **ESOPHAGUS-1**

ESOPHAGUS-1
  PART-OF-OBJECT HUMAN-1

Let us assume that the virtual patient does not have a memory of discomfort in the esophagus but does have a memory of some *symptom in its chest*.

SYMPTOM-1
  EXPERIENCER HUMAN-1
  LOCATION **CHEST-1**

CHEST-1
  PART-OF-HUMAN HUMAN-1

The components in boldface are the ones that must be matched. Is the LEIA's memory of a SYMPTOM in the CHEST close enough to the question about DISCOMFORT in the ESOPHAGUS? Assuming this LEIA knows what an esophagus is (which is not the case for all LEIAs), then it can recognize that

- DISCOMFORT is a child of SYMPTOM, forming a subsumption link of only one jump;
- ESOPHAGUS has a LOCATION of CHEST—information available in the ontology; and
- both body parts belong to the same human, the virtual patient, as indicated by the PART-OF-OBJECT property.

According to our matching algorithm—in conjunction with the fact that the virtual patient assumes sincerity conditions in its conversations with the physician—the virtual patient's memory of this event sufficiently matches the physician's question, so the virtual patient can respond affirmatively to the question: Yes, it has a memory of the symptom the physician is asking about.

There is one more aspect of ontological paraphrase that deserves mention: the paraphrasing that occurs as a result of a need to interpret signals from perception modes other than language. For example, in the MVP system, virtual patients are capable of interoception, which is interpreting the bodily signals generated by the simulation engine. We will not delve into the details of physiological simulation here (see chapter 8). Suffice it to say that physiological simulations are driven by ontological descriptions that capture how domain experts think about anatomy and physiology, employing concepts such as DYSPHAGIA (difficulty swallowing), PERISTALSIS (wavelike contractions of a series of muscles), and BOLUS (the contents of a single swallow—a chewed piece of food or a gulp of liquid). A given instance of a virtual patient may or may not know about such concepts—it might be endowed with no medical knowledge, some medical knowledge, or extensive knowledge (the latter being true of virtual patients who, by profession, are physicians). It is the job of the interoception engine to align what is generated by the simulation engine—for example, DYSPHAGIA—with the closest available concept(s) in the given patient's ontology, which might be DISCOMFORT in the CHEST after SWALLOW events, or PAIN near the STOMACH when food is stuck there, or various other things. Depending on the level of coverage of a particular agent's ontology, the results of the interpretation of interoception will be different and will constitute ontological paraphrases of the same event.

2.9 Further Exploration

1. Compare the knowledge bases used in OntoAgent with other available online resources:

- WordNet. Read about it at https://wordnet.princeton.edu and explore the resource at http://wordnetweb.princeton.edu/perl/webwn. There are also wordnets for languages other than English.
- FrameNet. Read about it at https://framenet.icsi.berkeley.edu/fndrupal/ and access the frame index—which is the easiest way to explore the resource—at https://framenet.icsi.berkeley.edu/fndrupal/frameIndex.
- Wordnik. Read about it at https://www.wordnik.com/about and search it at https://www.wordnik.com.

2. For each resource, think about/discuss both its potential utility and its limitations for NLP/NLU.

3. We mentioned in passing John McWhorter's take on Neo-Whorfianism. Read his book, *The Language Hoax*, or watch him lecture on the topic on YouTube—for example, at https://www.youtube.com/watch?v=yXBQrz_b-Ng&t=28s.

3

Pre-Semantic Analysis and Integration

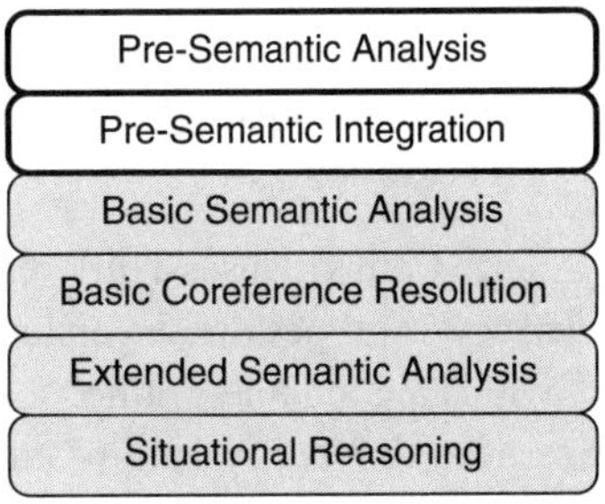

Before addressing semantics, the LEIA carries out two preparatory stages of analysis. The first one, *Pre-Semantic Analysis*, includes preprocessing and syntactic parsing, for which we use externally developed tools. The reasons why we have not developed knowledge-based alternatives are these: tools addressing the needed phenomena exist; they are freely available; they yield results that are acceptable for research purposes; Ontological Semantics makes no theoretical claims about pre-semantic aspects of language processing; and our approach to semantic analysis does not require pre-semantic heuristics to be complete or perfect. However, using externally developed tools comes at a price: their output must be integrated into the agent's knowledge environment. This is carried out at the stage called *Pre-Semantic Integration.*

This chapter first introduces the tool set LEIAs use for pre-semantic analysis and then describes the many functions needed to mold those results into the most useful heuristics to support semantic analysis. Although the specific examples cited in the narrative apply to a particular tool set at a particular stage of its development, there is a more important generalization at hand: building systems by combining independently developed processors will always require considerable work on integration—a reality that is insufficiently addressed in the literature describing systems that treat individual linguistic phenomena (see section 2.6).

3.1 Pre-Semantic Analysis

For pre-semantic analysis (preprocessing and syntactic parsing), LEIAs currently use the Stanford CoreNLP Natural Language Processing Toolkit (Manning et al., 2014). Although CoreNLP was trained on full-sentence inputs, its results for subsentential fragments are sufficient to support our work on incremental NLU.

For preprocessing, LEIAs use results from the following CoreNLP annotators:

- *ssplit*, which splits texts into sentences;
- *tokenize*, which breaks the input into individual tokens;

- *pos*, which carries out part-of-speech tagging;
- *lemma*, which returns the lemmas for tokens;
- *ner*, which carries out named-entity recognition; and
- *entitymentions*, which provides a list of the mentions identified by named-entity recognition.

Since CoreNLP uses a different inventory of grammatical labels than Ontological Semantics, several types of conversions are necessary, along with a battery of fix-up rules—all of which are too fine-grained for this description. We mention them only to emphasize the overhead that is involved when importing external resources and why it is infeasible to switch between different external resources each time a slight gain in the precision of one or another is reported.

For syntactic analysis, CoreNLP offers both a constituent parse and a dependency parse.[1] A constituent parse is composed of nested constituents in a tree structure, whereas a dependency parse links words according to their syntactic functions. Figures 3.1 and 3.2 show screenshots of the constituent and dependency parses for the sentence *A fox caught a rabbit*, generated by the online tool available at the website corenlp.run.[2]

Consider one example of the difference in information provided by these different parsing strategies. Whereas the constituency parse labels *a fox* and *a rabbit* as noun phrases and places them in their appropriate hierarchical positions in the tree structure, the dependency parse indicates that *fox* is the subject of *caught*, and *rabbit* is its direct object.

Both kinds of parses provide useful information for the upcoming semantic analysis. However, at the current state of the art, the results are error-prone, especially in less-formal speech genres such as dialogs. Therefore, rather than rely on either type of parse wholesale, the NLU system uses parsing output judiciously, as described below.

3.2 Pre-Semantic Integration

The Pre-Semantic Integration module adapts the outputs of preprocessing and parsing to the needs of semantic analysis. The subsections below describe the contentful (not bookkeeping-oriented) procedures developed for this purpose.

3.2.1 Syntactic Mapping: Basic Strategy

Syntactic mapping—or syn-mapping, for short—is the process by which a LEIA matches constituents of input with the syn-struc (syntactic structure) zones of word senses in the lexicon. This process answers the question, Syntactically speaking, what is the best combination of word senses to cover this input? Figure 3.3 illustrates the syn-mapping process for the input *He ate a sandwich*. It shows the relevant excerpts from two senses of *eat* (presented in section 2.2), one of which is syntactically suitable (eat-v1) and the other of which is not (eat-v2).

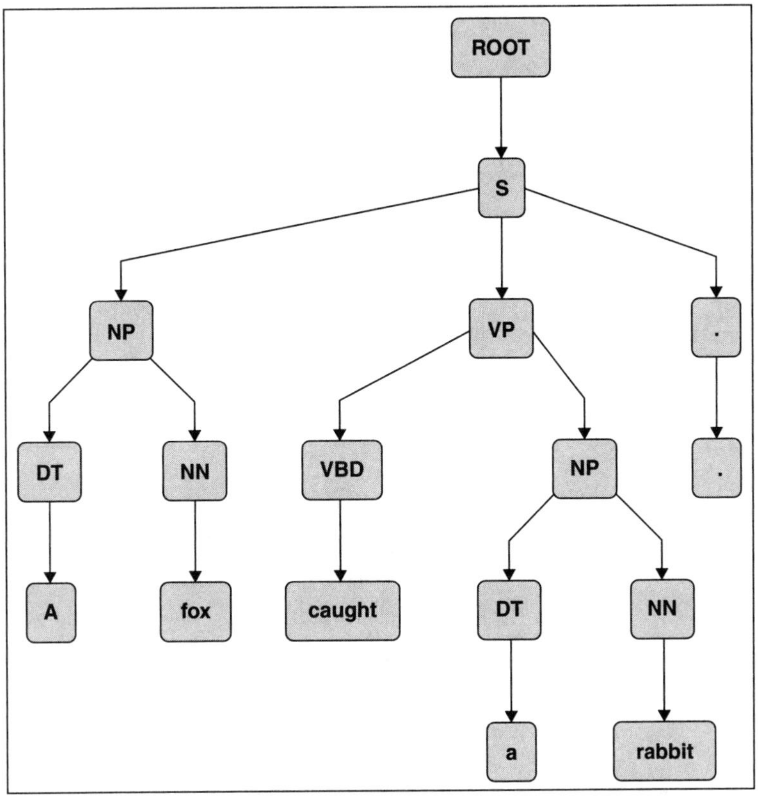

Figure 3.1
The constituency parse for *A fox caught a rabbit.*

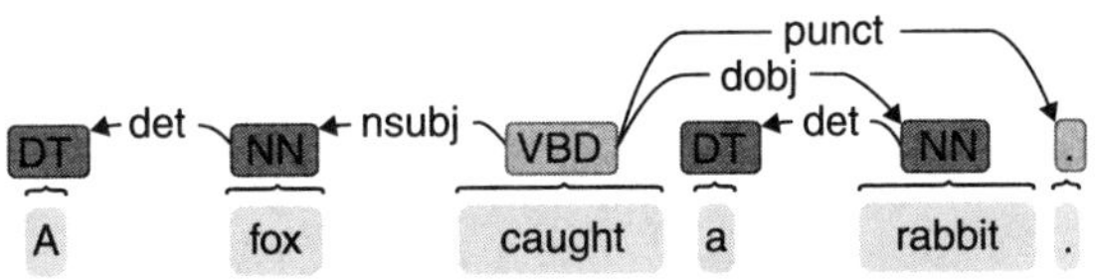

Figure 3.2
The dependency parse for *A fox caught a rabbit.*

Later on, during Basic Semantic Analysis, the LEIA will determine whether the meanings of the variables filling the subject and direct object slots of eat-v1 are appropriate fillers of the AGENT and THEME case roles of INGEST.[3]

Although the syn-mapping process looks easy for an example like *He ate a sandwich*, it gets complicated fast as inputs become more complex. In fact, it often happens that no syn-mapping works perfectly. There are many reasons for this, four of which we cite for illustration.

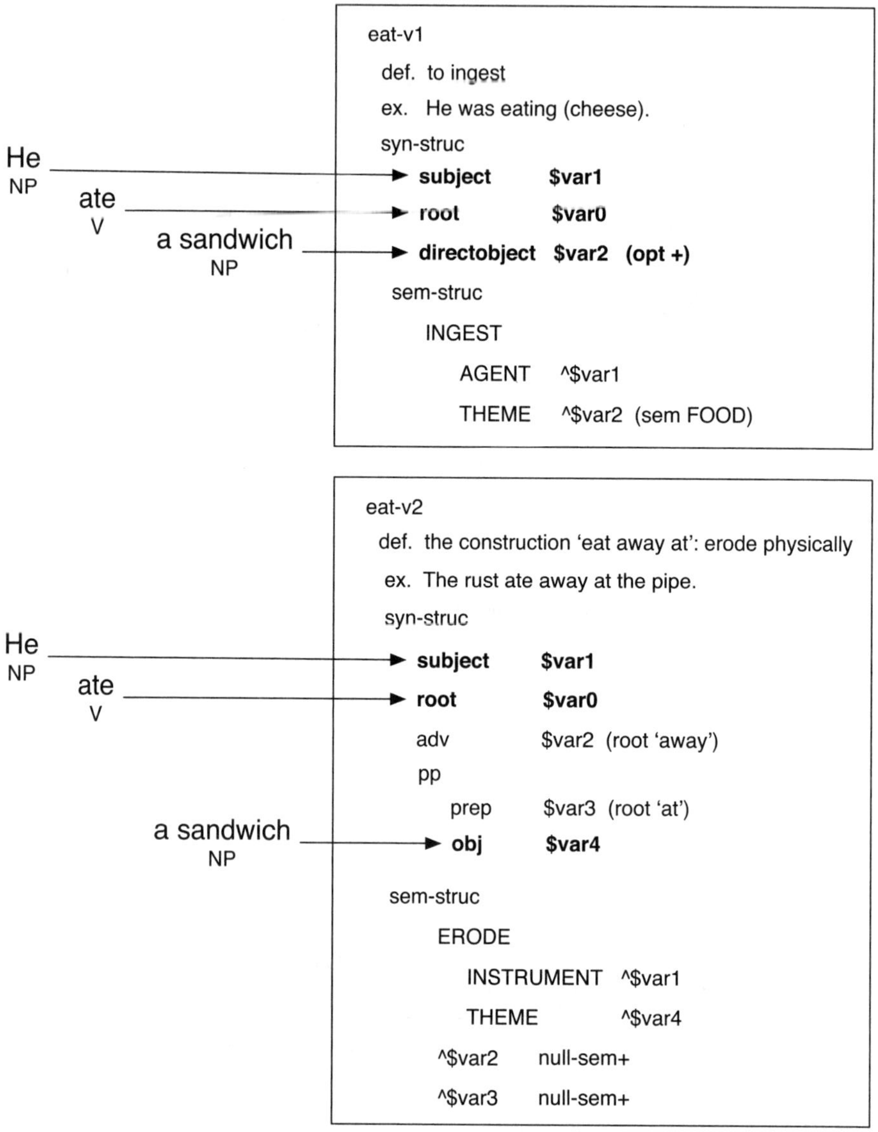

Figure 3.3
A visual representation of syn-mapping. For the input *He ate a sandwich*, eat-v1 is a good match because all syntactic expectations are satisfied by elements of input. Eat-v2 is not a good match because the required words *away* and *at* are not attested in the input.

1. Our syntactic theory does not completely align with that of CoreNLP: for example, our inventories of parts of speech and syntactic constituents are different from those used by CoreNLP. Although we have implemented default conversions, they are not correct in every case.
2. The parser is inconsistent in ways that cannot be anticipated linguistically. For example, the multiword expressions *Right at the light* and *Left at the light*—which, in the context of giving directions, mean *Turn right/left at the next traffic light*—are linguistically parallel, but the parser treats them differently, tagging *right* as an adverb but *left* as a verb. These kinds of inconsistencies are a good example of the challenges that arise when implementing models in systems. After all, the most natural and efficient way to prepare LEIAs to treat such expressions (i.e., the best modeling strategy) is to

    a. record the lexical construction "[direction] at the N", such that 'direction' can be filled by *right*, *left*, *hard right*, *hard left*, *slight right*, *slight left*, and N can indicate any physical object; and
    b. test the construction using a sample sentence to be sure that it is parsed as expected. However, when using a statistically trained parser, a correct parse for one example does not guarantee a correct parse for another structurally identical example. To generalize, any construction that includes variable elements can end up being parsed differently given different actual words of input.

3. The lexicon is incomplete. It can, for example, include one sense of a word requiring one syntactic construction but not another sense requiring a different syntactic construction. The question is, If an input uses a known word in an unexpected syntactic construction, should the system create a fuzzy match with an existing sense—and use that sense's semantic interpretation—or assume that a new sense needs to be learned? The answer: It depends. We illustrate the eventualities using simple examples that artificially impoverish our lexicon:

    a. Let us assume, for the sake of this example, that all of the verbal senses of *hit* in the lexicon are transitive—that is, they require a subject and a direct object. Let us assume further that the input is *He hit me up for 10 bucks*. Although any verb can accommodate an optional prepositional phrase (here: *for 10 bucks*), particles (here: *up*) cannot be freely added to any verb, so fuzzy matching would be the wrong solution. Instead, the agent needs to attempt to learn this new (idiomatic) word sense.
    b. By contrast, let us assume that the only available sense of the verb *try* requires its complement to be a progressive verb form, as in *Sebastian tried learning French*. Assume further that the input contains an infinitival complement: *Sebastian tried to learn French*. In this case, fuzzy matching of the syntactic structures would be correct since the same semantic analysis applies to both.

    So, when is fuzzy matching of syntactic structures appropriate and when isn't it?

Although our examples suggest a couple of rules of thumb (avoid fuzzy matching in the case of unexpected particles; do fuzzy matching given different realizations of verbal complements), the overall problem is larger and more complex.

4. Many inputs are actually noncanonical, reflecting production errors such as repetitions, disfluencies, self-corrections, and the like. These cannot, even in principle, be neatly syn-mapped.

Because of these complications, we have enabled agents to approach syn-mapping in two different ways, each one appropriate for different types of applications.

1. *Require a perfect syn-map.* Under this setting, if there is no perfect syn-map, the agent bypasses the typical syntax-informs-semantics NLU strategy and jumps directly to Situational Reasoning, where it attempts to compute the meaning of the input with minimal reliance on syntactic features (see section 7.2). This strategy is appropriate, for example, in applications involving informal, task-oriented dialogs because (a) they can contain extensive fragmentary utterances, and (b) the agent should have enough domain knowledge to make computing semantics with minimal syntax feasible.
2. *Optimize available syn-maps.* Under this setting, the agent must generate one or more syn-maps, no matter how far the parse diverges from the expectations recorded in the lexicon. These syn-maps feed the canonical syntax-informs-semantics approach to NLU. Optimizing imperfect syn-mapping is appropriate, for example, (a) in applications that operate over unrestricted corpora (since open-corpus applications cannot, in the current state of the art, expect full and perfect analysis of every sentence), (b) in applications for which confidently analyzing subsentential chunks of input can be sufficient (e.g., new word senses can be learned from cleanly parsed individual clauses, even if the full sentence containing them involves parsing irregularities), and (c) in lower-risk applications where the agent is expected to just do the best it can.

The processing flow involving syn-mapping is shown in figure 3.4.

Syn-mapping can work out perfectly even for subsentential fragments as long as they are valid beginnings of what might result in a canonical structure. Obviously, this is an important aspect of modeling incremental language understanding. For example, the inputs "The rust is eating" and "The rust is eating away" are both unfinished, but they will map perfectly to the syntactic expectations of eat-v2 presented earlier.

We already described how syn-mapping proceeds when everything works out well—that is, when the input aligns with the syntactic expectations recorded in the lexicon. Now we turn to cases in which it doesn't. Specifically (cf. figure 3.4), we will describe (a) the two recovery methods that attempt to normalize imperfect syn-maps and (b) the process of optimizing imperfect syn-maps when the input cannot be normalized.

One strategic detail is worth mentioning. When syn-mapping does not work perfectly, the agent waits until the end of the sentence to attempt recovery. That is, it does not attempt

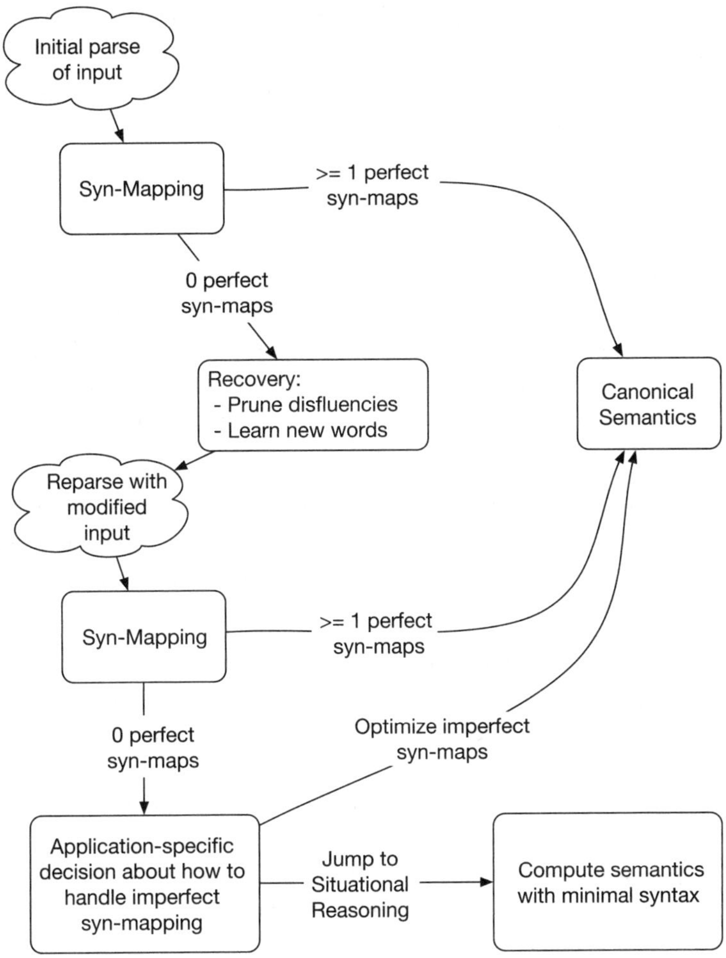

Figure 3.4
The processing flow involving syn-mapping. If the initial parse generates at least one perfect syn-map, then the agent proceeds along the normal course of analysis (stages 3–6: Basic Semantic Analysis, Basic Coreference Resolution, Extended Semantic Analysis, and Situational Reasoning). If it does not, then two recovery strategies are attempted, followed by reparsing. If the new parse is perfect, then the agent proceeds normally (stages 3–6). By contrast, if the new parse is also imperfect, the agent decides whether to optimize the available syn-maps and proceed normally (stages 3–6) or skip stages 3–5 and jump directly to stage 6, Situational Reasoning, where computing semantics with minimal syntax will be attempted.

recovery on sentence fragments during incremental analysis. This not only is a computationally expedient solution (the recovery programs need as much information as they can get) but also makes sense in terms of cognitive modeling, as it is unlikely that the high cognitive load of trying to reconstruct meaning out of nonnormative, subsentential inputs will be worthwhile.

3.2.2 Recovering from Production Errors

Noncanonical syntax—reflecting such things as disfluencies, repetitions, unfinished thoughts, and self-corrections—is remarkably common in unedited speech.[4] Even more remarkable is the fact that people, mercifully, tend not to even notice such lapses unless they look at written transcripts of informal dialogs. Consider, for example, the following excerpt from the Santa Barbara Corpus of Spoken American English, in which a student of equine science is talking about blacksmithing while engaged in some associated physical activity.

> we did a lot of stuff with the—like we had the, um, … the burners? you know, and you'd put the—you'd have—you started out with the straight … iron? … you know? and you'd stick it into the, … into the, … you know like, actual blacksmithing (DuBois et al. 2000–2005).[5]

Outside of context, and unsupported by the intonation of spoken language, this excerpt requires a lot of effort to understand.[6] Presumably, we get the gist by partially matching elements of input against the expectations in our mental grammar, lexicon, and ontology (i.e., we told you this was about blacksmithing).

The first method the LEIA uses to recover from these lapses is to strip the input of disfluencies (e.g., *um*, *uh*, *er*, *hmm*) and precisely repeated strings (e.g., *into the, ... into the*) and then attempt to parse the amended input. The stripping done at this stage only addresses the simplest, highest-confidence cases. The need for stripping is nicely illustrated by examples from the TRAINS corpus (Allen & Heeman, 1995).

(3.1) um so let's see where are there oranges $_{\text{(TRAINS)}}$

(3.2) there's let me let me summarize this and make sure I did everything correctly $_{\text{(TRAINS)}}$

If the new, stripped input results in a parse that can be successfully syntactically mapped, recovery was successful.

A research question is, Can more stripping methods be reliably carried out at this stage, or would this risk inadvertently removing meaningful elements before semantics had its say? For now, more sophisticated stripping methods are postponed until Situational Reasoning (chapter 7).

3.2.3 Learning New Words and Word Senses

The LEIA's lexicon currently contains about 30,000 word senses, which makes it sufficient for validating our microtheories but far from comprehensive. This means that LEIAs must be able to process both unknown words and unknown senses of known words. For example, if the lexicon happens to lack the word *grapefruit*, then the agent will have to undertake new-word learning when analyzing the seemingly mundane *I ate a grapefruit for breakfast*. The results of its learning will be similar to what many readers would conclude if faced with the input *Paul ate some cupuacu for breakfast this morning*: it must be some sort of food but it's unclear exactly what kind. (It is a fruit that grows wild in the Amazon rain forest.) The frequent need for new-word learning is actually beneficial for our program of work in developing LEIAs. After all, the holy grail of NLU is for agents to engage in lifelong learning of lexical, ontological, and episodic knowledge, so the more practice LEIAs get, and the more we troubleshoot the challenges they face, the better.

Since at this stage of the processing the agent is focusing exclusively on syntax, the only operation it carries out to handle unknown words is to posit a new template-like lexical sense that allows for syn-mapping to proceed in the normal way. To enable this, we have created templates for each typical dependency structure for open-class parts of speech. For example, the template for unknown transitive verbs is as follows. It will be used if the system encounters an input like *Esmeralda snarfed a hamburger*, which includes the colloquial verb *snarf*, meaning 'to eat greedily.'

```
new-transitive-verb-v1
   syn-struc
      subject          $var1
      v                $var0
      directobject     $var2
   sem-struc
      EVENT-#1
         CASE-ROLE     ^$var1
         CASE-ROLE     ^$var2
   meaning-procedures
      seek-specification EVENT-#1
```

Note that although the syntactic description is precise (exactly the same as for known transitive verb senses), the semantic description is maximally generic: an unspecified EVENT is supplied with two unspecified CASE-ROLE slots to accommodate the meanings of the attested arguments. (Note that the EVENT is indicated by EVENT-#1 in order to establish the necessary coreference between its use in the sem-struc and meaning-procedures zones.) Later on, during Basic Semantic Analysis (and, sometimes, Situational Reasoning), the agent will attempt to enhance the nascent lexicon entry by using the meanings of those case role fillers to (a) narrow down the meaning of the EVENT and (b) determine which case roles are appropriate. This future processing is put on agenda using the meaning-procedures zone, which contains a call to the procedural semantic routine called seek-specification, whose argument is the underspecified EVENT.

Whereas there is a single template for unknown transitive verbs, there are three templates for unknown nouns, since they can refer to an OBJECT (*meerkat*), EVENT (*hullabaloo*), or PROPERTY (*raunchiness*). The agent generates all three candidate analyses at this stage and waits until later—namely, Basic Semantic Analysis—to not only discard the unnecessary two but also attempt to narrow down the meaning of the selected one.[7] The following is the new noun template mapping to OBJECT.

```
new-noun-object-n1
    comments    This sense covers the eventuality that the noun refers to an OBJECT.
    syn-struc
        root $var0
    sem-struc
        OBJECT-#1
    meaning-procedures
        seek-specification OBJECT-#1
```

As for adjectives, such as *fidgety*, they are learned using the following template:

```
new-adj-adj1
    syn-struc
        mod    $var0
        n      $var1
    sem-struc
        PROPERTY-#1
            DOMAIN    ^$var1
    meaning-procedures
        seek-specification PROPERTY-#1
```

To reiterate, at this stage, the agent learns the syntax of new words and prepares for learning the associated semantics later on.

3.2.4 Optimizing Imperfect Syn-Maps

If, after attempting to normalize imperfect syn-maps (figure 3.4), there is still no perfect syn-map, the agent needs to make an application-oriented decision about whether to optimize imperfect syn-maps (essentially, do the best it can to push through the normal flow of processing) or skip to stage 6, where it can attempt to compute semantics with minimal syntax. Here we describe the default behavior: optimizing the imperfect syn-maps.

When the agent chooses to optimize imperfect syn-maps, it (a) generates all possible binding sets (i.e., mappings from elements of input to variables in lexical senses), (b) prioritizes them in a way that reflects their syntactic suitability, and (c) removes the ones that are highly implausible. This leaves a reasonable-sized subset of candidate mappings for the semantic analyzer to consider.[8] The need for this process is best illustrated by an example:

(3.3) Cake—no, *chocolate* cake—I'd eat every day.

CoreNLP generates an underspecified dependency parse for this input (see the online appendix at https://homepages.hass.rpi.edu/mcsham2/Linguistics-for-the-Age-of-AI.html for a screen shot of its output). Although the parse asserts that *I* is the subject of *eat* and *every day* is its modifier, it does not capture that *chocolate cake* is the direct object of *eat*. This leaves the syn-mapper little to go on in determining how to fill the case role slots of the available lexical senses of *eat* using all the available syntactic constituents.

The syn-mapper's approach to solving this problem is to assume, from the outset, that any non-verbal constituent can (a) fill any argument slot of the most proximate verb or (b) not fill any argument slot at all.[9] (Note that the latter is important in our example: the first instance of *cake* and the word *no* actually do not fill slots of *eat*.) The idea is to generate all candidate binding sets and then prune out the ones that seem too implausible to pass on to the semantic analyzer. Table 3.1 shows a small subset of the available binding sets if eat-v1 (the INGEST sense) is being considered as the analysis of *eat* in our sentence. Recall that eat-v1 is optionally transitive, which means that it does not require a direct object.

As a human, you might think that it makes no sense to even consider *every day* as the subject or direct object of *eat*, and that it makes no sense to leave both *chocolate cake* and *cake* unbound when they so obviously play a role in the eating being described. But making sense involves semantic analysis, and we haven't gotten there yet! What this *syntactic* analysis process sees is "NP—ADV, NP—NP AUX V NP," along with the parser's best guesses as to syntactic constituency and dependencies, but, as we have explained, its reliability decreases as input become more complex and/or less canonical.

The syn-mapper's algorithm for preferring some binding sets to others, and for establishing the plausibility cutoff for passing candidate binding sets to the semantic analyzer, involves a large inventory of considerations, most of which are too detailed for this exposition. But a few examples will illustrate the point.

Table 3.1
This is a subset of the binding sets that use *eat-v1* to analyze the input *Cake—no, chocolate cake—I'd eat every day*. The ellipses in the last row indicate that many more binding sets are actually generated, including even a set that leaves everything unbound, since this computational approach involves generating every possibility and then discarding all but the highest-scoring ones.

Subject	(Direct Object)	Unbound	Unbound	Unbound
Cake	chocolate cake	every day	I	
chocolate cake	Cake	every day	I	
I	Cake	every day	chocolate cake	
I	chocolate cake	every day	cake	
I	every day	chocolate cake	cake	
I	-	chocolate cake	cake	every day
chocolate cake	-	I	cake	every day
every day	I	chocolate cake	cake	
...	...	...	...	...

- Lexical senses for multiword expressions require particular lexemes; if those lexemes are not in the input, then the multiword senses are excluded. For example, the idiomatic sense of *eat* that covers *eat away at* (ERODE) will not be used to analyze the input *He ate a sandwich.*[10]
- If a lexical sense is mandatorily transitive, but the input has no direct object, then the given sense is strongly penalized. For example, the transitive sense of *walk* that is intended for contexts like *Miranda is walking the dog* will not be used to analyze inputs like *Miranda walks in the evenings.*
- If the lexical sense expects its internal argument to be an NP but the input contains a verbal complement, then that sense is strongly penalized. For example, the transitive sense of *like* that is intended for inputs such as *I like ice cream* will not be used for inputs such as *I like to ski.*
- In imperative clauses, the subject argument must not be bound. For example, when analyzing *Eat the cookies!* the syn-mapper will exclude the candidate set in which *the cookies* is used to fill the subject slot of the lexical sense eat-v1.

3.2.5 Reambiguating Certain Syntactic Decisions

No matter how the syn-mapping process proceeds—whether or not it involves recovery procedures, whether or not it generates perfect syn-maps—certain additional parse-modification procedures need to be carried out. This is because syntactic parsers are usually engineered to prefer yielding one result. However, they are not suited to making certain decisions in principle because the disambiguating heuristics are semantic in nature. Three syntactic phenomena that require parsers to make semantics-oriented guesses are prepositional phrase (PP) attachments, nominal compounds, and phrasal verbs.

- ***PP attachments.*** When a PP immediately follows a post-verbal NP, it can modify either the verb or the adjacent NP. A famous example is *I saw the man with the binoculars.*
    - If the binoculars are the instrument of seeing (they are used to see better), then the PP attaches to the verb: I [$_{VP}$ saw [$_{NP}$ the man] [$_{PP}$ with the binoculars]].
    - If the binoculars are associated with the man (he is holding or using them), then the PP attaches to the NP: I [$_{VP}$ saw [$_{NP}$ the man [$_{PP}$ with the binoculars]]].
- ***Nominal compounds.*** Nominal compounds containing more than two nouns have an internal structure that cannot be predicted syntactically; it requires semantic analysis. Compare:
    - [[kitchen floor] cleanser]
    - [kitchen [floor lamp]]

- ***Phrasal verbs.*** In English, many prepositions are homographous with (i.e., have the same spelling as) verbal particles. Consider the collocation *go after* + NP, which can have two different syntactic analyses associated with two different meanings:
  - [verb + particle + direct object] has the idiomatic meaning 'pursue, chase': The cops $went_{V}$ $after_{PARTICLE}$ the $criminal_{DIRECT\text{-}OBJECT}$.
  - [verb + preposition + object of preposition] has the compositional meaning 'do some activity after somebody else finishes their activity': The bassoonist $went_{V}$ $after_{PREP}$ the $cellist_{OBJECT\text{-}OF\text{-}PREP}$.

While there are clearly two syntactic analyses of *go after* that are associated with different meanings, and while there is often a default reading depending on the subject and object selected, it is impossible to confidently select one or the other interpretation outside of context. After all, *The cops went after the criminal* could mean that the cops provided testimony after the criminal finished doing so, and *The bassoonist went after the cellist* could mean that the former attacked the latter for having stepped on his last reed.

For all of these, LEIAs reambiguate the parse. That is, they always, as an across-the-board rule, create multiple candidates from the single one returned by the parser. Selecting among them is the job of the semantic analyzer at the next stage of analysis.

3.2.6 Handling Known Types of Parsing Errors

This book concentrates primarily on ideas—our theory of NLU, the rationale behind it, and how systems that implement it support the operation of intelligent agents. These ideas could be implemented using a wide range of engineering decisions, which are not without interest, and we have devoted significant effort to them. However, had we decided to discuss them in detail, this would have doubled the length of this book. Still, we will mention select engineering issues and solutions to emphasize that engineering must be a central concern for computational linguistics, being at the heart of the model-to-system transition (see section 2.6).

The engineering solution we consider here involves parsing errors. As mentioned earlier, syntactic parsing is far from a solved problem, so parsing errors are inevitable, even for inputs that are linguistically canonical. For example, our lexicon includes a ditransitive sense of *teach* intended to cover inputs like *Gina taught George math*. However, the parser we use incorrectly analyzes *George math* as a nominal compound.[11] How did we detect this error? Manually, as a part of testing and debugging. (The agent cannot independently recognize this particular error because the parse actually works out syntactically, the input being structurally parallel to *Gina taught social studies*, which does include a nominal compound and aligns with the transitive sense of *teach* in the lexicon.)

Rather than go down the rabbit hole of creating fix-up rules for the parser, we do the following: If inputs with the given structure are not crucial for a current application, we ignore the error and allow the associated inputs to be incorrectly analyzed. The agent then treats them as best it can despite the error. If, by contrast, such inputs *are* crucial for a current application—for example, if they must be featured in a robotic system demo tomorrow—then we use a recovery strategy that works as follows. We invent a sample sentence, run it through the parser, record its actual syntactic analysis, and then manually provide the necessary linking between syntactic and semantic variables. All of this information is stored in an adjunct database that does not corrupt the original lexicon.

Let us work through our *teach* example by way of illustration. Below is the canonical lexical sense of ditransitive *teach*, teach-v1.

```
teach-v1
    def.        to teach someone some subject matter
    ex.         Mary taught John physics.
    syn-struc
        subject           $var1
        v                 $var0
        indirectobject    $var3
        directobject      $var2
    sem-struc
        TEACH
            AGENT         ^$var1
            THEME         ^$var2
            BENEFICIARY   ^$var3
```

Compare this with the supplementary sense that accommodates the parsing error, teach-v101. This sense includes two traces that it is not canonical: it uses a special sense-numbering convention (100+), and it includes an associated comment in the comments field.

```
teach-v101
    def.        to teach someone some subject matter
    ex.         Mary taught John physics.
    comments    accommodates a parsing error
    syn-struc
        subject           $var1
        v                 $var0
        np
            n             $var2
            n             $var3
    sem-struc
        TEACH
            AGENT         ^$var1
            THEME         ^$var3
            BENEFICIARY   ^$var2
```

3.2.7 From Recovery Algorithm to Engineering Strategy

The recovery algorithm just described morphed into an available—although not default—engineering practice for adding new construction senses to the lexicon. This practice is used when acquirers either suspect or have evidence that a construction will not be treated by the parser in the way anticipated by our linguistic theory. The rationale for this practice is best explained by tracing what lexicon acquirers and programmers each want.

Lexicon acquirers want to record senses fast, with as few constraints on their expressive power as possible. They don't want to worry about the quirks of actual processors—or, more formally, about model-to-system misalignments. Programmers, for their part, want the available processors to output what the knowledge engineers are expecting. As it turns out, all of these desiderata can be met thanks to a program we developed for this purpose: the ExampleBindingInterpreter. The ExampleBindingInterpreter requires two types of input:

1. a lexical sense whose syn-struc is underspecified: it contains an ordered inventory of fixed and variable components, but the acquirer need not commit to the constituents' parts of speech or their internal structure; and
2. a sample sentence that indicates which words align with which syntactic components.

The ExampleBindingInterpreter does the rest. It creates a syn-map, no matter the actual parser output, allowing for subsequent semantic analysis to proceed in the normal way. We will illustrate the method using an example from an autonomous vehicle application.

The input is the command from the user to the agent "Turn right at the light," which is so frequent in this particular application that it merits being recorded explicitly in the lexicon. The semantic description of this input

a. includes the REQUEST-ACTION conveyed by the imperative verb form;
b. disambiguates the verb *turn* (which can also mean, e.g., 'rotate');
c. disambiguates the word *right* (which can also mean, e.g., 'correct');
d. disambiguates the word *at* (which can also, e.g., indicate a time);
e. concretizes the meaning of *the* as 'the next one' (we do not want the system to use general coreference procedures to try to identify an antecedent for *light*); and
f. disambiguates the word *light* (which can also mean, e.g., 'lamp').

If we were to record the syn-struc for this multiword expression in the usual way, it would look as follows.

```
turn-v12
   def.        The expression 'Turn right at the light.'
   ex.         Turn right at the light.
   syn-struc
      v            $var0 (form infinitive)
```

```
        adv       $var1 (root 'right')
        pp
            prep  $var2 (root 'at')
            obj   $var3 (root 'light')
    sem-struc
        ...
```

But consider all the mismatches that might occur during parsing: The parser might consider *right* an adjective or a verb rather than adverb; it might attach the PP to *right* rather than to *turn*; and this formalism does not readily allow for the explicit inclusion of the word *the*, which we actually want to treat specially in the sem-struc by blocking generic coreference procedures. Now compare this with the underspecified syn-struc in the lexicon entry shown in turn-v101 (which includes the sem-struc as well). The variables x, y, and z allow for the parser to tag the given words with any parts of speech.

```
turn-v101
    def.        The expression 'Turn right at the light.'
    ex.         Turn right at the light.
    comments    Syntactically underspecified; uses ExampleBindingInterpreter.
    syn-struc
        use-example-binding t
        v           $var0
        x           $var1 (root 'right')
        prep        $var2 (root 'at')
        y           $var3 (root 'the')
        z           $var4 (root 'light')
    sem-struc
        REQUEST-ACTION
            AGENT           HUMAN-#1 ("speaker")
            BENEFICIARY     HUMAN-#2 ("hearer")
            THEME           refsem1
        refsem1
            TURN-VEHICLE-RIGHT
                AGENT       HUMAN-#2 ("hearer")
                LOCATION    NEXT-TRAFFIC-LIGHT
        ^$var1    null-sem+
        ^$var2    null-sem+
        ^$var3    null-sem+
        ^$var4    null-sem+
    example-bindings turn-0 right-1 at-2 the-3 light-4
```

Let us work through the above entry from top to bottom.

- The fact that automatic variable binding will occur is indicated by "use-example-binding t" in the syn-struc ('t' means 'true').
- In writing this syn-struc, the acquirer needs to commit to only one part of speech: the one for the headword ($var0). All other parts of speech can be either asserted (if the

acquirer is confident of a correct analysis) or indicated by variables. In this example, the part of speech for *at* (prep) is asserted and the rest are left as variables.

- The actual root word for each category can be listed or left open: e.g., the actual noun (*light*) could have been left out, allowing the expression to cover any input matching "Turn right at the N."
- Optional elements can be indicated in the usual way: (opt+).
- Variations on an element can also be indicated. For example, the two expressions "Turn right at the light" and "Turn left at the light" can be covered by changing the description of X to (root 'right' 'left') as long as the parser assigns the same part of speech to all listed variations. (Recall that for the elliptical *Right/left at the light*, 'right' and 'left' were assigned different parts of speech, which would make sense bunching impossible in this case.)
- As described earlier, the sem-struc
  - asserts that this is a REQUEST-ACTION;
  - includes disambiguation decisions for all component words by indicating the concepts that describe their meaning; and
  - uses those concepts to fill case roles slots.
- The sem-struc indicates the roles of the speaker and hearer, which must be grounded in the application. We will use the convention "HUMAN-#1 ("speaker")" and "HUMAN-#2 ("hearer")" throughout as a shorthand to indicate the necessary grounding.
- The concepts TURN-VEHICLE-RIGHT and NEXT-TRAFFIC-LIGHT are, like all concepts, described by properties in the ontology—they are not vacuous labels in upper-case semantics. The reason they were promoted to the status of concepts is that, within our driving script—which was acquired to support a particular application—they play a central role. It is, therefore, more efficient to encapsulate them as concepts rather than to compositionally compute the elements of the expression on the fly every time. (Cf. the discussion of *eating hot liquids with a spoon* in section 2.8.2.)
- The null-semming of the variables reflects that their meanings have already been incorporated into the sem-struc. (For example, ^$var1, which corresponds to the word *right*, is null-semmed because its interpretation is folded into the choice of the concept TURN-VEHICLE-RIGHT. The other variables are null-semmed for analogous reasons.) If these variables were not null-semmed, then the analysis system—based on its global processing algorithm—would try to compositionally incorporate all available meanings of these words into the TMR, even though their meanings are already fully taken care of by the description in the sem-struc.
- The example-bindings field contains the sample sentence to be parsed, whose words are appended with the associated variable numbers from the syn-struc.

The reason for presenting this automatic syn-mapping process in such detail is to underscore that it addresses two core needs of NLU: (1) the need to populate the lexicon with constructions, since so much of language analysis is not purely compositional, and (2) the need to proactively manage the inevitable mismatches between idealized models and the actual results of actual processors that are available to be used in systems.

You might wonder: Aren't we losing something in terms of cognitive modeling by not recording the canonical linguistic structure of constructions like these? Yes. We are sacrificing modeling desiderata in service of making a particular system, which uses a particular parser, actually work. This is a trade-off. We are certainly not recommending that underspecifying the parts of speech should be a blanket answer to recording knowledge about constructions. Instead, it should be used judiciously, like all tools in the system-building toolbox.

3.3 Managing Combinatorial Complexity

Unfortunately, syn-mapping can result in many candidates for the semantic analyzer to work through. Because agents must function in real time, we need to address this problem of combinatorial explosions head-on, which we do with the microtheory of combinatorial complexity, to which we now turn.

The first thing to say about this microtheory is that it anticipates and attempts to circumvent the consequences of combinatorial complexity at the interface between syntactic and semantic analysis. That is, we can foresee which kinds of lexical items will predictably spawn combinatorial complexity, and we can reduce that complexity using specific types of knowledge engineering. Since there is no dedicated processing module corresponding to the syntax-semantics interface, architecturally, this microtheory best resides in Pre-Semantic Integration.

Combinatorial complexity arises because most words have multiple senses. If a sentence contains 10 words, each of which has 3 senses, then the agent must consider $3^{10} = 59{,}049$ candidate analyses. Since, in our lexicon, prepositions have many senses each, and common light verbs (such as *have*, *do*, *make*) have several dozen senses each, this means that even midlength sentences that contain even one preposition or light verb can quickly run into the tens of thousands of candidate analyses.

Consider the example in figure 3.5, which, although obviously cooked, is nicely illustrative: *Pirates attack animals with chairs in swamps.* (In a cartoon, this might even make sense.) If we consider even just two senses of each word, there will be 128 candidate interpretations.[12]

Not only does this example offer 128 candidate interpretations, but the semantic constraints available during Basic Semantic Analysis will be able to weed out only some of the interpretations. Many will remain equally viable, meaning that there will be extensive residual ambiguity. For example, both senses of *pirate* (pirate at sea and intellectual

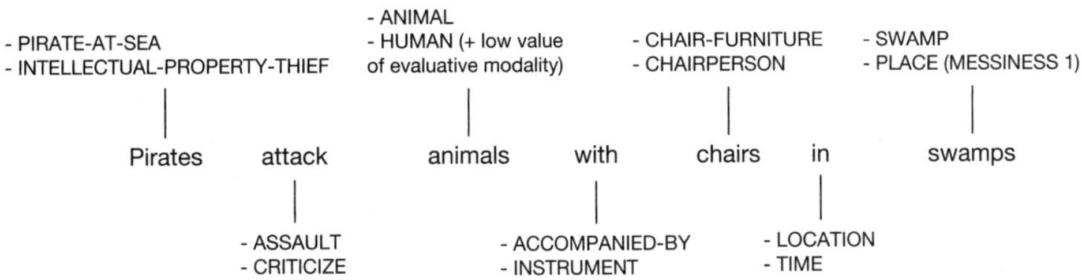

Figure 3.5
A subset of paired, syntactically identical senses.

property thief) are equally suitable as AGENTs of both senses of *attack* (physically attack and verbally assault); both senses of *swamp* (a bog and a messy place) are equally suitable as fillers of the LOCATION interpretation of *in*; both senses of *animal* (a living creature and a human viewed negatively) can be the THEME of the ASSAULT meaning of *attack*; and *with* can indicate the instrument the pirates use (chairs for sitting) or people accompanying the pirates in their actions (chairpersons).

Many instances of ambiguity are expected to remain unresolved at the stage of Basic Semantic Analysis, before more sophisticated reasoning has been leveraged. However, an interesting question arises at the interface of cognitive modeling and system engineering: Could candidate interpretations be bunched in a way that is both psychologically plausible and practically useful, in order to better manage the search space for the optimal analysis? The answer is yes, and it could be approached either through knowledge engineering or through dynamic reasoning.[13]

The knowledge-engineering solution (which we do not pursue) involves developing a hierarchical lexicon. Since both kinds of pirates are thieves, they could share the mapping THIEF, which could be used by default if the context failed to make clear which one was intended. Similarly, since both kinds of attack indicate a type of conflict, they could share the mapping CONFLICT-EVENT, which would correspondingly be used by default. There is much to like about this approach to knowledge engineering, not least of which is that it jibes with our intuitive knowledge of these word meanings. However, is developing a hierarchical lexicon—which is conceptually heavier and more time-consuming than developing a flat lexicon—really the best use of acquirer time, given that (a) many other types of knowledge are waiting to be acquired and (b) the underlying ontology is already hierarchical? We think not, which leads us to the more promising solution that involves runtime reasoning.

We have developed computational routines for dynamic sense bunching, with the following being the most useful so far:

- Bunching productive (i.e., not phrasal) prepositional senses into a generic RELATION.
- Bunching verb senses with identical syn-strucs and identical semantic constraints into their most local common ancestor. For example, *turn the steak* can mean INVERT (flip)

or ROTATE (let a different portion be over the hottest part of the grill), which have the common ancestor CHANGE-POSITION. Similarly, *distribute the seeds* can mean SPREAD-OUT or DISTRIBUTE (as to multiple people who bought them), whose common ancestor is CHANGE-LOCATION. In cases of literal and metaphorical sense pairs (e.g., *attack*), the common ancestor can be as imprecise as EVENT; however, even this is useful since it is a clue that the ambiguity could involve metaphorical usage.

- Bunching noun senses that refer to ANIMALS or HUMANS. For example, *pig* can refer to a barnyard animal, a messy person, or any animal who overeats (the latter two are described by multiconcept sem-strucs headed by HUMAN and ANIMAL, respectively).
- Bunching different senses of PHYSICAL-OBJECTS. These include, for example, the MACHINE and COMPUTER meanings of *machine* and the AUTOMOBILE and TRAIN-CAR meanings of *car*.

When senses are dynamically bunched, a procedural semantic routine is attached to the umbrella sense and recorded in the TMR. This offers the agent the option to attempt full disambiguation at a later stage of analysis.

Let us take as an example the preposition *in*. Below is one of the syntactically typical (nonphrasal) senses, followed by a list of some of the other syntactically identical senses.

```
in-prep1
   def.        refers to the physical location of an object or event
   ex.         The cat in the study is sleeping (the pp modifies 'cat').
               The cat is sleeping in the study (the pp modifies 'sleeping').
   syn-struc
      root    $var1 (cat (n v))
      pp
         prep   $var0
         obj    $var2
   sem-struc
      ^$var1
         LOCATION    ^$var2
```

Syntactically similar senses:

- in-prep14: DURING (*In the interview he said …*)
- in-prep17: TIME (*a meeting in January*)

Compare this with the umbrella sense below that bunches them. It links the meanings of $var1 and $var2 using the generic RELATION and includes a meaning procedure (seek-specification RELATION [in-prep1/14/17]) that points to the senses that can be consulted later for full disambiguation.

```
in-prep-501[14]
   def.        the umbrella sense for several senses of 'in'
   syn-struc
      root     $var1 (cat (n v))
      pp
         prep    $var0
         obj     $var2
   sem-struc
      ^$var1
         RELATION    ^$var2
   meaning-procedures
      seek-specification RELATION [in-prep1/14/17]
```

In many cases, the semantic analyzer will be able to select a winning sense. For example, *a meeting in January* will be analyzed confidently using in-prep17 since that sense requires the object of the preposition to refer to a temporal expression, such as MONTH, YEAR, or CENTURY. In such cases, sense bunching will clearly not be the final solution. However, in many cases it will be useful—for example, when speakers use prepositions noncanonically, which is a common type of performance error (see section 6.2.2), or when a multiword expression that would ideally be recorded as a lexical sense (e.g., *in good faith*) has not yet been acquired, thus necessitating a less precise analysis.

To give just one example of how much easier it is for people to read bunched outputs, consider the TMR for the sentence *A pirate was attacked by a security guard* in which the available analyses of *pirate* and *attack* are bunched (and *security guard* is unambiguous).

```
PHYSICAL-EVENT-1
   AGENT            SECURITY-GUARD-1
   THEME            HUMAN-1
   TIME             <find-anchor-time
   BUNCHED-FROM     ATTACK (attack-v1), CRITICIZE (attack-v2)

HUMAN-1
   BUNCHED-FROM     PIRATE-AT-SEA (pirate-n1),
                    INTELLECTUAL-PROPERTY-THIEF (pirate-n2)
```

The candidates this structure covers are as follows, in plain English:

- A pirate at sea was physically attacked by a security guard.
- A pirate at sea was yelled at by a security guard.
- An intellectual property thief was physically attacked by a security guard.
- An intellectual property thief was yelled at by a security guard.

Sense bunching can be applied in many ways: all available types of sense bunching can be carried out prior to runtime and employed across the board; select types of sense bunching (e.g., prepositions only) can be applied prior to runtime; or the agent can dynamically decide whether or not to bunch based on factors such as the number of candidate TMRs being too large or the extent to which the candidate TMRs do or do not fall within the agent's scope of interest. The actual strategy selected will depend on the requirements of the application system.

Of course, dynamic sense bunching is not the only way to deal with combinatorial complexity. In a particular application, the agent can opt to prefer domain-relevant interpretations from the outset, thereby reducing or even completely removing the problem of lexical disambiguation. (This is, in fact, what many developers of robotic systems routinely do, as this meets short-term goals.) We describe why we chose not to do this in the general case in chapter 7. Another option is to label a subset of senses as preferred, prototypical ones. But although this might seem like an easy type of knowledge acquisition at first glance, it quickly becomes complicated once we move beyond the relatively small set of simplest cases like *dog* defaulting to a canine companion. Ask people whether *cat* means a domesticated feline or a wild one, and the debate will be on! Moreover, even if we recorded knowledge to deal with most eventualities, there would still be residual ones, and one of the foci of our scientific investigation of NLU is to determine how we can best prepare a LEIA to deal with inputs that inevitably combine known and unknown information.

To conclude, although the repercussions of combinatorial complexity will not be encountered until later stages of processing, sense bunching can be incorporated into Pre-Semantic Integration to avoid at least some of those problems.

3.4 Taking Stock

This chapter has described the benefits and challenges of importing resources to carry out the pre-semantic stages of NLU; methods of preparing pre-semantic heuristics to best serve upcoming semantic analysis; the first stage of learning unknown words; and the process of dynamic sense bunching for dealing with combinatorial complexity.

In considering how much is involved in what we call Pre-Semantic Integration, one might ask, Why did we import external processors to begin with rather than developing our own? In fact, in the early days of Ontological Semantics, we *did* develop our own preprocessor along with a lexicalized parser that used a just-in-time parsing strategy (Beale et al., 2003). Although these processors were ideal for what they covered, they did not cover as many phenomena as the statistical preprocessors and parsers that were becoming available at the time. So we made the leap to import externally developed tools and invested extensive resources into integration. It was only once we had carried out the integration that we could assess how well the tools served our needs. As it turned out, there was a mixed bag of costs and benefits.

The original motivation for importing the tools was to save engineering time on pre-semantic issues. However, we did not foresee how much continued engineering effort would be needed (a) for integration (each new and improved version of the tool set can send ripples throughout our system) and (b) for developing methods to recover from unexpected results. In hindsight, it is unclear whether importing the tool set fostered or impeded our work on semantics and pragmatics. However, a clearly positive aspect of this decision is that it shows that we practice what we preach about science and engineering in AI: that systems need to actually work, and that no single group of individuals can solve the whole problem, so some sort of integration by different teams is ultimately inevitable. Consequently, developers must not shy away from making strategic decisions under uncertainty and incorporating the outcomes, whatever they may be, back into the overall program of R&D. In the case we describe, this has meant spending more time on recovering from unexpected syntactic parses than we could have anticipated a decade ago. But this led us down the path of paying particular attention to unexpected inputs overall, which is entirely to the good. We believe that herein lies a useful lesson for all practitioners in our field.

3.5 Further Exploration

1. Get acquainted with the Stanford CoreNLP parser using the online interface available at the website corenlp.run. To show the results of more than just the default annotators, click on the "Annotations" text field and select more options from the pull-down menu: for example, lemmas, coreference. In addition to grammatical sentences, try sentences that include production errors, such as repetitions (*Put the lamp on the on the table*) and highly colloquial ellipses (*Come on—that, over here, now!*). Even though utterances like these—and many more types of noncanonical formulations—are very common in real language use, they pose challenges to current parsing technologies.

2. Practice drawing parse trees using an online tree-drawing tool, such as the one at http://ironcreek.net/syntaxtree/. This will be useful because many aspects of the upcoming discussions assume that readers at least roughly understand the syntactic structure of sentences. If you need an introduction to, or refresher about, parse trees, you can look online (e.g., Wikipedia) or consult a textbook on linguistics or NLP, such as

- *Language Files: Materials for an Introduction to Language and Linguistics* (12th ed.), edited by Vedrana Mihalicek and Christin Wilson (The Ohio State University Press, 2011).
- *Natural Language Understanding* by James Allen (Pearson, 1994).

Avoid descriptions of syntactic trees within the theory of generative grammar since their X-bar structure reflects hypotheses about the human language faculty that are not followed by natural language parsing technologies.

3. Explore how PP-attachments work using the search function of the online COCA corpus (Davies, 2008–) at https://www.english-corpora.org/coca/. Use the search string *_nn with a _nn*, which searches for [any-noun + *with a* + any-noun]. Notice the variety of eventualities that have to be handled by semantic analysis. Learn to use the various search strategies available in the interface, since we will suggest more exercises using this corpus in upcoming chapters.

4

Basic Semantic Analysis

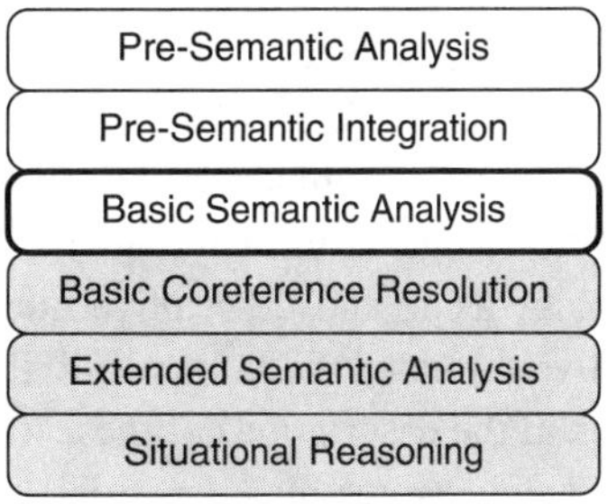

Chapter 2 introduced the basic principles of semantic analysis by LEIAs. As a refresher, the agent builds up the meaning of a sentence by

- syntactically parsing it to identify the verbal head of each clause and its arguments, which are usually noun phrases;
- identifying all available interpretations of the verbal head when used in the given syntactic environment;
- computing all available interpretations of the arguments, which might be multiword constituents (e.g., *a very big, green tree*);
- establishing one or more candidate semantic dependency structures by filling the case role slots (e.g., AGENT, THEME) of available EVENT interpretations with available interpretations of the arguments;
- integrating into the candidate meaning representation(s) the meanings of all other elements of input, such as sentence adverbs, modals, conjunctions, and so on; and
- generating one or more candidate text meaning representations (TMRs) that reflect available analyses; each of these is scored with respect to how well it aligns with the expectations recorded in the lexicon and ontology.

The status of these processes is as follows: They are necessary contributors to Basic Semantic Analysis. They are presented here at a conceptual level, not as an algorithm. In fact, in this book we will not present an algorithm for building a basic semantic analyzer (i.e., implementing stage 3 of analysis) using our knowledge bases and microtheories because (a) many different algorithms can implement this process and (b) such a description would be of interest exclusively to programmers, not the broader readership of this book. To understand why many algorithms can implement the process, consider the following comparison.

Building a basic semantic analyzer is much like putting together a jigsaw puzzle, except that a jigsaw puzzle allows for only one solution, whereas semantic analysis can result in multiple candidate solutions. The puzzle pieces for basic semantic analysis are the syntactic and semantic descriptions of word senses in the lexicon. Analyzing a sentence involves selecting the best combination of word senses, adjudged using various scoring criteria. Just as a jigsaw puzzle can be approached using many different algorithms—starting with the corners, the outer edges, a set of similarly colored pieces, and so on—so, too, can semantic analysis. In fact, over the years different software engineers in our group have implemented two completely different semantic analyzers based on the same theory, models, and knowledge bases. Although their algorithms organized in different ways how the recorded knowledge and descriptive microtheories were leveraged to analyze inputs, they effectively yielded the same result. For future NLU system developers, this book's main utility is in the descriptions of these knowledge bases and microtheories.[1] As history has shown, knowledge remains useful indefinitely, while system-implementation environments have a relatively short life. With this in mind, we concentrate our presentation around the content of the microtheory of Basic Semantic Analysis and its supporting knowledge bases.

Whereas chapter 2 described the general principles of Basic Semantic Analysis using only very simple examples, this chapter begins a progressively deeper exploration of the kinds of linguistic phenomena that make language at once rich and challenging. The Basic Semantic Analysis described in this chapter attempts to disambiguate the words of input and determine the semantic dependency structure of clauses (essentially, case roles and their fillers) using only the most local context as evidence. But, as this long chapter will reveal, many linguistic phenomena can be treated even at this early stage, when the agent is invoking only a subset of the heuristics and microtheories available to it.

Before proceeding to the phenomena themselves, a few general points deserve mention.

1. Roughly speaking, this stage computes what some linguists call *sentence semantics*: the meaning that can be understood from propositions presented in isolation. For example, the pronoun *he* refers to a male animal, but *which one* must be contextually specified. Similarly, the utterance *I won't!* means *I won't do something*, with that *something* requiring contextual specification. At this stage of processing, the basic analysis (i.e., *male animal* is (ANIMAL (GENDER male)); *do something* is (EVENT)) is recorded in the TMR, along with metadata that includes a call to a procedural semantic routine that can be run later to further specify the meaning.
2. In many cases, different *instances* of a given linguistic phenomenon, such as verb phrase ellipsis or nominal compounding, fall into different functional classes. Some instances can be fully analyzed at this stage, whereas others require heuristics or methods that only become available later.
3. For people, a lot of knowledge about language resides in their construction-rich lexicon; therefore, careful development of the lexicon is key to successful NLU by LEIAs. Any meaning that can be recorded in the lexicon—even using variable-rich constructions—can be processed as part of Basic Semantic Analysis.[2]

4. One reason for separating Basic Semantic Analysis from its various extensions is that, in many cases, the LEIA does not need a deeper analysis. That is, Basic Semantic Analysis often gives the agent the gist of what the utterance means—which is often enough for it to decide whether digging deeper is needed.

Basic Semantic Analysis covers a large number of linguistic phenomena. The sections below present a representative sampling of them, stopping far short of what needs to be mastered by knowledge engineers working in the environment. The goals of the chapter are

1. to show the benefits of acquiring lexicon and ontology based on how language is actually used, rather than a linguistically pure idealization;
2. to flesh out the notion of *Basic*—as contrasted with *Extended*—Semantic Analysis in the overall functioning of LEIAs; and
3. to underscore just how many semantic phenomena must be treated by natural language processing systems—that is, how much is left to do beyond the tasks most commonly undertaken by mainstream NLP systems.

In wrapping up this introduction, let us make one tactical suggestion, which derives from our experience in giving talks and teaching classes about this material. LEIA modeling is a practical endeavor with one of the key challenges being to balance the desire for human-level agent behavior with the reality that a large amount of rigorous linguistic description remains to be done. It is a fun pastime, but of little practical value, to think up exceptions to every useful generalization about language and the world. We recommend that readers not engage in this because it would distract attention from the considerable amount of information being presented and because fringe cases related both to language (e.g., garden-path sentences[3]) and to the world (e.g., humans who meow) are just not important enough to demand attention—not until we have gotten much further along in handling the very substantial mass of regular cases. With this in mind, let us move on to the linguistic phenomena handled during Basic Semantic Analysis.

4.1 Modification

The subsections below classify modifiers (primarily adjectives and adverbs, although other syntactic entities can function as modifiers as well) according to how they are described in lexical senses.[4] Lexical descriptions can include a combination of static descriptors and procedural semantic routines.

4.1.1 Recorded Property Values

The meanings of many modifiers can be described in their lexicon entries as the value of some PROPERTY recorded in the ontology. As background, a PROPERTY is a concept whose name is an ontological primitive. Each property is constrained by the values for its DOMAIN and RANGE slots. The DOMAIN indicates the types of OBJECTS and/or EVENTS the PROPERTY can apply to, whereas the RANGE indicates the set of its legal fillers.

Properties are divided into three classes based on the fillers for their RANGE slot. The fillers for the RANGE of SCALAR-ATTRIBUTES can be a number, a range of numbers, a point on the abstract scale {0,1}, or a range on that scale. For example,

```
HAPPINESS
    DOMAIN      ANIMAL
    RANGE       {0,1}
```

The fillers for the RANGE of LITERAL-ATTRIBUTES are literals—that is, they are not ontological concepts and are, therefore, not written in small caps. For example,

```
MARITAL-STATUS
    DOMAIN      HUMAN
    RANGE       single, married, divorced, widowed
```

The fillers for the RANGE of RELATIONS are ontological concepts. For example,

```
EXPERIENCER-OF
    DOMAIN      ANIMAL
    RANGE       EVENT
```

Since these ontological descriptions already exist, lexical senses that use properties to describe word meanings need to assert a particular value for the RANGE, but they do not need to repeat the understood constraints on the DOMAIN. For example, the meaning of *happy* is described in the lexicon as .8 on the abstract scale of HAPPINESS.

```
happy-adj1
    def.       experiencing a positive feeling
    ex.        a happy chipmunk
    syn-struc
        adj        $var0
        n          $var1
    sem-struc
        HAPPINESS
            DOMAIN      ^$var1
            RANGE       .8
```

The fact that $var1 must refer to an ANIMAL is not written in the lexical sense because the fact that HAPPINESS can apply only to ANIMALS is available in the ontology. If $var1 in a particular input does *not* refer to an ANIMAL, this lexical sense will not be used to analyze the input since the property HAPPINESS cannot apply to it.

Below are specific examples that illustrate how property values are used to describe the meanings of modifiers.

- SCALAR-ATTRIBUTES
  - a happy goat: GOAT (HAPPINESS .8)
  - a severe hailstorm: HAILSTORM (INTENSITY .9)
  - a 19-pound xylophone: XYLOPHONE (WEIGHT 19 (MEASURED-IN POUND))
- LITERAL-ATTRIBUTES
  - a divorcé: HUMAN (GENDER male) (MARITAL-STATUS divorced)
  - an accredited college: COLLEGE (ACCREDITED yes)
- RELATIONS
  - a fearful wolf: WOLF (EXPERIENCER-OF FEAR)
  - a wooden chair: CHAIR (MADE-OF WOOD)
  - to explore endoscopically: INVESTIGATE (INSTRUMENT ENDOSCOPE)
  - to analyze statistically: ANALYZE (INSTRUMENT STATISTICAL-ANALYSIS)
  - a house on fire: PRIVATE-HOME (THEME-OF BURN)

As the lexical sense for happy-adj1 showed, PROPERTY-based descriptions of word senses are recorded in the sem-struc zones of the lexicon. During NLU, they are copied directly into the TMR, meaning that such modifiers present no challenges beyond the potential for ambiguity. For example, *blue* can refer to feeling sad (HAPPINESS .2) or to a color (COLOR blue). In many cases, the meaning of the head noun helps to disambiguate: people and some animals can be sad, whereas it is mostly inanimate objects, as well as some birds and fish, that are colored blue. Of course, nonliteral language can require the opposite interpretations: a *blue person* might describe someone in a blue costume, and a *blue house* might describe a house that invokes feelings of sadness in people who drive by it. However, these examples of extended usage should not sidetrack us from the main point, which is that the meaning of the modified lexical item (such as a noun) tends to provide good disambiguating power for lexical items (such as adjectives) that modify it. During Basic Semantic Analysis, if multiple candidate analyses are possible, they are all generated and scored, and the LEIA waits until later in the process to select among them.[5]

4.1.2 Dynamically Computed Values for Scalar Attributes

All the modifiers discussed so far could be described using static meaning representations recorded in the sem-struc zone of lexical senses. However, the meaning of adverbs that modify scalar attributes—such as *very* and *extremely*—must be computed dynamically, taking into consideration the meaning of the particular adjective they are modifying. For example, *smart* and *dumb* are described using the scalar attribute INTELLIGENCE, with *smart* having a value of .8 and *dumb* .2. If *very* is added to *smart*, it raises the value from .8 to .9, whereas if *very* is added to *dumb* it decreases the value from .2 to .1. This dynamic modification of property values is carried out using a procedural semantic routine recorded in the lexical sense for *very*. Since this routine is simple, local, and not error-prone, it is carried out as part of Basic Semantic Analysis.[6]

By way of illustration, this section describes two procedural semantic routines involving scalar attributes:

1. *delimit-scale*, for adverbs that modify the abstract value of a scalar attribute: for example, *very experienced, not extremely smart*; and
2. *specify-approximation*, for adverbs that indicate the bidirectional stretching of a cited number: for example, *around 10:00, about 80 feet wide*.

Delimit-scale. Delimit-scale is the procedural semantic routine that calculates the modified value of a scalar attribute that is expressed as a point on the abstract {0,1} scale. It is called from lexical senses for words like *very, extremely, quite, moderately*, and *somewhat* when they modify a scalar attribute. The function takes three arguments:

- the meaning of the word modified: for example, *intelligent* is (INTELLIGENCE .8), *dim* is (INTELLIGENCE .2);
- which direction along the scale the value should be shifted—toward the mean (e.g., *relatively intelligent*) or toward the extreme (*very intelligent*); and
- the amount the value should be shifted (e.g., .1, .2, .3).

So, *very small* is calculated by taking the value of *small* (SIZE .2) and shifting it to the extreme by .1, returning the value (SIZE .1). Analogously, *moderately small* is calculated by taking the value of *small* (SIZE .2) and shifting it toward the mean by .1, returning the value (SIZE .3).

An interesting situation occurs if one modifies a scalar value such that it is off the scale. For example, *extremely* is defined as shifting the scalar value by .2 toward the extreme, so an *extremely, extremely expensive car* will be calculated as: *expensive* (COST .8) + *extremely* (.2) + *extremely* (.2) = (COST 1.2). The value 1.2 lies outside the {0,1} scale—but this is exactly what we want as an interpretation of *extremely extremely*! To understand why this is so, we must return to the nature of the ontology.

In the LEIA's ontology, properties are associated with different *facets*, including

- *sem*, which introduces the typical selectional restrictions;
- *default*, which introduces a more restricted, highly typical subset of *sem*; and
- *relaxable-to*, which represents an extended interpretation of *sem*.

Whereas the fillers for all these facets are explicitly listed for objects, events, and literal attributes, there is often no need to explicitly state the *relaxable-to* values for scalar attributes: they are simply outside the range of *sem*, with the degree of unexpectedness depending on how far outside the range they are. So, it is not impossible for a car to cost a million dollars, even though that would be far outside the expectations recorded in the *sem* facet of the property COST in the ontological description of AUTOMOBILE.

Returning to our example of *extremely, extremely expensive*, a perfectly valid value of 1.2 is being returned, which means that the value lies significantly beyond the expectations recorded in the *sem* facet of COST.

Let us reiterate the reason for specifying relative scalar values in the first place. If one text says that the president of France bought a *very expensive car* using government funds, and another text says that the president of Russia bought an *extremely, extremely expensive car* using government funds, a person—and, accordingly, a LEIA—should be able to answer the question, "Who bought a more expensive car, the president of France or the president of Russia?" The answer is the president of Russia—at least if the authors of the texts were reasonably consistent in their evaluations of cost.

Of course, scalar values between 0 and 1 have a concrete meaning only if one knows the objective range of values for the given property applied to the given object or event. That is, an *expensive car* implies a different amount of money than an *expensive jump rope* or an *expensive satellite*. Moreover, an expensive Kia is less expensive than an expensive BMW—something that the LEIA can reason about if the associated information is either recorded in its knowledge bases or is available in resources (e.g., on the internet) that it can consult at runtime.

Specify-approximation. Some lexical items indicate that the cited number is an approximation: for example, *around, about, approximately*. Approximating is a useful human skill, fortunately accompanied by related expressive means in language. When you say you will arrive at *around 3:00*, you are intentionally being vague. If pushed to estimate what you mean, if you are a typical person, you might say *give or take 10 minutes*, but if you are a habitually late person, you might reply *probably no earlier than 3:20*. The estimations described in this section are population-wide generalizations that do not incorporate the modeling of any particular individual—though that is an interesting problem that is worth pursuing in applications such as personal robotic assistants. The fact that calculated numerical values are estimates is reflected in TMR by appending the metaproperty value "approx. +" to a calculated property value. This feature allows the LEIA to reason as follows: If it hears *Louise came at around 3:00* it will resolve the time to *2:50–3:10 (approx. +)*. If it later hears that Louise came at 3:12, it will not flag a discrepancy; it will simply use the more precise time as the actual answer. Note that flagging involves extralinguistic reasoning that must be specially incorporated, if needed, into the reasoning engines of particular application systems.

As a default strategy, the LEIA approximates the implied range using a 7% rule, which works pretty well in many contexts. For example:

- About 5 gallons (5 * .07 =.35) resolves to between 4.65 and 5.35 gallons.
- About 150 lbs. (150 * .07 = 10.5) resolves to between 139.5 and 160.5 lbs.

This maximally simple rule would, of course, benefit from a follow-up rounding procedure; however, it would need to be rather sophisticated since something as simple as rounding up to the next whole number would clearly not work.

As mentioned earlier, all such calculations are appended with the marker "approx. +" in the TMR to make it clear that the text itself included an approximation.

The 7% rule is too coarse-grained on at least two counts. First, the actual number from which the approximation derives is important in terms of what the approximation actually means. In most cases, one approximates from numbers like 10, 25, 100, or 5,000,000. It's unusual to say *about 97 feet* or *around 8.24 pounds*. If, however, someone does use such a turn of phrase, its interpretation must be different from what is returned by the 7% rule. We have not yet pursued such pragmatically nonnormative cases.

The second way in which the 7% rule is too coarse-grained is that approximations work in idiosyncratic ways for certain semantic classes. Here we take just a few examples:

- *Heights of people*. Using the 7% rule, *about 6 feet tall* would give a range of over 5 inches on either side, which is far too broad. Instead, fixing the approximation at 1 or 2 inches either way seems closer to what people do.
- *Ages*. Interpreting the approximation of a person's age depends on how old the person is, with the 7% rule working poorly for children but better for adults. For example, the 7% rule would make a baby who is *about 5 days old* "5 days +/- 8.4 hours," and it would make a child who is *about 5 years old* "5 years +/- 3.5 months." Clearly "give or take a day" and "give or take a year" are better approximations. As a person gets older, the 7% rule works better: a person who is about 80 years old would be roughly 75–85, and a person who is about 50 years old would be 46.5–53.5. In this case, as for heights, it seems more direct to simply set the buffer for the approximation of given age ranges rather than try to force the 7% rule.
- *Clock time*. The 7% rule is reasonable for round clock times but much less so for more precise clock times. Rather than employ it, we are using a different approach to calculating approximate clock times:
    - around (hour, ½ hour, *noon*, or *midnight*) = +/- 10 minutes
    - around (:05, :10, :15, :20, etc.) = +/- 5 minutes
    - around (other) = +/- 2 minutes

Whereas special cases like these are easy to treat individually, creating a full inventory of special cases would be a tedious process of questionable utility. However, if it *were* deemed useful, it could be outsourced to anyone with the patience to work out the details. After all, this is not a matter of linguistics; it is one of world knowledge.

4.1.3 Modifiers Explained Using Combinations of Concepts

The meaning of some modifiers is best explained using combinations of concepts. Consider the meaning of *overboard* in *A sailor threw some food overboard.* As explained in the last chapter, *overboard* tells us that before the THROW event, the food was in some surface water vehicle, and after the THROW event, it is in the body of water in which that vehicle is floating. In short, *overboard* tells us about the SOURCE and DESTINATION of a MOTION-EVENT (THROW is an ontological descendant of MOTION-EVENT). Accordingly, the lexical entry for this physical sense of *overboard* specifies

- that the event being modified must be a MOTION-EVENT; if it is not, the usage is metaphorical (e.g., *She went overboard in decorating for the party*);
- that its SOURCE is a SURFACE-WATER-VEHICLE; and
- that its DESTINATION is a BODY-OF-WATER.

This lexical sense allows the agent to generate the following TMR for *A sailor threw some food overboard.* Note that the representation of *some food* is simplified to just *food* to avoid introducing set notation before its time.

THROW-1
    AGENT   SAILOR-1
    THEME   FOOD-1
    SOURCE   SURFACE-WATER-VEHICLE
    DESTINATION   BODY-OF-WATER
    TIME   *<find-anchor-time*

If the input indicates the source or destination of the MOTION-EVENT explicitly, as in *A sailor threw some food overboard into the Mediterranean Sea*, then the associated constraint is replaced by the actual value provided.

THROW-2
    AGENT   SAILOR-2
    THEME   FOOD-2
    SOURCE   SURFACE-WATER-VEHICLE
    DESTINATION   SEA-5
    TIME   *<find-anchor-time*

SEA-5
    HAS-NAME   'Mediterranean Sea'
    DESTINATION-OF   THROW-2

Although *overboard into* might sound less typical than *overboard* by itself, a destination is actually included in many examples in the COCA corpus: "overboard into the wind-whipped waves," "overboard into Ellicott Bay," "overboard into the sea," and others.

Overboard is not an unusually complex singleton. Plenty of modifiers are most productively described using multiple concepts. Consider several more examples:

- *Ad infinitum*, when used to modify a SPEECH-ACT (*Grizelda talks ad infinitum!*), adds two features to the SPEECH-ACT: (DURATION 1) and (EVALUATIVE .1).[7] Both of these are measured on the abstract scale {0,1}. The first means "for an extremely long time," and the second one reflects the speaker's negative evaluation of how long the person is talking.
- *Occasional*, when used to modify an event (*an occasional run*), adds two features: iterative aspect (ITERATION multiple) and a typical time interval that is pretty long (TIME-INTERVAL .7).
- *Ambulatory*, when used to modify a person, indicates that the person can walk. So *an ambulatory patient* is analyzed as MEDICAL-PATIENT (AGENT-OF WALK (POTENTIAL 1)). This meaning representation includes the highest value of potential modality on the abstract scale {0,1}, which indicates that the person is able to walk.
- *Argumentative*, when used of a person, indicates that the person argues regularly and that this is viewed negatively by the speaker. So *an argumentative coworker* is analyzed as COLLEAGUE (AGENT-OF ARGUE) (ITERATION multiple) (EVALUATIVE .2). Although arguing is not inherently negative, being argumentative is.
- *Abusive*, when used of a person, indicates that the person is the agent of repeating ABUSE events. So an *abusive spouse* is SPOUSE (AGENT-OF ABUSE (ITERATION multiple)). Here it is not necessary to indicate the negative evaluation because ABUSE is always negative and is ontologically specified as such.

In sum, describing modifiers using combinations of already-available concepts gives the LEIA more reasoning power than inventing endless properties for every modifier used in a language.

4.1.4 Dynamically Computed Values for Relative Text Components

Relative text components are indicated by expressions like *the preceding paragraph*, *this section's heading*, and *X and Y, respectively*. We will use the latter (*the latter*—a relative text expression!) for illustration. Consider the pair of sentences (4.1a) and (4.1b), which are a bit stilted but are useful because they generate simple and concise TMRs.[8]

(4.1) a. The doctor and the nurse ordered soup and spaghetti.
b. The doctor and the nurse ordered soup and spaghetti, respectively.

The first sentence says nothing about who ordered what. It says only that the set comprised of the doctor and the nurse ordered the set comprised of soup and spaghetti. It could be that both people ordered both dishes, that they jointly ordered one portion of each dish to share, or that each one ordered a different dish. (See section 4.1.5 for more on sets.) However, add the adverb *respectively* and everything changes: The doctor ordered the soup and the nurse ordered the spaghetti.

The procedural semantic routine for *respectively* first computes the meaning of the input without the modifier in order to determine whether the TMR shows the necessary types of parallelism. That is, if a set of two elements fills the AGENT case role, then a set of two elements must fill the THEME case role, as in our example.

The doctor and the nurse ordered soup and spaghetti.

```
ORDER-IN-RESTAURANT-1
    AGENT           SET-1
    THEME           SET-2

SET-1[9]
    ELEMENTS        PHYSICIAN-1, NURSE-1
    CARDINALITY     2

SET-2
    ELEMENTS        SOUP-1, SPAGHETTI-MEAL-1
    CARDINALITY     2
```

If there is no such parallelism, then further analysis of *respectively* is not undertaken as it will not work. Instead, the input will be left incompletely analyzed, the TMR will receive a low score, and the LEIA will have the opportunity later to attempt to recover from the failure—for example, by asking its human collaborator for clarification.

To incorporate the meaning of *respectively* into this TMR the LEIA must carry out the following five steps.

1. Verify that the EVENT to which this modifier is attached (here: ORDER-IN-RESTAURANT) has more than one property (here: AGENT and THEME), whose fillers are sets with the same cardinality, which must be greater than 1. If these conditions do not hold, the function stops because its reasoning will fail.

2. Create n new instances of the EVENT, such that n equals the cardinality of the sets, and remove the original single instance from the TMR, leaving

```
ORDER-IN-RESTAURANT-2
ORDER-IN-RESTAURANT-3
```

3. Add all case roles from the original instance, but without fillers:

```
ORDER-IN-RESTAURANT-2
    AGENT
    THEME

ORDER-IN-RESTAURANT-3
    AGENT
    THEME
```

4. Pair up the members of the sets across their respective case roles, following the word order of the input. This results in the following TMR.

```
ORDER-IN-RESTAURANT-2
    AGENT       PHYSICIAN-1
    THEME       SOUP-1

ORDER-IN-RESTAURANT-3
    AGENT       NURSE-1
    THEME       SPAGHETTI-MEAL-1
```

5. Copy any properties of the event that are not involved in the sets—e.g., adverbs of manner—into all new event frames. (Not applicable in our example.)

This is a good time to reiterate one of the benefits of the Ontological Semantics approach to language modeling. Once this meaning procedure has been implemented for one language, it can be ported to the lexicons and analysis systems of other languages, resulting in a substantial savings of time and effort. For example, although in Russian the adverb meaning *respectively*, *sootvetsvenno*, must be preposed rather than postposed—as shown by (4.2)—the basic TMR created without the adverb will be exactly the same as for English, so the procedural semantic routine to compute the adverb's meaning can be exactly the same as well.

(4.2)	Anna	i	Ivan	zakazali,	sootvetsvenno,	sup	i	spagetti.
	Anna	and	Ivan	ordered,	respectively,	soup	and	spaghetti.

4.1.5 Quantification and Sets

Although quantification has attracted a remarkable amount of scholarship within the field of formal semantics, for language understanding by LEIAs, it is no more important than dozens of other linguistic phenomena. Not having become a special priority to date, the associated microtheory—while covering the needs of our current application systems—is understood to require further development.

The notation used to describe sets and quantifiers is shown below. The comment after each semicolon indicates the valid types of fillers for the given property.

```
SET
    MEMBER-TYPE            ; an ontological concept
    ELEMENTS               ; a list of concept instances
    NOT-ELEMENTS           ; a list of concept instances or a pointer to another set
    QUANT                  ; 0–1, indicating a relative amount or percentage
    CARDINALITY            ; any integer
    RELATIVE-TO-NORM       ; 0–1 (used to describe words like 'too' in 'too many')
```

HAS-SUBSET	; a pointer to another set
SUBSET-OF	; a pointer to another set
COMPLETE	; yes or no
ORDINALITY	; any integer
SET-TYPE	; conjunctive or disjunctive
any property	; any valid filler for the given property

The following are examples of how sets are represented in TMRs.

Wolves

SET -1

MEMBER-TYPE	WOLF
CARDINALITY	>1

too many gray wolves

SET-2

MEMBER-TYPE	WOLF
CARDINALITY	>1
RELATIVE-TO-NORM	>0.6
COLOR	gray

two gray wolves

SET-3

MEMBER-TYPE	WOLF
CARDINALITY	2
COLOR	gray

both wolves

SET-4

MEMBER-TYPE	WOLF
CARDINALITY	2
COMPLETE	yes

80% of wolves

SET-5

MEMBER-TYPE	WOLF
QUANT	.8

a lot of wolves

SET-6

MEMBER-TYPE	WOLF
QUANT	.7< >.9

Two of the wolves are hungry

SET-7

MEMBER-TYPE	WOLF
CARDINALITY	>2
HAS-SUBSET	SET-8

SET-8

MEMBER-TYPE	WOLF
CARDINALITY	2
EXPERIENCER-OF	HUNGER-1
SUBSET-OF	SET-7

all of the wolves except for Whiskers

SET-9

MEMBER-TYPE	WOLF
QUANT	1
NOT-ELEMENTS	WOLF-1

WOLF-1

HAS-PERSONAL-NAME	'Whiskers'

Gray wolves are sleeping

SET-10

MEMBER-TYPE	WOLF
CARDINALITY	>1
COLOR	gray
EXPERIENCER-OF	SLEEP-1

A few of the properties used to describe sets deserve further comment.

- RELATIVE-TO-NORM indicates a state relative to what is considered optimal: for example, *too many wolves*. A value of 0.5 indicates the optimal state, with lower values indicating 'too little/few' and higher values 'too much/many.'
- COMPLETE indicates whether a set has been exhaustively described. It is overtly specified only if the value is 'yes.' It is used to described strings like *both* and *all three*.
- Any property defined in the ontology can be included in a set description. Our examples show COLOR and EXPERIENCER-OF.

The fact that any property, including RELATIONS, can be included in a set description opens up a much larger discussion regarding (a) what it means to fully interpret a set reference and (b) when and how this is best done by LEIAs.

Compare the inputs *Gray wolves are sleeping* and *Both gray wolves are sleeping*. Whereas *gray wolves* leaves the number of wolves unspecified, *both gray wolves* tells us that there are exactly two. The latter allows for a much more precise meaning representation to be generated: namely, the agent can generate and store in memory exactly two different instances of WOLF, each of which is engaging in a different instance of SLEEP. This provides more precise information for downstream reasoning.

The full analysis of *Both gray wolves are sleeping* is generated in two stages. The first stage records the meaning using the set notation just described, resulting in the following initial TMR:

```
SET-1
    MEMBER-TYPE        WOLF
    CARDINALITY        2
    COMPLETE           yes
    COLOR              gray
    EXPERIENCER-OF     SLEEP-1
```

The need for an enhanced analysis is triggered by the fact that the set has a specific cardinality.[10] In all such cases, the agent can generate the listed number of instances of the given MEMBER-TYPE and attribute to each one of them all of the properties applied to the set. (The fact that this type of set expansion might not be desirable for very large sets, especially if they are not described by any additional properties—e.g., *10,000 grains of sand*—is a separate decision that we will not explore here.)

The initial TMR above can be automatically expanded into the more explicit TMR below:

```
SET-1
    ELEMENTS           WOLF-1, WOLF-2
    CARDINALITY        2
    COMPLETE           yes

SLEEP-1
    EXPERIENCER        WOLF-1

WOLF-1
    MEMBER-OF          SET-1
    COLOR              gray
    EXPERIENCER-OF     SLEEP-1

SLEEP-2
    EXPERIENCER        WOLF-2

WOLF-2
    MEMBER-OF          SET-1
    COLOR              gray
    EXPERIENCER-OF     SLEEP-2
```

The functional difference between the basic and reasoning-enhanced set notation might not be immediately obvious because when you—a powerful human reasoning machine—read *Both gray wolves are sleeping*, you automatically see in your mind's eye two different gray wolves sleeping. The agent, by contrast, must carry out the reasoning processes we described to make this explicit. This is all done as part of Basic Semantic Analysis.

Additional aspects of set-based reasoning can also be handled during Basic Semantic Analysis, as long as they are prepared for by recording appropriate constructions in the lexicon. Below are two examples of lexical senses that record constructions involving sets. The nature of each construction is made clear in the definition and example slots. Each sense includes a call to a specific procedural-semantic routine that will compute construction-specific aspects of meaning. Whereas the procedural-semantic routine for the first example is rather simple, the procedural-semantic routine for the second is not. (Note that numbered instances of the variable *refsem* are used for certain instances of coreference within a lexical sense.)

```
out_of-prep2
   def.   The construction 'NUM out of NUM N'
   ex.    three out of four dentists
   syn-struc
      num          $var1
      prep         $var0
      num          $var2
      n            $var3
   sem-struc
      SET
         MEMBER-TYPE   ^$var3
         QUANT         refsem1
   meaning-procedures
      refsem1 = (value ^$var1 / value ^$var2)
```

The TMR for *Nine out of ten people* that is generated using the sense out_of-prep2 is

```
SET-1
   MEMBER-TYPE   HUMAN
   QUANT         .9
```

The second example involves a construction recorded as a lexical sense of *and.* Since the meaning procedure is complicated, we use subscripts to indicate how its elements align with the example sentence listed in and-conj31.

```
and-conj31
   def.   The construction 'the ORD and ORD most ADJ N'
   ex.    the fourth and fifth most important companies
   syn-struc
      det      $var1 (root 'the')
      num      $var2 (type ordinal)
      and      $var0
      num      $var3 (type ordinal)
      adv      $var4 (root 'most')
      adj      $var5
      N        $var6
   sem-struc
      SET
         MEMBER-TYPE   ^$var6
         ELEMENTS      refsem1, refsem2
      ^$var1   null-sem+
      ^$var4   null-sem+
```

```
meaning-procedures
   refsem1 =
      ^$var6-#1CORPORATION (^$var5IMPORTANCE  set-#1.ordinality.^$var2FOURTH)
   value refsem2 =
      ^$var6-#2CORPORATION (^$var5IMPORTANCE  set-#1.ordinality.^$var3FIFTH)
```

Let us trace how the example for this lexical sense, *the fourth and fifth most important companies*, will be analyzed. The sem-struc description generates the following TMR chunk:

```
SET
   MEMBER-TYPE    CORPORATION
   ELEMENTS       refsem1, refsem2
```

The refsem calculations, for their part, are as follows. They include a dot notation that is used in many ways throughout the LEIA's knowledge bases.

```
refsem1
CORPORATION-1
   IMPORTANCE     SET-1.ORDINALITY.4

refsem2
CORPORATION-2
   IMPORTANCE     SET-1.ORDINALITY.5
```

When these chunks are combined, they yield the final TMR:

```
SET-1
   MEMBER-TYPE    CORPORATION
   ELEMENTS       CORPORATION-1, CORPORATION-2

CORPORATION-1
   IMPORTANCE     SET-1.ORDINALITY.4

CORPORATION-2
   IMPORTANCE     SET-1.ORDINALITY.5
```

An additional example will illustrate how lexical senses like these accommodate many variations on a theme. The TMR for *the second and third most popular TV shows* that is generated using the lexical sense above is as follows:

```
SET-1
   MEMBER-TYPE    TELEVISION-PROGRAM
   ELEMENTS       TELEVISION-PROGRAM-1, TELEVISION-PROGRAM-2

TELEVISION-PROGRAM-1
   POPULARITY     SET-1.ORDINALITY.2

TELEVISION-PROGRAM-2
   POPULARITY     SET-2.ORDINALITY.3
```

Writing lexical senses for constructions like these, and implementing the associated procedural semantic routines, involves straightforward knowledge engineering. It is not difficult; it just takes time. By contrast, some aspects of set-related reasoning cannot be handled by lexical senses alone. For example, sometimes set-related information is distributed across multiple clauses, as in the following examples from the Cornell Natural Language Visual Reasoning corpus (CNLVR; Suhr et al., 2017):

(4.3) There are two towers with the same height but their base is not the same in color. $_{\text{(CNLVR)}}$

(4.4) There is a box with 4 items of which one is a blue item and 3 are yellow items. $_{\text{(CNLVR)}}$

The interpretation of such sentences must be distributed over multiple stages of processing to take care of coreference resolution across clauses as well as set-oriented reasoning beyond what can be recorded in lexical constructions. Preparing agents to analyze such inputs is challenging. For example, in (4.3) the singular noun phrase *their base* actually refers to two different bases, and *not the same in color* requires comparison of the values for COLOR for the bases of the two towers. Similarly, in (4.4), *of which* establishes a reference relation with the *4 items*, and the head nouns associated with *one* (item) and *3* (items) are elided.

Another challenge presented by sets involves interpreting the scope of modifiers in inputs such as "the ADJ N and N." In some cases, the modifier applies only to the first noun: in *the old men and children*, the children are clearly not old. In other cases, the intended meaning is ambiguous: in *the old men and women*, the women may or may not be old.

4.1.6 Indirect Modification

Indirect modification occurs when a modifier syntactically modifies one constituent but semantically modifies another. Most cases of indirect modification are best handled by dedicated lexical senses that reconstruct the intended meaning. For example:[11]

- When *married* modifies an event—e.g., *married sex, married conflicts*—it introduces into the context a married couple that is the agent or experiencer of the event.
- When *responsible* modifies an event—e.g., *responsible decision-making, responsible parenting*—it introduces into the context one or more human agents who are carrying out the given event responsibly.
- When *rural* modifies an abstract noun—e.g., *rural poverty, rural income*—it introduces into the context an unspecified set of people who live in a rural area. The semantic relationship between those people and the modified noun is not explicitly indicated and must be dynamically computed (just as with nominal compounds).
- When *sad* modifies anything but an animal, its meaning depends on the meaning of the noun it modifies. For example, when *sad* modifies a temporal expression—e.g., *sad time, sad year*—it introduces into the context one or more people who are sad during that time. When it modifies an abstract object—e.g., *sad song, sad news*—it means that

the object makes people feel sad. And when it modifies a person's body part or expression—e.g., *sad eyes, sad smile*—it means that the associated person is sad.

The meaning representations recorded in the sem-strucs of these senses include entities—most often, humans—that are not explicitly mentioned in the local dependency structure, though they might appear somewhere in the preceding context. Where things get interesting is with respect to specifying who, exactly, these entities are in contexts in which that information *is* provided. This can often be done using textual coreference procedures. For example:

- Interpreting *The committee is tasked with responsible decision-making* requires that the generic HUMAN(s) posited in the meaning representation for this sense of *responsible* be coreferred with the committee.
- Interpreting *Look into her sad eyes and you'll understand everything* requires that the generic HUMAN posited in the meaning representation for this sense of *sad* be coreferred with *her.*
- Interpreting *Married conflicts affect young couples' relationships* requires that the generic set of HUMANs posited in the meaning representation for this sense of *married* be coreferred with the *young couples* mentioned subsequently.

The procedural semantic routines needed to resolve these coreferences are recorded in the meaning-procedures zones of the associated lexical senses. They are called during the next stage of processing, Basic Coreference Resolution (chapter 5).

4.1.7 Recap of Modification

- Recorded property values
  - Scalar attributes—*accomplished person*: HUMAN (EXPERTISE-ATTRIBUTE .8)
  - Literal attributes—*divorced man*: HUMAN (GENDER male) (MARITAL-STATUS divorced)
  - Relations—*fearful fox*: FOX (EXPERIENCER-OF FEAR)
- Dynamically computed values for scalar attributes—*very smart*: *smart* (INTELLIGENCE .8) + *very* (increase value by .1) = (INTELLIGENCE .9)
- Modifiers explained using combinations of concepts—*argumentative person*: HUMAN (AGENT-OF ARGUE (ITERATION multiple) (EVALUATIVE .2))
- Dynamically computed values for relative text components—*The doctor and the nurse ordered soup and spaghetti, respectively* is effectively analyzed as the meaning of [*The doctor ordered soup*] and [*The nurse ordered spaghetti*]
- Quantification and sets—*Many wolves*: SET (MEMBER-TYPE WOLF) (QUANT .7<>.9)
- Indirect modification—*sad smile*: SMILE (AGENT HUMAN (HAPPINESS-ATTRIBUTE .2))

4.2 Proposition-Level Semantic Enhancements

The meaning of simple propositions can be enhanced in various ways. These include introducing values of modality or aspect; using a nonbasic (i.e., nondeclarative) mood, such as the imperative or interrogative; using the proposition as the complement of a non-modal, non-aspectual matrix verb; and combining any of these. Given the simple proposition *The firefighter swims*, whose bare-bones[12] TMR is

```
SWIM-1
    AGENT       FIREFIGHTER-1
    TIME        find-anchor-time
```

we can add various semantic enhancements like the following:

(4.5) [Add modality]
The firefighter *wants to* <*is trying to, can't, won't*>swim.

(4.6) [Add aspect]
The firefighter *is* swimming<is *starting to* swim, *finished* swimming>.

(4.7) [Add a non-modal, non-aspectual matrix verb]
The gardener *sees* the firefighter swimming.

(4.8) [Add the interrogative mood]
Who is swimming? *Is* the firefighter swimming?

(4.9) [Add the imperative mood]
Swim!

(4.10) [Combine several of the above]
Did the gardener *tell* the firefighter *to try to start* swimming?

We will consider each of these proposition-level enhancements in turn.

4.2.1 Modality

Ontological Semantics distinguishes ten types of modality, listed in table 4.1.[13] Each modal meaning is described in a MODALITY frame with the following properties:

- TYPE: effort, epistemic, and so on;
- VALUE: any value or range on the abstract scale {0,1};
- SCOPE: the ontological concept instance of the head of the proposition; and
- ATTRIBUTED-TO: indicates the individual responsible for reporting the modal meaning. By default, it is the speaker. For example, a speaker who says *Martina loves skydiving* is taking responsibility for Martina's positive evaluation of it. Martina might actually hate it.

In our firefighter example, two modality-enhanced versions are juxtaposed below.

Table 4.1
Types of modality used in Ontological Semantics

Modality type	Informal definition	Sample realizations
Effort	indicates effort expended	try to, not bother to
Epistemic	indicates factivity	did, didn't, will, won't
Epiteuctic	indicates success/failure	succeed in, fail to, barely manage to
Evaluative	indicates an assessment	be pleased with, disapprove of
Intentional	indicates intention	intend to, be planning to
Obligative	indicates obligation, requirement	must, have to, not be required to
Permissive	indicates permission	may, may not, can, can't
Potential	indicates ability, potential	can, can't, be unable to
Volitive	indicates want, desire	want do, be dying to, not have any desire to
Belief	indicates belief	believe, disbelieve

The firefighter wants to swim.

```
MODALITY-1
    TYPE            volitive
    VALUE           1
    SCOPE           SWIM-1
    ATTRIBUTED-TO   HUMAN-1 ("speaker")
    TIME            find-anchor-time

SWIM-1
    AGENT           FIREFIGHTER-1
```

The firefighter wants to try to swim.

```
MODALITY-1
    TYPE            volitive
    VALUE           1
    SCOPE           MODALITY-2
    ATTRIBUTED-TO   HUMAN-1 ("speaker")
    TIME            find-anchor-time

MODALITY-2
    TYPE            effort
    VALUE           1
    SCOPE           SWIM-1
    ATTRIBUTED-TO   HUMAN-1 ("speaker")

SWIM-1
    AGENT           FIREFIGHTER-1
```

4.2.2 Aspect

Aspect reflects specific time-related characteristics of an event. Formally, it scopes over a proposition, just like modality. It divides into PHASE and ITERATION, which have the following value sets:

- PHASE: *begin* (indicates the start of an event), *continue* (indicates its progression), *end* (indicates its completion), and *begin-continue-end* (represents the action viewed as a whole); and
- ITERATION: *single* (the event occurs once), *multiple* (it occurs multiple times), *any number or range of numbers.*

Aspect interacts with tense in complex ways that differ considerably across languages. Developing an overarching microtheory of tense and aspect—including time-tracking events throughout discourses—is a big job that we have not yet undertaken. When that microtheory comes on agenda, it will be informed by recent work on representing the relative times of events in corpus annotation (e.g., Mani et al., 2005).

Currently, TMRs explicitly show values of aspect in two cases: (a) when the value of PHASE can be unambiguously identified from verbal features—for example, present-progressive verb forms (*Paavo is* running) have "PHASE continue"—and (2) when they are instantiated by the lexical descriptions of specific words and phrases. For example, the verb *begin* is described in the lexicon as adding "PHASE begin" to the meaning of the event it scopes over. Similarly, the adverb *repeatedly* is described as adding "ITERATION multiple" to the meaning of the event it scopes over. Two more versions of our firefighter example illustrate the representation of aspect.

The firefighter is starting to swim.

```
ASPECT-1
    PHASE            begin
    SCOPE            SWIM-1
    TIME             find-anchor-time

SWIM-1
    AGENT            FIREFIGHTER-1
```

The firefighter wants to start swimming.

```
MODALITY-1
    TYPE             volitive
    VALUE            1
    SCOPE            ASPECT-1
    ATTRIBUTED-TO    HUMAN-1 ("speaker")
    TIME             find-anchor-time

ASPECT-1
    PHASE            begin
    SCOPE            SWIM-1

SWIM-1
    AGENT            FIREFIGHTER-1
```

4.2.3 Non-Modal, Non-Aspectual Matrix Verbs

Many verbs that convey meanings apart from modality and aspect take propositions as their complements. In such cases, the meaning of the matrix verb has a THEME slot that is filled by the meaning of the complement. This is illustrated below using the verbs *see* and *hear*.

The gardener saw the firefighter swimming.

INVOLUNTARY-VISUAL-EVENT-1	
EXPERIENCER	GARDENER-1
THEME	SWIM-1
TIME	*<find-anchor-time*

SWIM-1	
AGENT	FIREMAN-1

The gardener heard the boy playing the trumpet.

INVOLUNTARY-AUDITORY-EVENT-1	
EXPERIENCER	GARDENER-1
THEME	PLAY-MUSICAL-INSTRUMENT-1
TIME	*<find-anchor-time*

PLAY-MUSICAL-INSTRUMENT-1	
AGENT	HUMAN-1
THEME	TRUMPET-1

HUMAN-1	
GENDER	male
AGE	< 16 (MEASURED-IN YEAR)

4.2.4 Questions

There are many types of questions, including yes/no questions, *wh*-questions, choice questions, and tag questions. All questions set up a REQUEST-INFO frame whose THEME slot is filled by what is being asked. In many cases, representing what is asked requires the use of dot notation. Below are two examples of question frames that use this notation. The first example asks for the AGENT of SWIM, represented as SWIM-1.AGENT. The second example asks whether or not the statement is true—that is, it asks for the value of epistemic modality scoping over the proposition (if it is true, the value is 1; if it is false, the value is 0).

Who is swimming?

REQUEST-INFO-1	
AGENT	HUMAN-1 ("speaker")
THEME	SWIM-1.AGENT
BENEFICIARY	HUMAN-2 ("hearer")

SWIM-1	
TIME	*find-anchor-time*

Is the fireman swimming? or The fireman is swimming, isn't he?

REQUEST-INFO-1
    AGENT HUMAN-1 ("speaker")
    THEME MODALITY-1.VALUE
    BENEFICIARY HUMAN-2 ("hearer")

MODALITY-1
    TYPE epistemic
    SCOPE SWIM-1

SWIM-1
    AGENT FIREMAN-1
    TIME *find-anchor-time*

Most interrogative inputs are recognized as such during syntactic analysis, at which point the feature "interrogative+" is incorporated as metadata into the nascent TMR. Then, at this stage of Basic Semantic Analysis, that feature value is converted into an instance of the concept REQUEST-INFO. Special cases include tag questions and indirect questions. Tag questions are recorded in the lexicon as constructions whose syntactic components are illustrated in table 4.2. Indirect questions are discussed in section 4.4.

4.2.5 Commands

Commands, also called imperatives, are propositions in the imperative mood. They are recognized by the syntactic parser during Pre-Semantic Analysis. During Basic Semantic Analysis, the feature "imperative+" triggers the instantiation of a REQUEST-ACTION concept whose THEME is the action in question and whose main case role—usually AGENT—is the hearer, as illustrated below.[14]

Swim!

REQUEST-ACTION-1
    AGENT HUMAN-1 ("speaker")
    THEME SWIM-1
    BENEFICIARY HUMAN-2 ("hearer")

SWIM-1
    AGENT HUMAN-2 ("hearer")

Table 4.2
Examples of syntactic components of tag-question constructions

Clause	,	Aux	(neg)	pronoun	?
The Murphys won't go	,	will		they	?
Leslie can ski	,	can	't	she	?
Horatio should eat the nuts	,	should	n't	he	?

Eat two bananas, now!

REQUEST-ACTION-1	
AGENT	HUMAN-1 ("speaker")
THEME	INGEST-1
BENEFICIARY	HUMAN-2 ("hearer")

INGEST-1	
AGENT	HUMAN-2 ("hearer")
THEME	SET-1
URGENCY	1

SET-1	
MEMBER-TYPE	BANANA
CARDINALITY	2

4.2.6 Recap of Proposition-Level Semantic Enhancements

- Modality: The firefighter *wants to* <*is trying to, can't, won't*>swim.
- Aspect: The firefighter *is starting* to swim <*finished* swimming>.
- Non-modal, non-aspectual matrix verbs: The gardener *sees* the firefighter swimming.
- Questions: *Is* the firefighter swimming? *Who* is swimming?
- Commands: *Swim!*
- Combinations: The firefighter *wants* to *start* swimming.

4.3 Multicomponent Entities Recorded as Lexical Constructions

Knowledge of a language is, in large part, knowledge of how words fit together to express particular meanings. Even memorizing 10,000 individual words of English would not prepare a non-English speaker to express in a normal, idiomatic way ideas like, "Seconds, anyone?" "Are we there yet?" "Time's up." or "I will if you will." In fact, the lion's share of the work in learning a new language is memorizing tens of thousands of instances of how things are said.[15] Whereas some of these are fixed expressions, many are constructions—that is, templates containing a combination of particular words and slots for variables. If LEIAs are to ever achieve human-level language capabilities, they need a very large, construction-packed lexicon that mirrors the lexical knowledge possessed by native speakers.

Unfortunately, the word *construction* comes with baggage, so we need to take a short detour to address terminology. Recently, *construction grammar* has become a popular subfield of theoretical linguistics (Hoffman & Trousdale, 2013; see section 1.4.3.1). In keeping with its theoretical status, it attempts to account for certain aspects of human language processing—in particular, the form-to-meaning mapping of linguistic structures. We couldn't agree more that the latter is key to the study of language. And, as we have already demonstrated, the LEIA's lexicon records exactly such form-to-meaning mappings. However, where we part ways with construction grammarians is in the specific interpretation of the word *construction*.[16] For them, even non-argument-taking words are constructions in the sense that the syntactic form maps to the semantic interpretation. For us, by contrast, constructions must contain multiple constituents.

The constituents in a construction can be any combination of specific words and variable categories. Examples include:

- Idiomatic expressions: for example, *SUBJ kick the bucket*, *SUBJ spill the beans*
- *Verb +particle* collocations: *SUBJ give in* <*give out, give up*>
- Common expressions whose meanings are presumably remembered by people, not recomputed each time: *Have a nice day. Are we there yet? Salt, please.*
- Proverbs and sayings: *Nothing ventured, nothing gained. It takes one to know one.*
- Semantically non-compositional nominal compounds: *training data, trial balloon*
- Semantically constrained nominal compounds: FISH +*fishing* (i.e., any type of FISH followed by the word *fishing*) means a FISHING-EVENT whose THEME is that type of fish, for example, *trout fishing*
- Elliptical constructions: for example, NP V $\text{ADV}_{\text{COMPAR}}$ *than* NP is recorded as a sense of *than*, which covers inputs like *Betty jumps higher than Lucy* __. The lexical sense includes a procedural semantic routine that copies the meaning of V into the elided slot, effectively rendering our example, *Betty jumps higher than Lucy jumps*.
- All argument-taking words: adjectives, adverbs, verbs, prepositions, and so on.

The last class might come as a terminological surprise: Why would regular argument-taking words be considered constructions? Because the way they are recorded in the LEIA's lexicon, they fulfill our definition of construction: that is, they are multipart entities defined by a combination of required words and variable slots. Consider again the first verbal sense of *eat*, which was initially presented in section 2.2.

```
eat-v1
   def.        to ingest
   ex.         He was eating (cheese).
   syn-struc
      subject        $var1
      v              $var0
      directobject   $var2 (opt+)
   sem-struc
      INGEST
         AGENT       ^$var1
         THEME       ^$var2 (sem FOOD)
```

The verb *eat* is a required word in this construction, though it can appear in any inflectional form. Syntactically, it requires a subject and permits an optional direct object. Semantically, the subject must be a valid AGENT of INGEST, which the ontology indicates must be some type of ANIMAL. The direct object, if selected, must be some type of FOOD. Stated differently, this sense of *eat* is not treated as an isolated word mapped to the concept INGEST. Instead, it is described with its expected syntactic dependents, its expected set of case roles, and the semantic constraints on the fillers of those case roles. For this reason, it is a construction.

We further divide constructions into *lexical constructions* and *non-lexical constructions*. Lexical constructions must contain at least one specific word (which can, however, be used in different inflectional forms), which anchors the construction in the lexicon. Lexical constructions can contain any number of other required words and/or syntactic constituents.[17] All of the examples above are of lexical constructions.

Non-lexical constructions, by contrast, contain only category types. For example, the syntactic construction called *object fronting* places the NP serving as the direct object in the sentence-initial position. This allows for *This I like!* to be used as an emphatic alternative to *I like this!* Similarly, the ontological construction FRUIT + TREE → TREE (HAS-OBJECT-AS-PART FRUIT) allows the nominal compound *apple tree* to be analyzed as TREE (HAS-OBJECT-AS-PART APPLE) (see section 6.3.1). Non-lexical constructions must be recorded in separate knowledge bases since they have no required word to serve as an anchor in the lexicon.

Lexical constructions, as we define them, are a supercategory of what descriptive linguists and practitioners of NLP call *multiword expressions*.[18] Multiword expressions require multiple specific words, possibly with some variable slot(s) as well. For example, the idiomatic verb phrase *kick the bucket* is considered a multiword expression whose subject slot is a variable.

When LEIAs process constructions, their goal is the same as for any input: to compute the input's full contextual meaning. This has little to do with the most popular threads of work on multiword expressions over the past thirty years by descriptive linguists and practitioners of NLP. Descriptive linguists have primarily pursued classification, including analyzing the degree to which multiword expressions are fixed versus variable. Practitioners of NLP, for their part, have pursued the automatic detection of multiword expressions and their translation in machine translation systems. It is noteworthy that neither detecting nor translating multiword expressions directly addresses their meaning since even a correct translation achieved using statistical methods does not imply that the expression has been understood in a way that would support reasoning by intelligent agents.[19]

Let us return to lexical constructions as recorded in the LEIA's lexicon. The top-level distinction is between constructions that can be treated as lexemes with white spaces and those that cannot. The first category, which includes entities like *vice president*, *stock market*, and *nothing ventured, nothing gained*, is trivial. The components must occur in the listed order, they do not permit modifiers or other elements to intervene between them, and only the last word, if any, is subject to inflection. We record such entities in the lexicon as multipart head words with an underscore indicating each white space. This approach is simple and works nearly perfectly—only *nearly* because in rare cases an expletive, speaker correction, or interruption might occur between the elements. (This can also, by the way, happen in the middle of regular words: *decon (ouch!) struction*.) As with all unexpected input, such deviations must be handled by recovery procedures, which amount to a sequence of attempts to relax certain constraints, such as the expectation that only a blank space can intervene between components of a multiword head entry.

Other constructions, as we have said, can have any combination of particular words and variable constituents. Consider the idiom *X pays homage to Y*. It is recorded in the lexicon as a sense of the verb *pay*.

```
pay-v16
   def.        The construction 'pay homage to', which maps to PRAISE
   ex.         They paid homage to the king.
   syn-struc
      subject         $var1
      v               $var0
      directobject    $var2 (root 'homage')
      pp
         prep         $var3 (root 'to')
         obj          $var4
   sem-struc
      PRAISE
         AGENT        ^$var1
         THEME        ^$var4
      ^$var2     null-sem+
      ^$var3     null-sem+
```

Syntactically, this sense permits an NP of any legal form to fill the subject and object-of-preposition slots, but the direct object must be the noun *homage* and the preposition must be *to*. These constraints are indicated by appending the feature "root 'homage'" to the direct object slot and the feature "root 'to'" to the preposition slot. Semantically, *pay homage to* is interpreted using the ontological concept PRAISE. The subject of *pay homage to* fills the AGENT slot of PRAISE and the object of the preposition fills the THEME slot. The constraints on the AGENT and THEME are drawn from the ontological description of PRAISE.

Much more could be said about the expressive power of the metalanguage used to record lexical constructions, but those details are more appropriate for knowledge engineers than for the general reader. Suffice it to reiterate that (a) constructions can be composed of any sequence of lexical, syntactic, or ontological categories, which can be constrained by any of the morphological, syntactic, or semantic features used in the system, and (b) *lexical* constructions require that at least one particular word be fixed so that it can anchor the construction in the lexicon.

4.3.1 Semantically Null Components of Constructions

Some components of some constructions do not carry independent meaning, such as *the bucket* in the idiom *kick the bucket*. Such components are marked with the feature "null-sem+" in the sem-struc zone of the associated construction so that they are not analyzed compositionally.

Although null-semming might seem to be a perfect solution for getting rid of non-compositional elements, it has one complication. Occasionally, a null-semmed element is modified, and its meaning must somehow be attached to the meaning of the construction overall. For example, in (4.11) the modifier *goddamned* is not modifying *the bucket*; it is expressing the speaker's frustration at the person's having died before paying back the money.

(4.11) My neighbor kicked the goddamned bucket before he paid me back the money he owed me!

There *is* a solution to this modification problem, which we will turn to once we have described typical uses of null-semming in the LEIA's lexicon.

4.3.2 Typical Uses of Null-Semming

The most obvious use of null-semming is in canonical idioms like *kick the bucket*, but this mechanism has broader uses as well. For example, it can remove wordy reformulations of simpler meanings (as in examples (4.12)–(4.15) below), turns of phrase that primarily serve a discourse (rather than a semantic) function (4.16), and aspects of meaning that are so fine-grained or difficult to explain that they are, for the time being, not being chased down (4.17). In the examples, ^ indicates "the meaning of the given string" (we borrow this convention from the sem-struc zones of lexical senses).

(4.12) ~~The fact that~~ the guy's religious will have nothing to do with it. (COCA)
^((the guy's religious) will have nothing to do with it)

(4.13) But ~~the thing is, is that~~ it was relaxed. (COCA)
^(but it was relaxed)

(4.14) ~~It's just that~~ he'll lose in a heartbeat. (COCA)
^(he'll lose in a heartbeat)

(4.15) And here we are in ~~the month of~~ April. (COCA)
^(And here we are in April)

(4.16) ~~It turned out that~~ he had a small baggie of marijuana. (COCA)
^(he had a small baggie of marijuana)

(4.17) She ~~couldn't help but~~ laugh. (COCA)
^(She laughed)

Let us linger for a moment on the last example: What *is* the meaning of *She couldn't help but laugh*? Perhaps something like: "Irrespective of whether or not she wanted to laugh, she did so because, given some unspecified properties of the situation, it would have been too difficult for her not to laugh." However, even if this were deemed a reasonable analysis, it is still only an English paraphrase, which is a far cry from a formal representation that could support useful automatic reasoning. The formal representation would be quite complex and it is not clear what goal it would serve for LEIAs in the foreseeable future. So spending time trying to specify it would be a poor use of limited resources.

4.3.3 Modification of Null-Semmed Constituents

As mentioned above, although null-semming can be a useful, elegant solution to handling some relatively superfluous text elements, it has its downside: in some cases, the null-semmed elements can be the target of modification. From the COCA corpus and other sources, we found quite a variety of examples:

- *Pay homage to* means PRAISE, but the nature of the homage can be specified as *silent*, *deliberate*, or *the strictest*.
- *Put a spell on* means BEWITCH, but that spell can be specified as *protective*, *love*, or *magic*.
- *Raise eyebrows* means SURPRISE, but "raised some serious eyebrows" appears multiple times on the internet.

If we null-sem the elements *homage*, *spell*, and *eyebrows* in these constructions, their modifiers will be left hanging. The solution is to anticipate such eventualities.

We will consider two cases by way of example: the previously mentioned *kick the goddamned bucket*, which illustrates proposition-level modification, and *put a ADJ spell on*, which illustrates modification of a meaningful element in a semi-idiomatic turn of phrase.

Kick the bucket is fully idiomatic since there is neither kicking nor a bucket involved. The only modifiers that are typically permitted in this expression are *proverbial* and expletives like *bloody* or *goddamned*. *Proverbial* can be included as an optional modifier in the *kick the bucket* sense with no semantic modifications needed. As for the expletives, they are best handled in a separate sense of the construction that requires an expletive adjective. The associated semantic description indicates that a very low value of evaluative modality (namely, .1 on the {0,1} scale), attributed to the speaker, scopes over the DIE event. In other words, the speaker who is reporting the event of dying is very unhappy about it.

One might ask, Isn't there a generalization that expletives can syntactically modify non-compositional elements of all idioms? And, if so, shouldn't the LEIA have a general rule to this effect, rather than having to rely on additional, expletive-inclusive senses of individual idioms? Perhaps, but there is a practical problem with implementing such a rule:

Where would it live in the system? Developing language-understanding capabilities for LEIAs is a practical endeavor. The system needs to work, the knowledge bases need to be inspectable, and knowledge engineers need to be able to trace what is happening and why. The program that uses lexical senses to build TMRs expects certain kinds of information in those senses and should not rely on fix-up rules with a high potential of getting lost in the code—even if one believes that such rules might exist in the minds of humans. Of course, specific decisions about practicality and engineering can change in different project configurations. One might invent an architecture that elegantly houses all kinds of fix-up rules. The approach just described for dealing with expletives in constructions reflects our current preference, which is based on real-world experience with organizing knowledge bases and engines that employ them.

Now we turn to *cast/put a spell on/over* as an example of a semi-idiomatic turn of phrase that can include modification. This construction means BEWITCH (remember, this is an ontological concept, not an English word). The idiomatic aspect of the construction is the verb choice—it must be *put* or *cast*. *Spell*, by contrast, is used in one of its canonical meanings. If we only ever expected the unmodified input *X cast a spell on Y*, then null-semming *a spell* would be a fast and sufficient solution since the ontology already tells us that the INSTRUMENT of BEWITCH is SPELL. However, since *spell* can, in fact, be modified, we choose to leave it as an explicit INSTRUMENT of BEWITCH in the sem-struc of the construction sense, as shown below.

```
cast-v2
  def.        The construction 'cast/put a spell on someone'
  ex.         The witch cast a spell on the frog.
  comments    Not null-semming $var2 means that it can be modified in the normal way: 'The witch
              cast an evil/magical spell on the frog.'
  syn-struc
    subject         $var1
    root            $var0
    directobject    $var2 (root 'spell')
    pp
      prep          $var3 (root 'on')
      obj           $var4
  sem-struc
    BEWITCH
      AGENT         ^$var1
      THEME         ^$var4
      INSTRUMENT    ^$var2
    ^$var3    null-sem+
  synonyms    put
```

If *spell* is not modified, then this representation is simply redundant: both the lexicon and the ontology tell us that the INSTRUMENT is SPELL—not a problem. But if *spell* is modified, then the modification can be applied to ^$var2 in the normal way.

Preparing for modification-oriented eventualities in constructions involves extra work, and that work may or may not be a priority at a given time in system development. This raises the question of what happens if modification is not accounted for in a construction sense but occurs in an input. In this case, overall processing will not fail, but the meaning of the modifier will be lost since the head that it modifies will be ignored.

One additional detail is worth noting. Analysis of the COCA corpus revealed that the constructions discussed above are overwhelmingly used in their base forms—not an unexpected outcome for essentially fixed expressions:

- The collocation *paid* followed by *homage* (within a distance of four words) yielded 138 hits, of which 123 were exactly *paid homage*, 11 inserted a modifier, and 4 included a pronominal indirect object (i.e., *paid him/her homage*).
- The collocation *kicked the* followed by *bucket* (within a distance of four words) yielded 34 hits, all of them exactly *kicked the bucket*.
- The collocation *put a* followed by *spell* (within a distance of three words) yielded 60 hits, 57 of which were exactly *put a spell*.

In conclusion, it is useful to be aware of the potential for modification of construction components and to prepare for them to the degree that resources permit. The fact that they are usually used in their base forms is informative when determining how to best allocate knowledge acquisition resources.

4.3.4 Utterance-Level Constructions

Some full utterances are recorded as senses in the lexicon. There are both theoretically and practically motivated reasons for this, which we consider in turn.

Theoretical motivations. The modeling of LEIAs is inspired by our hypotheses about human cognition. It seems likely that after we have encountered utterances some number of times, we simply remember their analyses: *When are we gonna be there? I'm hungry! Salt, please*. Such utterances do not have to be idiomatic. Any utterance that is frequent for a given person (possibly, in a given physical or social context) likely has a remembered analysis. For example, a running coach might begin every long-distance run with the elliptical reminder "Water?" meaning "Have you brought water to drink during the run?" The mapping between the word "Water?" and the full proposition is likely to be remembered, rather than recomputed, by the runners after the first practice or two. As concerns modeling LEIAs, it makes no difference whether these remembered analyses are recorded in the LEIA's lexicon or in a separate repository of remembered text meaning representations (see section 6.1.6 for a discussion of the latter). Either way, the LEIA has the associated text-to-meaning mappings in its knowledge substrate.

Practical motivations. On the practical front, recording the meaning of full utterances can provide immediate language support for agent reasoning and action before all of the component linguistic phenomena can be handled using general-purpose methods. For example, say a robotic agent is capable of visually locating, picking up, and passing to its human collaborator a hammer. A typical way to ask for this to happen is by saying, "Hammer!" So, the elliptical "Hammer!" must be understood as "Give me a hammer."

If "Hammer!" were to be analyzed from first principles, the agent would need to use goal- and plan-based reasoning supported by domain knowledge (see chapter 7). Specifically, it would need to recognize that the object HAMMER is not a proposition, so there must be an implied EVENT. The agent must determine which hammer-oriented events it is capable of carrying out and, of those, which is the most relevant at the moment of speech. Depending on the domain, this might be complicated and error-prone. However, if "Hammer!" is typically used to mean "Give me a hammer," then it merits having a lexical sense that records exactly this meaning when this word is used as an independent utterance.

```
hammer-n2
   def.          'Hammer!' or 'Hammer.' used to mean 'Give me a hammer.'
   ex.           'Hammer!' 'Hammer.'
   comments      The punctuation mark indicates that this is an utterance, not a word
   syn-struc
      root       $var0
      punct      $var1 (root *period* *exclamation-point*)
   sem-struc
      REQUEST-ACTION
         AGENT          HUMAN-#1 ("speaker")
         THEME          TRANSFER-POSSESSION-#1
         BENEFICIARY    HUMAN-#2 ("hearer")
      TRANSFER-POSSESSION-#1
         AGENT          HUMAN-#2
         THEME          HAMMER
         BENEFICIARY    HUMAN-#1
      ^$var1     null-sem+
```

Different domains have analogous elliptical structures, as in this example about surgery.

(4.18) "Everyone ready? Let's go. Knife." The nurse pops the handle of the scalpel into my palm. (COCA)

Recording analyses for the elliptical uses of *Hammer*, *Knife*, and others is not only a convenient strategy, it also models the fact that, in certain application domains, members of human-robotic teams will likely encounter these elliptical utterances so often that they will simply remember the intended meanings, without having to dynamically resolve the ellipsis every time.

4.3.5 Additional Knowledge Representation Requirements

As microtheories go, the microtheory of constructions is quite advanced, meaning that it covers a lot of eventualities. However, language never ceases to surprise, and a recent evaluation study (McShane, Nirenburg, & Beale, 2015) pointed out the need for more precise knowledge acquisition with respect to constructions along three lines.

1. Exclusion criteria can be needed to avoid false positive construction matches. For example, *X can tell that Y* means UNDERSTAND, as shown by (4.19). However, this idiom does not permit an indirect object. If there *is* an indirect object, then the idiom conveys emphasis, as shown by (4.20).

(4.19) I can tell that you like this film. (COCA)

(4.20) Well, I can tell you that the markets are on edge right now. (COCA)

Two different constructions, recorded in two different lexical senses, are needed to cover these different usages.

2. Constructions are more ambiguous than might be expected, which makes treating polysemy as much a priority for constructions as it is for simple lexemes. Pairs of literal and metaphorical meanings are particularly common, as illustrated by *take a bath* in (4.21) and (4.22) and *take a look at* in (4.23) and (4.24).

(4.21) I scrubbed the tub and took a bath. (COCA)

(4.22) Yes, they took a bath in the stock market, "but not as badly as some people," Dottie says. (COCA)

(4.23) I took a look at his shoes: winterized Air Jordan Six-Rings, gleaming black. (COCA)

(4.24) So as part of that investigation we took a look at her finances … (COCA)

3. In some cases, constructions overlap. For example, (4.25) uses the construction *let it go*, which means to stop unproductively attending to something. By contrast, (4.26) uses the construction *let it <something>go to hell (in a handbasket)*, which means to let something deteriorate by not attending to it.

(4.25) How can you be so judgmental? Life's life. Let it go. (COCA)

(4.26) Doggone it, this is our—this is our community. And we're not going to let it go to hell in a handbasket. (COCA)

The best strategy for handling this last eventuality is for the LEIA to select the largest textspan that matches a recorded construction. That is, if one construction completely subsumes another, then the longer one should be selected if all of its elements are attested in the input.

The reason for pointing out these complications is to emphasize that although constructions do represent relatively stable islands of confidence for language understanding, they are not without their challenges. This is important because, in reading the knowledge-lean NLP literature on constructions (or, more specifically, the subset that they call *multiword expressions*), one can get the impression that the main challenge is automatically detecting them. In reality, that is just the beginning.

4.3.6 Recap of Constructions

- Idioms: Moira has kicked the bucket. (COCA)
- Idioms with indirect modification: And you people tell me the Communists are running rampant in the outlying provinces and that if Mikaso kicks the damned bucket we could lose all ties to the Philippines … (COCA)
- Wordy formulations: But ~~the thing is, is that~~ it was relaxed. (COCA)
- Frequent locutions: Signature, please. (COCA)
- Application-specific frequent locutions: Knife. (COCA)
- Additional knowledge representation requirements involving (a) exclusion criteria, (b) ambiguity, and (c) overlapping constructions.

4.4 Indirect Speech Acts, Lexicalized

Often, questions and commands are expressed using indirect speech acts. An indirect speech act differs from a direct speech act in that its form (the locutionary act) does not align with its function (the illocutionary act). Starting from the basics, an alignment between form and function occurs when

- An assertion is used to make a statement: The cake is fantastic. (COCA)
- A question is used to ask a question: How is the game? (COCA)
- A command is used to issue a command: Get me a coffee. (COCA)

Various types of misalignments can occur for good pragmatic reasons, such as being polite or conveying emphasis, as shown by the following set of examples.

(4.27) [An assertion used to request information]
"I need to know what you're talking about." (COCA)

(4.28) [An assertion used to request action]
"… Quinn, it would be great if you'd bring your own card." (COCA)

(4.29) [An assertion used as a command]
"But now you had better be off home." (COCA)

(4.30) [A question used to request action]
"I have a lost kid here. Can you help us find his parents?" (COCA)

(4.31) [A question used as an assertion. *Are you kidding me* expresses nonagreement and/or frustration]
These people are the moral exemplars of the 21st century, are you kidding me? (COCA)

(4.32) [A command used as an assertion. *Dream on!* indicates that the hope, plan, or wish just expressed is unrealistic, will not happen]
Damn her, she knew exactly what he was trying to do, and would she help him? Dream on! (COCA)

(4.33) [A command used as a threat]
"Come on, big boy," he yelled. "Make my day!" (COCA)

A convenient aspect of indirect speech acts is that many of them are formulaic and can be recorded in the lexicon as constructions. For example, *I need to know X* and *You need to tell me X* are both typical ways of asking for information about X. By recording such constructions in the lexicon, we give LEIAs the same knowledge of their formulaic nature as a person has.

Of course, every formula that can serve as an indirect speech act also has a direct meaning. For example, you can say, *I need to know Mary's address* to a person who you think can give you that information or to a person who couldn't possibly know it by way of explaining why you are shuffling through your address book.

LEIAs treat speech-act ambiguity in the same way as any other type of ambiguity: by recognizing all available interpretations during Basic Semantic Analysis and then waiting until Situational Reasoning to select the contextually appropriate one.

Of course, not all indirect speech acts use well-known conventions. You can say, *The mail just came* with the implication that you want the interlocutor to go fetch it, despite the lack of any linguistic flag to indicate the veiled request. Indirect speech acts of this type are more difficult to detect, and configuring LEIAs to seek them out must be approached judiciously since we would not be well served by paranoid agents who assume that every utterance is a call to action.

4.5 Nominal Compounds, Lexicalized

A nominal compound (hereafter, NN) is a sequence of two or more nouns in which one modifies the other(s): e.g., *glass bank*. Although NNs are often subsumed under general discussions of constructions, they pose sufficiently idiosyncratic issues to merit a separate computational microtheory.[20]

Most computational work on NNs has focused exclusively on establishing the implied relationship between the nouns without disambiguating the nouns themselves. However, the latter is actually more challenging. For example, *glass bank* can mean a coin storage unit made of glass, a slope made of glass, a storage unit for glass, a financial institution with a prominent architectural feature made of glass, and more. And even though some NNs might seem unambiguous at first reading—e.g., *pilot program* (feasibility study) and *home life* (private life, how one lives at home)—they actually have other available readings: *pilot program* could mean a program for the benefit of airplane pilots, and *home life* could refer to the length of time that a dwelling is suitable to be lived in (by analogy with *battery life*).

A LEIA's analysis of NNs involves both the contextual disambiguation of the nouns and the establishment of the semantic relationship between them. We start by discussing the treatment of two-noun compounds for which the relevant senses of both nouns are available in the lexicon. Compounds with more than two nouns and/or unknown words are discussed at the end of the section.

As a prelude to describing LEIAs' approach to NN analysis, let us briefly consider best-case NN analysis results reported by others. Some examples are shown in table 4.3. Columns 3 and 4 juxtapose optimal results of LEIA processing with optimal results from three other paradigms: Tratz and Hovy (2010) (T); Rosario and Hearst (2001) (R); and Levi (1979) (L).

The points below explain why LEIA analysis is semantically more comprehensive—albeit much more expensive to operationalize—than the relation-selection approach undertaken by the others.

1. The LEIA analyses include disambiguation of the component nouns along with identification of the relation between them, whereas relation-selection approaches address only the relation itself.
2. The LEIA analyses are written in an unambiguous, ontologically grounded metalanguage (a reminder: strings in small caps are concepts, not English words), whereas the relation-selection approaches use ambiguous English words and phrases.

Table 4.3
Comparison of best-case analyses of NNs across paradigms

	Example	**Full NN analysis by OntoAgent**	**Relation selection from an inventory**
1	cooking pot	POT (INSTRUMENT-OF COOK)	perform/engage_in (T)
2	eye surgery	PERFORM-SURGERY (THEME EYE)	modify/process/change (T)
3	cat food	FOOD (THEME-OF INGEST (AGENT CAT))	consumer + consumed (T)
4	shrimp boat	BOAT (INSTRUMENT-OF CATCH-FISH (THEME SHRIMP))	obtain/access/seek (T)
5	plastic bag	BAG (MADE-OF PLASTIC)	substance/material/ingredient + whole (T)
6	court order	ORDER (AGENT LEGAL-COURT)	communicator of communication (T)
7	gene mutation	MUTATE (THEME GENE)	defect (R)
8	papilloma growth	CHANGE-EVENT (THEME PAPILLOMA) (PRECONDITION SIZE (< SIZE.EFFECT))	change (R)
9	headache onset	HEADACHE (PHASE begin)	beginning of activity (R)
10	pet spray	LIQUID-SPRAY (THEME-OF APPLY (BENEFICIARY PET))	for (L)

3. In 2, 7, 8, and 9 of table 4.3, the meaning of the "relation" in relation-selection approaches is actually not a relation at all but, rather, the meaning of the second noun or its hypernym: for example, *growth is-a change*. By contrast, since the LEIA's treatment involves full analysis of all aspects of the compound, the meaning of each of the nouns is more naturally incorporated into the analysis.
4. The relation-selection approach can merge relations into supersets that are not independently motivated, such as (T)'s *obtain/access/seek*.[21] For LEIAs, by contrast, every relation available in the independently developed ontology is available for use in compounding analysis—there is no predetermined list of compounding relations. This harks back to an early observation that practically any relationship could be expressed by a nominal compound (Finin, 1980).
5. Relation-selection approaches can include unbalanced sets of relations: for example, *consumer+consumed* (T) has been promoted to the status of a separate relation, but many other analogous correlations have not.
6. Relation-selection approaches occlude the semantic identity between paraphrases. By contrast, LEIAs generate the same meaning representation whether the input is *headache onset, the onset/beginning/start of a headache*, or *someone's headache began/started.*

In sum, for LEIAs there is no isolated NN task that exists outside the overall semantic analysis of a text. LEIAs need to compute the full meaning of compounds along with the full meaning of everything else in the input, with the same set of challenges encountered at every turn. For example:

- *Processing the elided relations in NN compounds is similar to processing lexically underspecified ones.* In NNs (e.g., *physician meeting*), the relation holding between the nouns is elided and must be inferred. However, in paraphrases that contain a preposition (e.g., *meeting of physicians*), the preposition can be so polysemous that it provides little guidance for interpretation anyway. Both of these formulations require the same reasoning by LEIAs to determine the intended meaning—here, that there is a MEETING whose AGENT is a set of PHYSICIANS.
- *Unknown words are always possible.* Encountering out-of-lexicon words is a constant challenge for agents, and it can be addressed using the same types of learning processes in all cases (see chapter 8).
- *Many word combinations are lexically idiosyncratic.* Although past work has considered compounds like *drug trial* and *coffee cup* to be analyzable as the sum of their parts, they are actually semantically idiosyncratic: they represent specific elements of a person's world model whose full meaning cannot be arrived at by compositional analysis. Ter Stal and van der Vet (1993) are correct that much more lexical listing is called for in treating compounds than the community at large acknowledges; here is a short excerpt from their discussion:

In natural science, (room temperature) means precisely 25 degrees Centigrade. A process able to infer this meaning would have to make deductions involving a concept for room, its more specific interpretation of room in a laboratory, and the subsequent standardisation that has led to the precise meaning given above. All these concepts play no role whatsoever in the rest of the system. That is a high price to pay for the capacity to infer the meaning of (room temperature) from the meanings of *room* and *temperature*. Thus, (room temperature) is lexicalized.

- *Analyses should not introduce semantic ellipsis*. The relation-selection method often introduces semantic ellipsis. For example, (T) analyze *tea cup* as *cup* with the *purpose* of *tea*. But objects cannot have purposes; only events can. So this analysis introduces ellipsis of the event *drink*. Similarly, *shrimp boat* is cited in (T) as an example of *obtain/access/seek*, but the boat is not doing any of this; it is the *instrument* of the *fisherman* who is doing this. If intelligent agents are to be equipped to reason about the world like people do, then they need to be furnished with nonelliptical analyses or else configured to dynamically recover the meaning of those ellipses.

In sum, viewing NN compounds within the context of broad-scale semantic analysis is a different task from what has been pursued to date in descriptive and NLP approaches. Let us turn now to how we are preparing LEIAs to undertake that task.

When a LEIA encounters an NN, it calls a confidence-ordered sequence of analysis functions. The first method that successfully analyzes the NN is accepted as long as it is contextually appropriate—a point we return to at the end of the section.[22] There are two lexically supported strategies for analyzing NN compounds, and one default strategy for cases that are not covered by either of these.

The first lexicon-oriented strategy is to record the NN as a head entry: *abnormal psychology, gas pedal, finish line, coffee cup*. As just mentioned, this is actually necessary since a lot of compounds have incompletely predictable meanings. A *coffee cup* is not a cup that can only contain coffee or that does, at the moment, contain coffee; instead, it is a particular type of object that can contain water, coins, or a plant.

The second lexicon-oriented strategy is to use one of the words in a two-word NN as the key for a lexical construction whose other component is a variable. For example:

- One sense of the noun *fishing* expects an NN structure in which *fishing* is N2 and some type of FISH is N1. It covers NNs such as *trout fishing, bass fishing*, and *salmon fishing*, analyzing them as

```
FISHING-EVENT
   THEME        TROUT/BASS-FISH/SALMON
```

- In compounds of the structure *N detective*, if N is a kind of CRIMINAL-ACTIVITY, then the overall meaning is that the DETECTIVE is the AGENT-OF an INVESTIGATE event whose THEME is N. So, *homicide detective* is analyzed as

```
DETECTIVE
    AGENT-OF        INVESTIGATE-#1
INVESTIGATE-#1
    THEME           MURDER
```

If, by contrast, the input were *university detective*, then this construction would *not* match, and the NN would be passed on for other types of processing.

If a given NN is not covered by a construction recorded in the lexicon, or if it *is* covered but the associated analysis does not work within the overall semantic dependency structure of the clause, then the agent opts for an underspecified analysis. Specifically, it links all available meanings of N1 with all available meanings of N2 using the most generic relation, RELATION. These candidates are subjected to deeper analysis during Extended Semantic Analysis (section 6.3.1).

4.6 Metaphors, Lexicalized

Metaphors are a frequent occurrence in natural language, and they are essential to people's ability to understand abstract ideas.[23] As Lakoff and Johnson (1980, p. 3) explain, "Our ordinary conceptual system, in terms of which we both think and act, is fundamentally metaphorical in nature." Bowdle and Gentner (2005, p. 193) concur: "A growing body of linguistic evidence further suggests that metaphors are important for communicating about, and perhaps even reasoning with, abstract concepts such as time and emotion.… Indeed, studies of scientific writing support the notion that far from being mere rhetorical flourishes, metaphors are often used to invent, organize, and illuminate theoretical constructs." As regards the frequency of metaphors, the latter report: "In an analysis of television programs, Graesser et al. (1989) found that speakers used approximately one unique metaphor for every 25 words" (p. 193).

Since LEIAs treat some metaphors (conventional metaphors) during Basic Semantic Analysis and postpone others (novel metaphors) until Extended Semantic Analysis, the discussion of metaphors is divided between this chapter and chapter 6. But before we turn to LEIA-specific issues, let us begin with a short overview of past work on metaphor.

4.6.1 Past Work on Metaphor

Metaphor has been addressed from a broad variety of premises and in different contexts: in rhetoric since Aristotle, literary criticism (e.g., Skulsky, 1986), semiotics (e.g., Eco, 1979), a variety of schools in linguistics (e.g., Lakoff & Johnson, 1980; Steen, 2017), psychology (e.g., Bowdle & Gentner, 2005), psycholinguistics (e.g., Glucksberg, 2003), philosophy (e.g., Bailer-Jones, 2009; Lepore & Stone, 2010), and neuroscience (e.g., Goldstein et al., 2012).[24] The distinction between conventional and novel metaphors has been firmly established in linguistics (e.g., Nunberg, 1987) and psychology (e.g., Gibbs, 1984). Bowdle and Gentner (2005) view the novel-to-conventional metaphor continuum in an etymological perspective and argue that metaphors conventionalize and diachronically lose their metaphoricity. Most metaphors discussed within the popular conceptual metaphor theory (e.g., Lakoff, 1993) are actually conventional and, therefore, presumably exist in a native speaker's lexicon. Even if the early AI approaches to metaphor do not state it overtly, their underlying motivation was to use metaphor processing as a means of bypassing the need for lexical and conceptual knowledge acquisition.

Theorists go beyond the novel/conventional distinction. Steen (2011) introduces a distinction between deliberate and nondeliberate metaphors: "A metaphor is deliberate when addressees must pay attention to the source domain as an independent conceptual domain (or space or category) that they are instructed to use to think about the target of the metaphor" (p. 84). But as he concedes, "the processes leading up to the product of metaphor comprehension … are largely immaterial to the question of whether their product counts as a deliberate metaphor or not" (p. 85). This corroborates our position: To successfully process input containing conventional metaphors, the hearer does not need to realize that a metaphor is present. Conventional metaphor qua metaphor may be of interest to scholars or as the subject of an entertaining etymological parlor game. But to understand *ballpark* in *ballpark figure*, it is not necessary to know that it is a baseball metaphor.

We hypothesize that people usually process novel and deliberate metaphors in the same manner in which they process unknown lexical units that are not metaphorical—by learning their meanings over time from their use in text and dialog and recording those meanings in their lexicons for later use. In other words, the novel (nonmetaphorical) senses of *pocket* and *bank* in *He pocketed the ball by banking it off two rails* will be learned with the help of knowledge of the domain (billiards) and general knowledge of what can typically be done with a billiard ball. By the same token, the meaning of *albatross* in *"But O'Malley's heaviest albatross is the state of his state"*[25] will also be understood based on the hearer's knowledge of the overall context, with no need for the hearer to have read, or even know about the existence of, Coleridge's *The Rime of the Ancient Mariner.*

Of course, building an agent that models an etymologist is a potentially interesting research direction, but it is much more important in agent systems to cover conventional metaphors. We believe that the best way to do this is to view the task as a routine part of the lifelong enhancement of an agent's knowledge resources. An agent will fail to register the aesthetic contribution of an extended metaphor like the following, but this is equally true about many people—after all, not everybody knows about baseball: (A team leader cajoling a team member) *Eric, the bases are loaded; tomorrow's demo is crucial. Please stop grandstanding and playing hardball, step up to the plate, join the effort, and lead off with a ballpark figure.*

We discussed metaphorical language in a paper entitled *Slashing Metaphor with Occam's Razor* (Nirenburg & McShane, 2016b). Initially, some readers might not have fully understood the meaning of the title. However, most everyone will have guessed that we intended to say something negative about the study of metaphor. Some readers will also have understood that we would justify this attitude on the grounds that the study of metaphor is unnecessary from some point of view. Having read on, readers who still remember the title would realize what it intended to convey: that separating metaphor detection and interpretation from the treatment of other types of figurative language and other semantic anomalies violates the dictum *entities must not be multiplied beyond necessity.*

Now, readers (such as LEIAs) with no training in philosophy may have recognized Occam as a named entity without realizing that *Occam's razor* refers to the above dictum. Such readers would fully understand this paper's title only after having read the previous sentence. The above observations further motivate our contention that delayed interpretation of input is a viable and potentially effort-saving strategy for agents. Some readers will also appreciate the double entendre in the paper's title: the metaphorical use of an action (*slashing*) associated with the physical tool (*razor*) that once served as the source of the metaphor to describe the mental tool (*Occam's razor*). While recognizing this may be a nice bonus, it is not essential for understanding the main argument of the paper. This observation illustrates and motivates our contention that agents can often function well without understanding all of an input. Anybody who has ever communicated in a foreign language can vouch for this. Sometimes incomplete understanding can lead to misunderstandings or embarrassment but, more often than not, it works well enough to achieve success in communication. Of course, the $64,000 question is how to teach LEIAs to determine what, if any, parts of an input they can disregard with impunity. This is a direction of ongoing work for our team.

4.6.2 Conventional Metaphors

Conventional metaphors are represented as regular word senses in the LEIA's lexicon. Lakoff and Johnson (1980) propose an inventory of conventional-metaphor templates, which are associated with a large number of linguistic realizations that can be recorded in the lexicon as senses of words and phrases. Below are examples of the Ontological Semantic treatments of some of Lakoff and Johnson's templates. Note that one of the challenges in meaning representation is that some of these meanings are vague—which is a knowledge representation challenge not specific to metaphorical language. Our goal here, as in all knowledge engineering, is to provide the agent with an analysis that will support its reasoning about action.

Template: Argument is war

someone attacks someone

CRITICIZE-1

AGENT	HUMAN-1
THEME	HUMAN-2
INTENSITY	>.8

an idea is on target

IDEA-1

ACCURACY	1

someone shoots down a proposal

ACCEPT-1

AGENT	HUMAN-1
THEME	PROPOSAL-1
SCOPE-OF	MODALITY-1

MODALITY-1

TYPE	epistemic	
VALUE	0	; indicates negation
SCOPE	ACCEPT-1	

Template: Time is money

an activity cost someone an hour (this has a negative implication, in contrast to *took* someone an hour)

EVENT-1

AGENT	HUMAN-1
DURATION	1 (MEASURED-IN HOUR)
SCOPE-OF	MODALITY-1

MODALITY-1

TYPE	evaluative
VALUE	<.5
SCOPE	EVENT-1

Template: Happy is up, sad is down

someone is in high spirits

HUMAN-1

HAPPINESS	.8

someone falls into a depression

CHANGE-EVENT-1

EXPERIENCER	HUMAN-1
PRECONDITION	HAPPINESS-1
EFFECT	HAPPINESS-2

HAPPINESS-1
  DOMAIN HUMAN-1
  RANGE >.5

HAPPINESS-2
  DOMAIN HUMAN-1
  RANGE <.2

Template: The mind is a machine

someone grinds out a solution to a problem (analyzed as 'someone thinks very hard and succeeds in solving a problem')

SOLVE-1
  AGENT HUMAN-1
  THEME PROBLEM-1
  PRECONDITION THINK-1
  SCOPE-OF MODALITY-1

THINK-1
  AGENT HUMAN-1
  SCOPE-OF MODALITY-2

MODALITY-1
  TYPE epiteuctic
  VALUE 1
  SCOPE SOLVE-1

MODALITY-2
  TYPE effort
  VALUE 1
  SCOPE THINK-1

someone runs out of steam

ANIMAL-1
  EXPERIENCER-OF FATIGUE-1

FATIGUE-1
  INTENSITY >.8

In sum, conventional metaphors pose the exact same inventory of meaning representation challenges and opportunities as any other lexemes or phrases that happen to not have what linguists would consider a historical source in metaphor.

4.6.3 Copular Metaphors

Many creative (not conventionalized) metaphors are of the form NP1 *is* NP2. Lakoff and Johnson (1980, p. 139) say that creative metaphors "are capable of giving us a new understanding of our experience." Their example: *Love is a collaborative work of art* (p. 139).

We describe this class using the syntactic term *copular metaphors* because their defining feature is, in fact, syntactic: they use the copular verb *be*.

We prepare LEIAs to analyze copular metaphors using a lexical sense of the verb *be* that

1. expects the syntactic structure NP1 *is* NP2;
2. puts no semantic constraints on the meanings of the NPs;
3. links the meanings of the NPs using the almost vacuous concept JUXTAPOSITION; and
4. includes a meaning procedure that, if called during Extended Semantic Analysis, will attempt to identify the most salient properties of the second NP and apply them to the first NP (see section 6.3.3).

Note that this sense of *be* is just one of many senses of this verb that links two noun phrases. It is the least-specified one semantically, meaning that others will be preferred if their semantic constraints are satisfied. For example, *John is a doctor* will be handled by a sense of *be* that requires NP2 to indicate a SOCIAL-ROLE. That sense will be used to analyze this input as follows:

```
HUMAN-1
    HAS-PERSONAL-NAME   'John'
    HAS-SOCIAL-ROLE     PHYSICIAN
```

The sense used for novel metaphors, which has no semantic constraints, will be used as a fallback when more narrowly constrained senses like this one are not applicable.

4.6.4 Recap of Metaphors

- Conventionalized metaphors are recorded as lexical senses.
- Nonconventionalized metaphors in copular constructions (*Life is a garden*) are recognized as potential metaphors, analyzed using the property JUXTAPOSITION, and flagged with a procedural semantic routine that will be run during Extended Semantic Analysis.
- Metaphorical meanings not treated by these methods will often give rise to a low-scoring TMR, indicating that something is wrong. That "something" will be explored during Extended Semantic Analysis.

4.7 Metonymies, Lexicalized

Like metaphors, many word senses that can historically be analyzed as metonymies are, in synchronic terms, regular lexical senses that are recorded in the LEIA's lexicon, such as *get a pink slip* (get fired) and *red tape* (excessive bureaucratic requirements). There are also ontological classes of metonymies: for example, a piece of clothing can be used to refer to the person wearing it (*Give the red shirt a glass of milk*). The latter are not treated at this stage. Instead, semantic analysis results in an incongruity, reflected by a low score for the TMR (there is no sense of *give* that expects a SHIRT as the BENEFICIARY). This low score is a flag to track down the source of the incongruity during Extended Semantic Analysis (section 6.2).

4.8 Ellipsis

Of all the topics treated in this chapter on Basic Semantic Analysis, ellipsis is likely to be the most surprising. After all, ellipsis is not only an aspect of reference resolution—which is a prime example of a *pragmatic* phenomenon—it is a particularly difficult aspect of it. However, if one dispenses with preconceived notions about linguistic modularity and, instead, considers the twin challenges of detecting and resolving ellipsis, it turns out that much of the work can be neatly subsumed under Basic Semantic Analysis. In fact, in some cases *all* of the work can be, as the subsections below will explain.

4.8.1 Verb Phrase Ellipsis

Modal and aspectual verbs can be used either with or without an overt complement, as shown by the juxtaposed (a) and (b) versions of (4.34) and (4.35).

(4.34) a. "You have to get up."
b. "I know you don't feel like getting up, but you have to __." (COCA)

(4.35) a. "I just started playing."
b. "They've been playing at least five years, maybe three years, and I just started __." (COCA)

Structures whose verbal complements are not overt are said to contain verb phrase (VP) ellipsis.

We prepare LEIAs to detect and resolve VP ellipsis by recording lexical senses of modal and aspectual verbs in pairs. Whereas one member of the pair expects an overt VP, the other expects VP ellipsis. The elliptical senses posit a placeholder EVENT along with a meaning procedure indicating that it requires downstream coreference resolution (section 5.5). In other words, an input like "I just started" will be analyzed as "I just started EVENT", and the nature of the EVENT will be flagged as needing to be tracked down later.

By explicitly preparing for VP ellipsis during lexical acquisition, we solve two problems at once. First, we account for the fact that we, as people, do understand—even outside of context—that sentences with VP ellipsis imply some event; so, too, should LEIAs. Second, we do not need a special process for detecting VP ellipsis. When an elliptical input is processed, it simply uses the lexical sense that expects the ellipsis.

Consider the example *John washed his car yesterday but Jane didn't __*. Its basic TMR—shown in two parts below for readability's sake—indicates that, prior to forthcoming coreference procedures, all the LEIA knows about Jane from this utterance is that she didn't *do something* (EVENT-1).

John washed his car yesterday but

```
WASH-1
    AGENT               HUMAN-1
    THEME               AUTOMOBILE-1
    TIME                (combine-time -1 day)    ; "yesterday"

HUMAN-1
    HAS-PERSONAL-NAME   'John'
    GENDER              male
    AGENT-OF            WASH-1

AUTOMOBILE-1
    RELATION            HUMAN-1                  ; "his"
    THEME-OF            WASH-1

CONTRAST-1                                       ; "but"
    DOMAIN              WASH-1
    RANGE               EVENT-1
```

Jane didn't

```
EVENT-1
    CASE-ROLE           HUMAN-2
    SCOPE-OF            MODALITY-1
    COREF               seek-specification

HUMAN-2
    HAS-PERSONAL-NAME   'Jane'
    GENDER              female
    AGENT-OF            EVENT-1

MODALITY-1
    TYPE                epistemic
    VALUE               0                        ; negation
    SCOPE               EVENT-1
```

To reiterate, the point of this example is to show that during Basic Semantic Analysis the agent detects the VP ellipsis in "Jane didn't ___" and provisionally resolves it as an unspecified EVENT whose precise meaning will be sought during the next stage of processing, Basic Coreference Resolution.

4.8.2 Verb Phrase Ellipsis Constructions

Identifying the sponsor for an elided VP can be quite difficult, which is why, in the general case, it is postponed until the dedicated Basic Coreference Resolution module (see section 5.5). However, there exist elliptical constructions in which identifying the elided meaning is quite straightforward. We record these in the lexicon, and the associated lexical senses allow for the ellipsis to be fully resolved right now, during Basic Semantic Analysis. The example in (4.36) shows a construction that indicates that the agent applies maximum effort to carrying out the action.

(4.36) [Subj$_1$ V as ADV as Pronoun$_1$ can/could]
Agatha wrote back as fast as she could ___. (COCA)

In this example, the modal verb *could* is used without its VP complement, making it an elliptical structure. But the ellipsis is resolved by copying the meaning of the same verb that this expression modifies: that is, Agatha wrote back as fast as she could *write back*. This means that no discourse-level reasoning is needed to resolve the ellipsis; the answer lies in the construction itself.

Deciding how many ellipsis-oriented constructions to record, and how to balance literal and variable elements in them, represents a microcosm of knowledge engineering overall. For example, the input *Boris gives his children as many gifts as he wants to* can be covered by either of the constructions shown in table 4.4. However, the second, more generic one also covers examples like *Boris takes as few as he can.*

In general, the more narrowly defined the construction, the more likely it will give rise to a unique and correct semantic analysis. But recording constructions takes time, and more narrowly specified constructions offer less coverage. So knowledge acquirers must find the sweet spot between the generic and the specific.

Below are some VP ellipsis constructions, presented using an informal notation, along with examples that were automatically extracted from the COCA corpus. The elliptical gaps are indicated by underscores, and the antecedents for the elided categories are underlined.

(4.37) [V <all, everything, everything and anything, anything and everything, whatever> Pronoun (ADV) AUX]
a. My biggest focus right now is just learning all I can ___. (COCA)
b. The government gobbled up whatever it could ___. (COCA)

(4.38) [VP as ADV as Pronoun AUX]
Yeah, he wanted me to come pick him up as quickly as I could ___. (COCA)

Table 4.4.
VP ellipsis constructions

Boris	**gives**	**his children**	**as many**	**gifts**	**as**	**he**	**wants to**
Subj	V	(NP)	as many	N	as	Pronoun	wants to
Subj	V	(NP)	QUANT	(N)	as	Pronoun	MODAL
Boris	**takes**		**as few**		**as**	**he**	**can**

(4.39) [VP as ADV/ADJ as Pronoun AUX]

a. He said, 'It's going to hit me, so I'm going to enjoy life as best as I can__.' (COCA)

b. On this picture, I feel like I got my way as much as I could __. (COCA)

To conclude this section, certain cases of VP ellipsis can be fully treated—both detected and resolved—during Basic Semantic Analysis thanks to constructions recorded in the lexicon.

4.8.3 Event Ellipsis: Aspectual + NPOBJECT

Another type of event ellipsis occurs when aspectual verbs take an NP complement that refers to an OBJECT rather than an EVENT. Such clauses must involve ellipsis because aspectual meanings can only ever apply to events. Consider the following pair of examples.

(4.40) He boldly went up to her and started a conversation … (COCA)

(4.41) He wrote and directed plays, started __ a book about the Yiddish language with his grandfather … (COCA)

Example (4.40) refers to starting a conversation, which is an EVENT, so there is no ellipsis. By contrast, (4.41) refers to starting a book, which is an OBJECT; what is meant is that he started *writing* a book.

We prepare LEIAs to treat *aspectual +NPOBJECT* cases the same way as VP ellipsis: by creating a lexical sense of each aspectual verb that expects its complement to refer to an OBJECT. The semantic description of such constructions includes an underspecified EVENT whose THEME is that OBJECT. So, his starting a book in (4.41) will be analyzed as his being the AGENT of an EVENT (scoped over by "PHASE begin") whose THEME is BOOK. This is all the text says—and that is the very definition of Basic Semantic Analysis. The rest requires nonlinguistic reasoning that the agent will pursue, if it chooses to, during Extended Semantic Analysis and/or Situational Reasoning (chapters 6 and 7).

4.8.4 Event Ellipsis: Lexically Idiosyncratic

Certain words, when used in certain constructions, always imply a particular kind of EVENT. For example, when someone invites someone else to some place, there is always an implied MOTION-EVENT.

(4.42) Marino brought the young artist into his sphere, secured several commissions for him, and eventually invited him to Rome. (COCA)

The TMR for the clause *he invited him to Rome* is as follows.

```
INVITE-1
    AGENT          HUMAN-1
    THEME          MOTION-EVENT-1
    BENEFICIARY    HUMAN-2
    TIME           <find-anchor-time
```

MOTION-EVENT-1
  AGENT HUMAN-2
  DESTINATION CITY-1

HUMAN-1
  GENDER male

HUMAN-2
  GENDER male

CITY-1
  HAS-NAME 'Rome'

The lexical sense of *invite* that covers this construction explicitly lists the MOTION-EVENT, which explains how it turns up in the TMR. Similarly, when one *forgets* some PHYSICAL-OBJECT that can be carried (i.e., it is an ontologically licensed THEME of a CARRY event), the elided event is, by default, TAKE. There is a lexical sense of *forget* that anticipates, and resolves, this ellipsis.

As discussed earlier with respect to VP ellipsis constructions, the challenge in writing such lexical senses is determining the sweet spot for coverage versus precision. Let us continue with the example of forgetting an OBJECT. Above, we suggested that the object in question must be able to be carried in order for the elided verb to be understood as CARRY. This works for forgetting one's keys, notebook, lunch, and many other objects. It does not cover forgetting one's car or one's file cabinet. But what does it mean to forget one's car or one's file cabinet? It is impossible to say without context since there is no high-confidence default interpretation. Forgetting one's car might mean forgetting to move it to the other side of the street according to alternate-side-of-the-street parking rules, or it might mean forgetting to drive it to school, rather than ride one's bike, in order to help transport something after class. Forgetting one's file cabinet might mean forgetting to look there for a lost object or forgetting to have it transported when moving from one office to another. Since we know that nonholdable objects can occur as the direct object of *forget*, we need to write another lexical sense that anticipates them. This sense, like the aspectual senses discussed earlier, will initially underspecify the event—listing it as simply EVENT—and call a procedural semantic routine that will later attempt to reason more precisely about what it might mean.

4.8.5 Event Ellipsis: Conditions of Change

Events and states cause other events and states.

(4.43) [An event causes an event; 'rain(s)' is analyzed as RAIN-EVENT]
Heavy rains caused the river and its tributaries to flood … (COCA)

(4.44) [A state causes an event]
"There's room for two," Sophie called out. Her excitement made Mr. Hannon laugh. (COCA)

(4.45) [An event causes a state]
... The disappearance of the valuables made people nervous ... (COCA)

(4.46) [A state causes a state]
The deaths and the publicity about the state's raging rivers have taken a toll on commercial rafters' business. The conditions have made people nervous ... (COCA)

Events and states cannot be caused by objects. However, language permits us to express situations as if objects could cause events and states.

(4.47) [An object is said to cause an event]
Investigators want to closely inspect the engines to figure out how exactly the birds caused the plane to fail so badly and so fast. (COCA)

(4.48) [An object is said to cause a state]
And I knew ice cream was something that made people happy. (COCA)

In such cases, the named object participates in an event or state that is the actual cause of another event or state. In our examples, the accident happened because of something the birds did, and ice cream plays some role in an event (eating it) that makes people happy.

We prepare LEIAs to detect elided events in causal clauses using special lexical senses of words and phrases that express causation—e.g., *cause (sth.)*, *make (sth.) happen*, *bring (sth.) about*. The given senses expect the named cause to be, ontologically speaking, an OBJECT, and they explicitly posit an EVENT for which that OBJECT serves as a case role. For example, the sense of *make* that expects the construction NP_{OBJECT} *makes NP V* has a semantic representation that will generate the following TMR for *The onion made her cry.*

CRY-1
- EXPERIENCER HUMAN-1
- CAUSED-BY EVENT-1
- TIME *<find-anchor-time*

EVENT-1
- INSTRUMENT ONION-1
- COREFER *seek-specification*

HUMAN-1
- GENDER female

This analysis posits an EVENT without specifying its nature.[26] The fact that the event is underspecified is reflected in the TMR by the call to the meaning procedure *seek-specification*. If, later on, the LEIA has reason to believe that the nature of this event is important, it can attempt to track it down—though it will be successful only if it has a sufficient amount of domain and situational knowledge to support the analysis (see chapter 7).

4.8.6 Gapping

Gapping is a type of verbal ellipsis that occurs in structurally parallel coordinate and comparative structures. The following examples illustrate two common gapping constructions, presented informally.

(4.49) [Subj_1 V DO_1(,) and Subj_2, DO_2]
Of course, thoughts influence actions, and actions, thoughts. (COCA)

(4.50) [Subj_1 V IO_1 DO_1(,) and Subj_2, DO_2]
The plumber charged us $200, and the electrician, $650.

Although gapping is not used all that commonly in English, it makes sense to cover the most frequent eventualities using constructions like those illustrated above, which are anchored in the lexicon on the keyword *and.* These senses are supplied with meaning procedures that can be run right away, during Basic Semantic Analysis. They semantically analyze the overt verb and then reconstruct the elided one as a different instance of the same ontological type. This means that gapping can be detected and fully resolved at this stage of processing for any input that matches a recorded gapping construction.

4.8.7 Head Noun Ellipsis

In English, ellipsis of the head noun in noun phrases is permitted when the head noun follows a number (4.51), follows a quantifier (4.52 and 4.53), or participates in constructions like *someone's own* (4.54).

(4.51) He had a number of offspring but only two __ were considered worthy contenders for the ducal crown. (COCA)

(4.52) "Tea, Mr. Smith?" Tracy asked. "Yes, I'd love some __." (COCA)

(4.53) Who are the people who worked for him? Unless he didn't have any __? (COCA)

(4.54) The voice was not my own __. (COCA)

The lexicon includes a special sense of each applicable word and phrase that anticipates head noun ellipsis. These senses include a call to a procedural semantic routine that guides the LEIA in resolving the ellipsis by attempting to identify the most recent mention of an entity that matches the selectional constraints of the clause's main verb.

4.8.8 Recap of Ellipsis

- Verb phrase (VP) ellipsis: Our fondness for sweetness was designed for an ancestral environment in which fruit existed but candy didn't __. (COCA)
- VP ellipsis constructions: She hides her true identity as long as she can __. (COCA)
- Event ellipsis—Aspectual + $\text{NP}_{\text{OBJECT}}$: "Started __ the book yet?" (COCA)

- Event ellipsis—Lexically idiosyncratic: She made friends with a French girl, who invited her __ to Paris. (COCA)
- Event ellipsis—Conditions of change: The acid in vinegar caused the iron in the steel to combine rapidly with oxygen from the air. (COCA)
- Gapping: Lou likes Coke, and Sherry __, Pepsi.
- Head noun ellipsis: He was good, getting up to eight skips. At best Annabel got three __. (COCA)

4.9 Fragmentary Utterances

We define *fragmentary utterances* as nonpropositional, freestanding utterances—in contrast to the midsentence fragments that are analyzed as a matter of course during incremental NLU. Examples of fragmentary utterances are "Large latte" and "Not yet." During Basic Semantic Analysis, all of the available analyses of such utterances are posited as candidates. Selecting the intended meaning and incorporating it into the larger context is the shared responsibility of Extended Semantic Analysis (section 6.4) and, if needed, Situational Reasoning (chapter 7).

4.10 Nonselection of Optional Direct Objects

Some verbs, such as *read* and *paint*, are optionally transitive. This means that they *can* select a direct object but do not require one. Nonselection of a direct object is not ellipsis but it occasionally presents an interesting problem: the unexpressed object can be needed to interpret a subsequent referring expression, as in (4.55).

(4.55) They won't be doing any hiring this year apart from replacing those __ who leave.

The elided noun in the noun phrase [$_{NP}$ those __ who leave] is *employees*. This concept was implicitly introduced into the context as the filler for the THEME of HIRE. Engineering details aside, the point is this: During Basic Semantic Analysis, if the direct object of an optionally transitive verb is unselected (i.e., not explicit in the input), a special slot for it is created in the TMR. That slot is filled by the generic constraint found in the ontology (here: HIRE (THEME EMPLOYEE)) and is available for later coreference as needed. This is an excellent example of why it is risky to split language processing tasks finely across systems and developers. If one does so, then phenomena like this will more than likely fall between the cracks as nobody's responsibility.

4.11 Unknown Words

As mentioned earlier, the LEIA's lexicon contains about 30,000 word senses, making it substantial but far from comprehensive. This means that LEIAs must be able to process both unknown words and unknown senses of known words. We already described the first stage of treating unknown words: during Pre-Semantic Integration, LEIAs posit syntactically specific, but semantically underspecified, lexical senses for unknown words. Now, during Basic Semantic Analysis, they try to narrow down the meaning of the newly learned word senses using ontological search. The specifics of the process vary depending on (a) the part of speech of the newly learned word sense and (b) whether what is being learned is a completely new string or a known string in a different part of speech. In the latter case, the meaning being learned might, though need not, be related to the known meaning.

4.11.1 Completely Unknown Words

New-word learning focuses on open-class words—currently nouns, adjectives, and verbs, which we consider in turn.

Unknown nouns. Syntactically, simple nouns take no arguments. Semantically, they can refer to an OBJECT, EVENT, or PROPERTY. During Pre-Semantic Integration, the LEIA generates three candidate senses for each unknown noun, one for each of these semantic mappings. Each candidate sense is then evaluated at this stage with two goals: (a) to choose the best of these mappings for the context and (b) if possible, to narrow down that interpretation to a more specific concept in the given branch of the ontology. Consider example (4.56), which contains the unknown noun *hobo.*

(4.56) A hobo came to the door. (COCA)

The sense of *come* that best fits the context has the syntactic structure "Subject V PP" and the semantic analysis "COME (AGENT ^Subject) (DESTINATION ^PP-obj)". Since the meaning of the subject must fill the AGENT slot, the best interpretation of *hobo* is OBJECT (rather than EVENT or PROPERTY). However, based on the fact that this OBJECT has to fill the AGENT slot of COME, the agent can narrow it down to ANIMAL, since that is the *sem* filler of this property recorded in the ontology. This results in the following TMR for *A hobo came to the door.*

COME-1
- AGENT ANIMAL-1
- DESTINATION DOOR-1
- TIME *<find-anchor-time*

ANIMAL-1
- *from-sense* *new-noun-object-n1*

The metadata in this TMR (i.e., the *from-sense* slot) carries a trace that unknown-word processing was carried out, should the LEIA decide to pursue a more fine-grained analysis of this word through learning by reading (Nirenburg et al., 2007) or by interacting with a human collaborator. Now consider example (4.57), in which the only unknown word is *tripe.*

(4.57) Jane was eating tripe with a knife.

The verb *eat* has several senses in our lexicon, all but one of which cover idiomatic constructions that are rejected on lexico-syntactic grounds (e.g., *eat away at*). So the LEIA can immediately narrow down the choice space to the main sense of *eat* (eat-v1), which is optionally transitive. It maps to an INGEST event whose case roles are AGENT and THEME. The THEME is specified as FOOD, which is more constrained than the ontologically listed disjunctive set [FOOD, BEVERAGE, INGESTIBLE-MEDICATION] (i.e., one does not eat a beverage or medication). This leads to the following TMR for *Jane was eating tripe with a knife*.

ASPECT-1	
PHASE	continue
SCOPE	INGEST-1
INGEST-1	
AGENT	HUMAN-1
INSTRUMENT	KNIFE-1
THEME	FOOD-1
SCOPE-OF	ASPECT-1
TIME	<*find-anchor-time*
HUMAN-1	
AGENT-OF	INGEST-1
HAS-PERSONAL-NAME	'Jane'
GENDER	female
FOOD-1	
THEME-OF	INGEST-1
from-sense	*new-noun-object-n1*

Unknown adjectives. Syntactically, adjectives modify nouns. Semantically, they map to a PROPERTY, and the meaning of the noun they modify fills the DOMAIN slot of that property. The RANGE, however, depends on the meaning of the adjective itself. For example, the subject noun phrase in (4.58) includes the unknown adjective *confrontational.*

(4.58) A confrontational security guard was yelling.

The LEIA's analysis of *confrontational security guard* is

SECURITY-GUARD-1	
DOMAIN-OF	PROPERTY-1

PROPERTY-1 is an instance of the most underspecified ontological property (the root of the PROPERTY subtree). If asked to, the LEIA can create the entire set of properties for which SECURITY-GUARD is a semantically acceptable filler for DOMAIN. This would narrow the interpretation from 'any property' to 'one of a listed set of properties'. However, in many cases—like this one—this set will still be too large to be of much more utility than the generic PROPERTY interpretation.

Unknown verbs. Verbs can take various numbers and types of arguments and complements, which can realize various semantic relations. We will use a transitive verb as our example. Transitive verbs most often use the subject to express the AGENT and the direct object to express the THEME. If the subject cannot semantically fill the AGENT role, then the next case role in line is INSTRUMENT. We can see how these case role preferences play out in example (4.59), where the unknown verb is *nicked.*

(4.59) The truck nicked the tree.

The agent already mapped *nicked* to EVENT during Pre-Semantic Integration. Now it tries to determine the case roles of its arguments using the abovementioned preferences. *The truck* cannot be the AGENT since it is inanimate, but it can be the INSTRUMENT. *The tree*, for its part, can be a THEME. This results in the following TMR:

```
EVENT-1
   THEME         TREE-1
   INSTRUMENT    TRUCK-1
   COREF         seek-specification
```

The question is, can the agent usefully constrain the interpretation of EVENT on the basis of what the ontology says about this combination of case roles and fillers? Not too much, so the EVENT remains underspecified at this stage and can be analyzed more deeply, if the agent chooses to, during Situational Reasoning.

4.11.2 Known Words in a Different Part of Speech

It is not unusual for the lexicon to contain a needed word but not in the part of speech needed for the input. For example, it might have the noun *heat* but not the verb *to heat.* The first thing to say about such situations is that there are many eventualities:

- The lexicon might contain exactly one sense, which luckily is semantically related to the new-part-of-speech sense.
- The lexicon might contain multiple senses, exactly one of which is related to the new-part-of-speech sense.
- The lexicon might contain any number of senses, none of which is related to the new-part-of-speech sense.

This complexity is just one manifestation of the open-world problem. No matter how many words and phrases an agent knows, an input might contain a new one.

To keep useful bounds on this discussion, we will work through one example,

(4.60) A large radiator was heating the room.

with a number of simplifying assumptions:

1. The lexicon contains exactly one nominal sense of *heat*, which refers to temperature.
2. That sense is the one needed to learn the meaning of the verb *to heat*.
3. All the other words in the sentence have just one sense in the lexicon.

If (1) or (3) did not hold (i.e., if there was lexical ambiguity), then the process would iterate over all possibilities and generate multiple candidate analyses. This is not a problem; it is just inconvenient to present. If (2) did not hold, the agent would get the answer wrong and would suspect it only if all TMR candidates got low confidence scores, suggesting some problem in combining semantic heads with their arguments. What is interesting about this example is that the noun *heat* maps not to an OBJECT or EVENT but to an ontological PROPERTY—namely, TEMPERATURE.

Consider the TMR for *A large radiator was heating the room*, which the agent generates using the analysis process described below.

```
CHANGE-EVENT-1
    THEME            ROOM-1
    INSTRUMENT       HEATER-1
    PRECONDITION     TEMPERATURE-1
    EFFECT           TEMPERATURE-2
    TIME             <find-anchor-time
    SCOPE-OF         ASPECT-1

TEMPERATURE-1
    DOMAIN           ROOM-1
    RANGE            <TEMPERATURE-2.RANGE

TEMPERATURE-2
    DOMAIN           ROOM-1
    RANGE            >TEMPERATURE-1.RANGE

HEATER-1
    SIZE             .7
    INSTRUMENT-OF    CHANGE-EVENT-1

ASPECT-1
    PHASE            continue
    SCOPE            CHANGE-EVENT-1
```

1. The agent looks up the word *heat* and finds only a nominal sense, described as TEMPERATURE (RANGE (> .7)).
2. From its inventory of methods to treat different eventualities of new-word-sense learning, it selects the one aimed at learning new verb senses from noun senses that map to a PROPERTY value.

3. The chosen learning method directs the agent to hypothesize that the verb refers to a CHANGE-EVENT involving this property. CHANGE-EVENT is an ontological concept used to describe events whose meaning is best captured by comparing the value of some property in the event's PRECONDITION and EFFECT slots. For example, *speed up* is described as a CHANGE-EVENT whose value of SPEED is lower in the PRECONDITION than in the EFFECT. Analogous treatments are provided in the lexicon for *grow taller* (HEIGHT), *quiet down* (LOUDNESS), *go on sale* (COST), and countless other words and phrases (see McShane, Nirenburg, & Beale, 2008, for details).
4. The agent creates a CHANGE-EVENT TMR frame, along with its PRECONDITION and EFFECT slots.
5. It hypothesizes the direction of change (i.e., the comparison between the range of TEMPERATURE in the PRECONDITION and EFFECT) on the basis of the RANGE of the property in the nominal sense: if the nominal sense has a high value (like the .7 listed in the nominal sense of 'heat'), then it assumes that the direction of change is increase; if the nominal sense has a low value, then it assumes that the direction of change is decrease.
6. It interprets the THEME of the CHANGE-EVENT (here, ROOM-1) as the DOMAIN of the TEMPERATURE frames.
7. It links the meaning of *heater* to the CHANGE-EVENT using the case role INSTRUMENT, since the default case role for subjects (AGENT) cannot apply to inanimates like *heater*.
8. It deals with routine semantic analysis needs, such as the analysis of tense and aspect.

To reiterate, this algorithm works for the case of unknown verbs for which an available nominal sense maps to a SCALAR-ATTRIBUTE. This is only one of many eventualities, all of which need to be fully fleshed out algorithmically and then tested against corpus evidence. Our expectation is that, along with some impressive automatic results, we will encounter many false positives—that is, cases in which the meaning of a word will not be predictable from a morphologically related word form. That, in turn, will motivate further enhancements to the learning algorithms. This is, however, an envelope that we must push hard because automating lexical knowledge acquisition is a high priority for knowledge-based agent systems.

4.12 Wrapping Up Basic Semantic Analysis

Even after all this processing has been carried out, a lot of loose ends remain, such as residual lexical and referential ambiguity, low-scoring TMRs resulting from incongruities, procedural semantic routines that have been posited in the TMR but have not yet been run, and fragmentary utterances that have not yet been incorporated into the discourse structure. All of these can be pursued by a LEIA if it chooses to do so in later stages of processing—but it might not. By this point, the LEIA has an idea—or several competing ideas—of what the input is about, and the given topic may or may not be relevant to its operation. For example, if the LEIA is a mechanic's assistant, but the topic is football, and there are multiple humans involved in the conversation, there is no reason for the agent to ponder which meaning of football is intended (soccer or American football), and it should certainly not pester the speaker with clarification questions about it. In short, the conclusion of Basic Semantic Analysis is an important decision point for the LEIA with respect to its language- and task-related reasoning.

4.13 Further Exploration

1. Explore inventories and classifications of metaphors available online, using search terms such as "Lakoff and Johnson metaphor," "conceptual metaphor," "structural metaphor," and "English metaphor list." When analyzing individual metaphors, think about questions such as the following:

- Does it feel metaphorical to you, or does it seem more like a fixed expression at a distance from its metaphorical roots?
- Do you know the etymology?
- Do you think that knowing the etymology helps you to understand the intended meaning?
- Do you fully understand its intended meaning?
- Is that meaning precise or vague?
- Does the metaphor sound normal/everyday or creative/flowery?
- Can you quickly think of a nonmetaphorical paraphrase, or is the metaphor the default way of expressing the given idea?

2. Use the online version of the COCA corpus (https://www.english-corpora.org/coca/) to explore the distribution of speech acts (direct vs. indirect) for constructions that can canonically indicate an indirect speech act. Can you identify any heuristics to predict whether the direct or indirect meaning is intended? For example,

- When does *I need to know* indicate a direct speech act (simply that the speaker needs to know something) versus an indirect speech act (i.e., "Tell me")?
- When does *Can you …* indicate a direct speech act (asking about the hearer's ability to do something) versus an indirect speech act ("Please do it")?
- When does *It would be great if* indicate a direct speech act (the expression of a desire) versus an indirect speech act ("Please make this happen")?

3. Use the online version of the COCA corpus (https://www.english-corpora.org/coca/) to explore nominal compounds. There's a challenge, however: the interface does not allow you to search for patterns that are as unconstrained as "any noun followed by any noun." Invent search strategies that give you some insights into how nominal compounding works within the constraints of the search engine. For example, think about classes of nouns for which a particular word can serve as an example to seed exploration. For example, *professor _nn* can be used to investigate "social-role + any-noun" (e.g., *professor rank*). How do the hits compare with *carpenter _nn*? *nurse _nn*? *nn _nn chef* (e.g., *genius pastry chef*)?

4. Looking just at the table of contents at the beginning of the book, try to reconstruct what was discussed in each section of chapter 4 and recall or invent examples of each phenomenon.

5

Basic Coreference Resolution

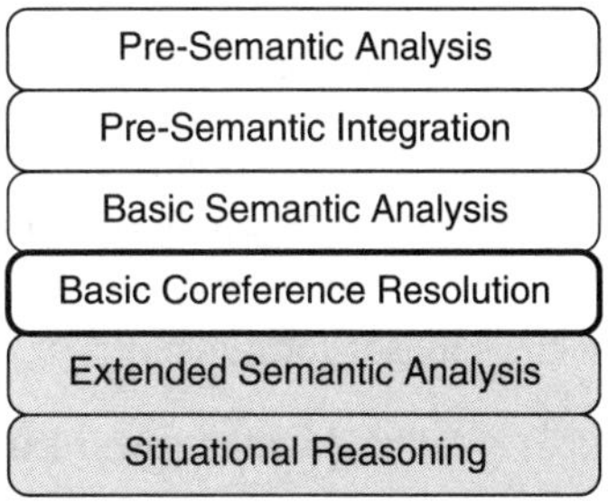

The process of Basic Semantic Analysis described in the previous chapter attempts to disambiguate the words of input and establish their semantic dependencies. For some inputs, like *A squirrel is eating a nut*, this process will result in a complete, unambiguous meaning representation. But for most inputs, it leaves loose ends, such as residual ambiguity (multiple analyses are possible), incongruity (no analyses work out well), and underspecification (an imprecise analysis has been posited in anticipation of a more precise one).

This chapter considers one class of outstanding issues: underspecifications resulting from the need for textual coreference. *Textual coreference* refers to linking different mentions of the same entity in a so-called *chain of coreference* within a language input. For example, the constituents with matching subscripts in (5.1) are coreferential.

(5.1) I handed the card$_1$ to Nina$_2$. She$_2$ read it$_1$ silently and then aloud. $_{\text{(COCA)}}$

Textual coreference is just one part of the much larger enterprise of *reference resolution*, which involves linking actual or implied mentions of objects and events to their anchors in an agent's memory. Agents undertake full reference resolution during Situational Reasoning (chapter 7), but they carry out many prerequisites to that during this stage of Basic Coreference Resolution. By contrast, most current NLP systems attempt only textual coreference—moreover, only select aspects of it (see section 1.6.8).

5.1 A Nontechnical Introduction to Reference Resolution

Since only linguists who already engage in the study of reference are likely to be familiar with the full problem space, we will start with an extended, example-based introduction. This is just a warm-up, intended to convey how much work needs to be done. It does not yet organize the phenomena or their treatment into a model—that will come later.

Some readers might find the introduction sufficient to satisfy their curiosity about the main topic of this chapter—textual coreference. The model of textual coreference is quite detailed, as it must cover many disparate phenomena; for the casual reader, the whole chapter beyond the introduction might read like a deep dive. There is no getting around this complexity, but there is a choice: one can study, casually read, or briefly skim the description of the model. Any of these should suffice as preparation for later chapters.

5.1.1 Definitions

A *referring expression* (RefEx) is a word or phrase that has referential function: that is, it points to some object or event in the world. An entity that helps to resolve a RefEx (that is, to specify its meaning) is called its *sponsor.* In (5.2), he_{1A} is the sponsor for he_{1B}, and he_{1B} is the sponsor for himself_{1C}.

(5.2) As he_{1A} walks toward the exit, he_{1B} admires himself_{1C} in the mirror behind the bar. (COCA)

We use the term *sponsor* (following Byron, 2004) rather than the more common *antecedent,* because not every sponsor is, in fact, an antecedent:

1. Whereas an antecedent must come before the RefEx in a speech stream or text, a sponsor can come either before or after it.
2. Whereas an antecedent must be part of the linguistic context, a sponsor can also be in the real-world context—as when something is pointed to and then visually perceived by the interlocutor.
3. Whereas an antecedent is assumed to have a strict coreference relationship with a RefEx, a sponsor can stand in various semantic relationships with it. For example, in (5.3) *a couple* is the sponsor of—though not coreferential with—*the husband*: it introduces two people into the context, only one of whom is coreferential with *the husband.*

(5.3) A voice came from across the dining room, where a couple was finishing their meal. "It happened to us too!" the husband volunteered. (COCA)

Entities that are evaluated as potential sponsors are called *candidate sponsors.* In (5.4), both *Einstein* and *Bohm* are candidate sponsors for the RefEx *he.*

(5.4) Einstein told Bohm that he had never seen quantum theory presented so clearly as in Bohm's new book … (COCA)

The *window of coreference* is the span of text/discourse where the sponsor is sought. Ideally, it is the most local segment of the discourse that is about the given topic. Stated differently, the window of coreference should not extend back into a portion of the discourse that is about something else. Unfortunately, it is currently beyond the state of the art to reliably, automatically analyze discourse structure. Therefore, most systems—including ours, for the moment—use a fixed window of coreference, usually a couple of sentences.

As mentioned earlier, coreference resolution involves linking textual elements that refer to the same thing. This is different from *reference* resolution, which involves linking the meanings of text strings—or things perceived using other channels of perception, such as vision—to their *anchors* in a person's or agent's memory.[1] If there is no existing anchor—that is, if the entity or event is new to the agent—a new anchor must be created. Textual coreference is often a necessary step toward full reference resolution, but it is never the end stage. The end stage always involves modifying the LEIA's memory. This happens during Situational Reasoning.

Many *words* in language inputs do not refer and, therefore, are excluded from reference resolution procedures. These include such things as pleonastic *it* (*It is raining*) and non-compositional components of idioms (*He kicked the bucket*[2]). Basic Semantic Analysis, which is carried out before reference resolution, determines whether an entity is referential. Stated briefly:

- All referring expressions, and *only* referring expressions, end up as concept instances in the TMR for an input. In English, nouns (including pronouns) and verbs can be referring expressions.
- Only concept instances are subject to reference resolution.
- By specifying which concept instances comprise a TMR, Basic Semantic Analysis solves the problem of detecting which entities in texts are referring expressions, which is a well-known challenge for knowledge-lean approaches.
- The TMR does retain, as part of its metadata, a trace of the form of the referring expression in the input: for example, *a fox* versus *the fox* versus *this fox* versus *it*. These linguistic clues influence which coreference procedures are run.

5.1.2 An Example-Based Introduction

For readers new to this topic, the number of reference-oriented reasoning challenges that can pile up in a short text might come as a surprise.[3] Consider the following excerpt from a piece by Sabrina Tavernise called "Buying on Credit Is the Latest Rage in Russia" (*New York Times,* 2003).

> New advertisements are appearing on Moscow's streets and subways. Comic-book-style stories portray the new quandaries of the Russian middle class. "If we buy the car, we can't afford to remodel the apartment," says a woman with a knitted brow, in one ad. Then comes the happy ending. Her husband replies, smiling: "We can do both! If we don't have enough, we'll take a loan!"

To fully interpret this excerpt, a person or LEIA must understand the following.

1. The meaning of *new advertisements*, *comic-book-style stories*, *one ad*, *a loan*, and *a woman* are among the mentions of new entities that must generate new anchors in memory.

2. *A knitted brow* must generate a new anchor and must also be linked, using the PART-OF-OBJECT relation, to the anchor for the woman whose brow it is.
3. *Her husband* must generate a new anchor and must also be linked to the anchor for the woman, using the relation HAS-SPOUSE.
4. The interpretation of *we* involves combining into a set the anchors for the woman and her husband, which do not form a syntactic constituent (i.e., they appear in different portions of the text). Note that the first mention of *we* comes before the mention of the woman and her husband, thus requiring coreference resolution using a *post*cedent, not an *ante*cedent.
5. Understanding the meaning of *Moscow's streets and subways* requires linking the anchor for some unspecified set of streets and subways to the anchor for the city, Moscow.
6. The *Moscow* in question must be understood as the one in Russia. This can be inferred from the title or by reasoning that only the most well-known *Moscow* is likely to be written about. Both readers and intelligent agents (who could have access to very large lists of geographical place names), might realize that there are many towns and cities called *Moscow* worldwide.
7. Although noun phrases with *the*—which are called *definite descriptions* since they use the definite article—often refer to entities previously mentioned in the text, this is far from always the case. There are conditions under which *the* should not trigger the search for a textual sponsor; this occurs, for example,
    a. when the entity has restrictive postmodification: *the new quandaries of the Russian middle class* (*of the Russian middle class* is the postmodifier that licenses the use of *the* with *quandaries*);
    b. when the entity includes a proper name modifier: *the Russian middle class*;
    c. in clichés and idioms: *the happy ending*;
    d. when the meaning of the entity is generic: *the personal check*;
    e. when there is semantic ellipsis—that is, the omission of words/meanings that are necessary to fully understand the text but are not syntactically obligatory: *the car* and *the apartment* do not refer to just any car and any apartment—*the car* is the one the couple is thinking of buying and *the apartment* is the one they already own or rent.
8. All cases of ellipsis (i.e., missing elements) must be detected and resolved. In *If we don't have enough*, the meaning *money* must be understood as what is lacking. Moreover, that lacking must be attributed to the couple who needs the money to buy a car and remodel their apartment.

9. The meaning of all EVENTS—most often realized as verbs in text—must undergo reference resolution, just like OBJECTS. Events in this excerpt are conveyed by the words *appearing, portray, buy, afford, remodel, says, comes, replies, smiling, do, have, take.* Although most of these do not have a textual coreferent (instead, they directly create new anchors in memory), *do (both)* does have a textual coreferent: the set comprised of the event instances *buy (the car)* and *remodel (the apartment).*
10. *The happy ending*, which must be a new anchor in memory, must be linked to this particular comic-book-style story, not stories in general. In addition, *the happy ending* must be understood as a cliché, which, in this context, conveys that the entire passage is a spoof.

Although this example does not exhaust or systematically organize all types of overt and elided RefExes, it is sufficient to illustrate the overall magnitude of the reference problem and the extent to which reference decisions must be integrated with overall semantic analysis and reasoning about language and the world.

5.1.3 A Dozen Challenges

As the next step in our big-picture overview, let us consider some issues that are more challenging than they might appear at first blush.[4]

Challenge 1. *Detecting RefExes can be difficult.* A given text string can be referential in some contexts and nonreferential in others, as shown by the contrastive examples in table 5.1.

Challenge 2. *The surface features (number, gender, animacy) of RefExes and their sponsors need not match, although they most commonly do.* For example, in English, ships can be referred to as *she*, and a beloved inanimate object can be referred to as *he* or *she.*

(5.5) Do not touch my car—she is only ever to be driven by me!

Likewise, *they* is making fast inroads as a gender-neutral singular pronoun.

(5.6) If someone got hurt, they would blame it on some outside force … (COCA)

Table 5.1
Referential and nonreferential uses of the same types of categories

RefEx type	Referential use	Nonreferential use
Definite description	Look at *the boat*!	On *the one hand*, …; on *the other hand*, ….[5]
Indefinite description	*A bug* just landed on your head.	Danny is *a plumber.*[6]
Pronoun	Take the vase and put *it* there.	*It* is not a good idea to lie.
Verb	George *has* a red car.	George *has*[7] finished painting the house.

Challenge 3. *Two noun phrases with the same head (the same main noun) do not necessarily corefer.* Although head-matching, like surface-feature matching, is a useful heuristic for identifying coreferential categories, it is quite normal for different entities of the same type, like the guards in (5.7), to be referred to by NPs with the same head noun.[8]

(5.7) The guard flipped a switch in his booth and a strip of spikes hinged up from the asphalt. The other guard set to circling the truck with the dog. (COCA)

Challenge 4. *Some personal pronouns can have a specific, a generalized, or a hybrid referent.* For example, in English the pronoun *you* can refer to one or more specific animate entities (5.8), people in general (5.9), or a nonspecific hybrid of those two meanings, implying "you and anyone else in the same position" (5.10).[9]

(5.8) If you eat another piece of pizza you'll explode.

(5.9) It's tough to live in the suburbs if you don't drive.

(5.10) If you speed you risk getting a ticket.

Similarly, *they* can refer to specific or nonspecific individuals, the latter illustrated by (5.11).

(5.11) They say it will rain tomorrow.

Challenge 5. *Fully interpreting referring expressions can require making implicatures.* For example, if a sixty-year-old married woman says, *We're going to Italy this summer,* the normal implication is that she and her husband are going. But if she were thirty, had kids, and was going to Disneyland, the kids would be implied as well.

Challenge 6. *Coreferential entities need not be of the same syntactic category.* For example: an NP can corefer with a verb (5.12), a predicate-adjective construction (5.13), a modality expressed by a verb (5.14), a set of NPs that must be dynamically combined (5.15), or a span of text (5.16).

(5.12) [Both expressions instantiate the concept INVADE]
If you have been invaded and that invasion is an accomplished fact several years down the line, you can not ignore it. (COCA)

(5.13) [All three expressions instantiate the concept AESTHETIC-ATTRIBUTE with a value of 1]
[Ramona:] And just remember how beautiful it is out here. It's really gorgeous. [Elam:] The beauty of the Flint Hills is especially apparent when you drive the ranch loop with Jane. (COCA)

(5.14) [Both expressions instantiate (MODALITY (TYPE EFFORT) (value 1))]
They had never said much to each other, she and her father, and when they tried now, their efforts felt strained and pointless. (COCA)

(5.15) [A set must be dynamically composed from *the Falcons* and *the Bills*]
The Falcons started with the Bills in Atlanta in Week 4. The two teams were tied at 17–17 in the fourth quarter, but the Bills pulled it out 23–17. (COCA)

(5.16) [The first clause, minus the discourse adverb *well*, corefers with *that*]
Well by the 1970s the risk factors has caused the Chesapeake Bay to loose [*sic*] 99 percent of its native oyster population. That's bad. (COCA)

Challenge 7. *Referentially related entities are not always coreferential.* Instead, they can express semantic relationships other than coreference, such as a set-member relationship (5.17), an instance-type relationship (5.18), different instances of a given type (5.19), and so-called bridging references,[10] in which the mention of one entity implicitly introduces another into the frame of discourse (5.20–5.23).[11]

(5.17) [Set-member relationship]
And I recently talked to a couple. And the wife, she was a nurse. (COCA)

(5.18) [Instance-type relationship]
The repairman canceled again. These people are so unreliable!

(5.19) [Different instances of a given type]
Every time I see you and Marshall together I think there's a happy marriage. I want one too. (COCA)

(5.20) [Bridging involving an event and its subevent]
For the second straight game, Ryan Hartman committed two penalties in the second period … (COCA)

(5.21) [Bridging involving an object and its part]
Who started the ridiculous big-grille trend, anyway? Audi, maybe? Not only is it an ugly look, but it's also exponentially absurd because most cars get their engine air through vents below the bumper, not through the grille. (COCA)

(5.22) [Bridging involving an event and one of its participants]
… We counted eight playoff series in which the home team won the first game but lost the last game. (COCA)

(5.23) [Bridging involving an event and one of its nonhuman props]
He remembers the electricity in the locker room before the game, the stadium's muddy field, the game's overall frustration … (COCA)

Challenge 8. *The sponsor for a RefEx can be vague.* For example, a speaker—let's call her Alice—might conclude an hour-long speech about the dangers of smoking by saying, "And *that's* why you shouldn't smoke!" Given that the whole speech was about smoking, when Alice gets to "that's why," which part of what she said is she actually referring to?

Clearly, she is including whatever reasons she gave not to smoke, and clearly, she is not including jokes or references to the temperature of the room. But if she presented vignettes about particular smokers' experiences, would those be among the reasons to avoid smoking? Could a typical listener even remember all the reasons or definitively tease apart the reasons from other speech content? Most likely not.

Challenge 9. *Even if a RefEx does not require a textual sponsor, it might have one—and the agent should track it down.* Some RefExes, such as well-known proper names and so-called *universally known* entities (e.g., *the universe*), do not require a textual sponsor. However, even if a RefEx does not require a textual sponsor, there are three good reasons for seeking one out anyway:

1. Universally known entities are lexically ambiguous. In addition to their universally known meanings—for example, *the atmosphere* referring to Earth's atmosphere—they have additional generic meanings. *The atmosphere* might refer to the atmosphere of some other planet or, metaphorically, to the collective mood of a group of people. In the best case, the meaning will be clear on first mention of the entity in a discourse, and all other mentions in the chain of coreference will take on the same interpretation.[12]
2. Proper names can have a large number of real-world referents. For example, *the CIA* can refer to the Central Intelligence Agency or the Culinary Institute of America, and the names *Henry*, *Dr. Adams*, and *Professor Tanenbaum* can refer to any number of real-world individuals. Ideally, the identity of any of these will be clear on first mention in a text, and all subsequent mentions will belong to the same chain of coreference.
3. Chains of coreference can help to compute discourse structure, which, as we said earlier, is a difficult and as-yet unsolved problem.[13] The idea is, if coreference links obtain between sentences, those sentences probably belong to the same discourse segment.[14] By contrast, if sentences do not contain any coreference links, there is likely to be a discourse boundary between them (i.e., the topic has shifted). Detecting discourse boundaries can be useful, for example, for dynamically establishing the window of coreference: one should look for sponsors only within the chunk of text that is about that particular thing.

Challenge 10. *RefExes and/or their sponsors can be elided.* Some types of ellipsis, like verb phrase ellipsis, are typical in any genre of English (5.24), while others, like subject ellipsis, are used only in stylistically marked contexts (5.25) or with the support of the nonlinguistic context (5.26).

(5.24) I like to wear running shoes if I can ___. (COCA)

(5.25) She felt her heart beat. ___ Felt the pulse of it against her face. (COCA)

(5.26) "It was an accident. I was aiming for that boy over there," she said and pointed to the ball field. (COCA)

Challenge 11. *There can be benign ambiguity with respect to the sponsor.* Benign ambiguity is ambiguity that does not impede reasoning. For example, in (5.27), the second *it* might be understood as the rose, the glass, the glass of water, or the glass of water containing the rose—all of which are functionally equivalent since, by the time we get to *it* in the sentence, the rose and the glass of water are already functioning as a unit.[15]

(5.27) She would also take her rose back to the office, put it in a glass of water, and place it on the windowsill. (COCA)

Challenge 12. *Anchoring RefExes in memory can be tricky.* As explained earlier, when a RefEx is encountered in a text, it might refer to something the agent already knows about (a known anchor), or it might refer to something new (and therefore require that a new anchor be created). The most obvious situation in which a RefEx should create a new anchor in memory is when an NP is introduced with an indefinite article (5.28) or certain quantifiers (5.29).

(5.28) Through the window, the more silent sounds of night. <u>A dog</u> barks somewhere. (COCA)

(5.29) Rachel was going over the company phone bill and saw calls to a number she didn't recognize, so she dialed it. <u>Some girl</u> answers. (COCA)

The LEIA knows that such expressions trigger the creation of a new anchor because the lexical senses for *a/an* and this meaning of *some* include a call to a procedural semantic routine called *block-coreference*. This instructs the agent to *not* seek a textual sponsor but, instead, to directly create a new anchor in memory during Situational Reasoning.

Note, however, that there are contexts in which an entity that is at first interpreted as new turns out to be already known, as in (5.30).

(5.30) I know <u>someone</u> was here and I know it was <u>the same person who vandalized my car</u> … (COCA)

We model coreference and memory management for such contexts analogously to a person's reasoning. When the LEIA encounters the string *someone*, it creates a new anchor in memory because this entity is being presented as new and unknown. The fact that it later becomes clear that it is actually known means that two different anchors need to be coreferred—or merged, depending on the memory management strategy. Thus, no special look-ahead strategy need be implemented to account for the possibility that an originally unrecognized entity is later recognized.

With many other types of referring expressions, it can be difficult to know when to link new information to existing memories. Say a LEIA encounters the following sentence:

(5.31) When Susan and Tom Jones met their 1930s Colonial, it wasn't exactly love at first sight. (COCA)

Although the name Tom Jones might immediately evoke the hero of a Henry Fielding novel—information that a well-read LEIA should know—this can't be the intended referent since the dates don't match up. And, in fact, the name Tom Jones is so common that a LEIA with even moderate real-world experience in text processing will likely have encountered multiple people named Tom Jones. For each person, the LEIA might know some distinguishing feature values—maybe age for one, profession for another, and the wife's name for a third—but it will certainly not know the full set of features that might be needed to decide whether incoming information relates to a known or new person with this name. In fact, even knowing that Tom Jones's wife is named Mary is of modest help since Mary, too, is a very common name, and there are many Tom and Mary Joneses in the world. In short, deciding which property values to take into account when determining whether a referent is known or new is a big challenge—one that is undertaken later, during Situational Reasoning.

Summary: To reiterate—and allow readers to test whether they can recall relevant examples—the dozen challenges presented above are as follows:

1. Detecting RefExes can be difficult.
2. The surface features of RefExes and their sponsors need not match.
3. Head-matching NPs do not necessarily corefer.
4. Some personal pronouns (e.g., *you*, *they*) can have a specific, generalized, or hybrid referent.
5. Fully interpreting RefExes can require making implicatures.
6. Coreferential categories need not be of the same syntactic category.
7. Referentially related entities need not be coreferential.
8. A RefEx's sponsor can be vague.
9. Even RefExes that don't require a sponsor might have one—and it should be sought out.
10. RefExes and/or their sponsors can be elided.
11. There can be benign ambiguity with respect to the sponsor of RefExes.
12. Anchoring RefExes in memory can be tricky.

5.1.4 Special Considerations about Ellipsis

Ellipsis is one method of realizing a referring expression. Its broadest definition is uncontroversial: it is the nonexpression of some meaning that can be recovered from the linguistic or real-world context. However, distinguishing ellipsis from related phenomena—such as fragments and telegraphic writing—is less clear-cut, as is classifying subtypes of ellipsis.

Our classification derives from a combination of linguistic principles, hypotheses about how people detect and resolve ellipsis, and the practical needs of building agent systems. We begin with some definitions.

Syntactic ellipsis is the grammatically licensed absence of an otherwise mandatory syntactic constituent. *Mandatory* is defined either by an argument-taking word's selectional constraints[16] (5.32) or by the grammar overall (5.33).

(5.32) [The multiword auxiliary *do not* requires a complement but permits its omission] Both mammals (synapsids) and birds, reptiles and crocodiles (sauropsids) exhibit some form of mother-caring behavior, but amphibians do not __. (COCA)

(5.33) [The so-called *gapping* construction allows for ellipsis of the main verb in certain types of highly parallel coordinate structures] Yet these forms invite the limpers to judge the runners; non-readers __, the readers; the inarticulate __, the articulate; and non-writers __, writers. (COCA)

Whereas syntactic ellipsis involves a constituent missing from a complete structure, a *fragment* is a subsentential constituent occurring in isolation. For example, in question-answer pairs like (5.34), the question sets up specific expectations about the answer, and the answer can be a fragmentary utterance that fills those expectations.

(5.34) "Where you going?" "North Dakota." (COCA)

Noncanonical syntax subsumes phenomena such as unfinished thoughts, interruptions, spurious repetitions, and self-corrections. It is not a type of ellipsis. Whereas both syntactic ellipsis and fragments are rule-abiding phenomena, noncanonical syntax is not. LEIAs treat it either by recovery procedures triggered by failures of syntactic analysis (during Pre-Semantic Integration) or by circumventing syntactic analysis and attempting meaning composition using an *ordered bag of concepts* methodology (during Situational Reasoning).

There are two subtasks in treating ellipsis: detection (identifying what's missing) and resolution (determining what it means). Both of these can be quite challenging, which explains why ellipsis has, to date, not received adequate attention in NLP systems, despite the extensive ink devoted to it in the theoretical literature.[17]

5.1.5 Wrapping Up the Introduction

If your impression at this juncture is, "There is more going on with reference than I had imagined," that is not surprising. One of the main challenges in treating the full spectrum of reference phenomena is organizing the treatment of all these phenomena across analysis modules in a way that (a) makes sense both theoretically and practically and (b) lends itself to enhancement in well-understood ways over time.

Readers have already seen the first stage of reference treatment: during Basic Semantic Analysis, all overt referring expressions, as well as many types of ellipsis, have been detected and at least partially semantically analyzed. For example, *he* is understood as HUMAN (GENDER male), and an elided verb phrase (licensed by a modal or aspectual verb; e.g., *You didn't __?*) is detected and provisionally resolved as an underspecified EVENT. What remains is to ground these initial semantic interpretations in the discourse context. Often, this involves identifying their textual sponsors.

The rest of the chapter works through the textual-coreference procedures that the LEIA carries out at this stage, which rely on the linguistic, lexical, and ontological knowledge that it possesses for general domains. Later, during Situational Reasoning, the agent will have access to more types of heuristic evidence (e.g., the results of vision processing) and more types of reasoning (e.g., reasoning about its role in the task at hand). These will help it to finalize coreference decisions and anchor referring expressions in its memory.

As we said, there is no getting around the sheer number of phenomena to be treated, their complexity, or the reality that achieving high-confidence results for all of them will require a lot of time and work. Accordingly, our claims about the current state of this microtheory are relatively modest:

1. It organizes the treatment of a large number of phenomena in a way that is cognitively motivated and practically useful.
2. It significantly advances the treatment of several important phenomena, such as anaphoric (including elliptical) event coreference, broad referring expressions, and bridging constructions.
3. It integrates coreference resolution with lexical disambiguation. That is, once the agent knows, for example, what a pronoun refers to, it can use that information to help disambiguate the other words in its clause.

The microtheory has proven useful even in its current state. We estimate that dedicating one linguist-year to each of around twenty well-defined problems would advance this microtheory to a seriously operational level for English-speaking LEIAs.

Each section of the upcoming narrative addresses a different type of referring expression: personal pronouns (5.2), pronominal broad referring expressions (5.3), definite descriptions (5.4), anaphoric event coreference (5.5), other types of elided and underspecified events (5.6), and overt event references (5.7). The more detailed sections conclude with a recap that extracts the main points and examples. These can be skimmed as a memory refresher and road map.

5.2 Personal Pronouns

LEIAs use a three-stage process to analyze personal pronouns like *he* and *her*. First, they import the results of an externally developed coreference engine (5.2.1). Then they run an internally developed personal pronoun–coreference function, which offers higher-confidence resolution for certain types of examples (5.2.2). And finally, they evaluate all posited coreference links for their semantic suitability (5.2.3).

5.2.1 Resolving Personal Pronouns Using an Externally Developed Engine

When LEIAs use the Stanford CoreNLP tool set for preprocessing and syntactic analysis, they also run its coreference resolver (hereafter, CoreNLPCoref)[18] and store the results for use at this stage. CoreNLPCoref is comprised of precision-ordered sieves that identify sponsors for certain *instances* of certain *types* of referring expressions. It offers state-of-the-art results within the limitations of knowledge-lean methods. The details of CoreNLP-Coref are described in Lee at al. (2013).[19]

For many types of referring expressions, CoreNLPCoref treats only a subset of instances. For example, although it treats the pronoun *it*, it does not treat examples in which the sponsor is one or more clauses (5.35), and although it treats the pronoun *they*, it does not treat examples in which the sponsor must be composed from different constituents on the fly (5.36).[20]

(5.35) But I'm glad she's happy! It makes me happy![21] $_{(COCA)}$

(5.36) A month or two later, she met another man. They fell in love and got married. $_{(COCA)}$

There are also coreference phenomena that CoreNLPCoref treats infrequently if at all, such as ellipsis and demonstrative pronouns (*this*, *that*).

CoreNLPCoref does quite well at resolving reflexive pronouns, proper names, definite descriptions, and first- and second-person pronouns. However, its precision on third-person pronouns is much lower. This is not unexpected since the latter's interpretation often requires real-world knowledge and reasoning. In fact, the Winograd Schema Challenge (Levesque et al., 2012) posits that the interpretation of third-person pronouns in pairs of sentences like (5.37a,b) can be used to gauge the intelligence of artificial intelligence systems.

(5.37) a. Joan made sure to thank Susan for all the help she$_{SUSAN}$ had given.
b. Joan made sure to thank Susan for all the help she$_{JOAN}$ had received.

Since LEIAs need a confidence estimate for each language-processing decision, we convert the precision scores for each sieve reported in Lee et al. (2013) into estimates of confidence. For example, the precision of the Exact String Match sieve serves as an estimate of the LEIA's confidence in resolving each RefEx treated by that sieve.

5.2.2 Resolving Personal Pronouns Using Lexico-Syntactic Constructions

As just mentioned, the weakest aspect of CoreNLPCoref (apart from not treating some types and instances of referring expressions at all) is resolving third-person pronouns. Although one might think that the only next step is to invoke semantics and pragmatics, it turns out that we can squeeze a bit more predictive power out of surface (lexico-syntactic) features by bunching them into constructions. These constructions reflect the fact that language is quite formulaic, and that phenomena such as parallelism have strong effects throughout the language system.[22]

Below we describe six constructions that we developed and tested to assess the utility of knowledge engineering in this domain. It is important to note that the published results of this evaluation (McShane & Babkin, 2016a) addressed exclusively the more difficult personal pronouns (*he*, *him*, *she*, *her*, *they*, and *them*), not taking credit for the fact that the same constructions would also resolve the easier ones (*I*, *me*, *you*, *we*, *us*).[23]

In the descriptions of each construction, *feature matching* means having the same value for person, number, and gender. In all cases, the pronoun and its sponsor are at the same level of quotation—that is, they are both either within or outside a direct quotation. *Sequential* implies that there are no other categories of the given type intervening. Square brackets indicate the number of examples evaluated and the percentage correct in the evaluation.[24] Each construction is illustrated by an example from the English Gigaword corpus (Graff & Cieri, 2003).

Construction 1. Sequential string-matching subjects of coordinated clauses [20 examples; 100% correct].

(5.38) The Warwickshire all-rounder Roger Twose has been named in the New Zealand squad to tour India beginning in October. Now he has taken the decision to make his life in New Zealand and he goes with our blessing and best wishes. (Gigaword)

Construction 2. Sequential feature-matching subjects of speech act verbs [20 examples; 90% correct].

(5.39) Established Zulu actors were used in the dubbing process, which took more than a month, he said. He said senior politicians, among them PWV provincial premier Tokyo Sexwale, and leading movie personalities had been invited to the gala … (Gigaword)

Construction 3. Sequential string-matching subjects in a [main-clause + subordinate-clause] structure [32 examples; 100% correct].

(5.40) Rabin also accused Iran of controlling the Islamic fundamentalist group Hezbollah, which has been blamed for several terrorist attacks. But he said he believed the weapons flow through Syria had slowed in recent months. (Gigaword)

Construction 4. Sequential feature-matching subjects in a [main-clause + subordinate-clause] structure [20 examples; 85% correct].

(5.41) President Bill Clinton warned Saturday that he would veto any attempt by Republicans to scrap plans to put 100,000 additional police on US streets in line with his prized crime-fighting package. (Gigaword)

Construction 5. Sequential feature-matching subjects of identical verbs [20 examples; 90% correct].

(5.42) The survivors of the family live under one roof. They live frugally on rice and beans distributed by the church. (Gigaword)

Construction 6. Direct objects of sequential coordinate clauses [20 examples; 85% correct].

(5.43) In addition, some 170 US soldiers will go to Saudi Arabia to take two Patriot missile batteries out of storage and transfer them to Kuwait, the Pentagon said. (Gigaword)

To be clear, we are not saying that CoreNLPCoref would get these wrong. However, even if its answers were correct, it would have no way of knowing that these resolutions were more confident than the corpus-wide average for third-person pronouns. We are also not saying that the results of this small-scale evaluation are definitive: after all, one can come up with counterexamples for all these generalizations. What we *are* saying is that constructions are a useful tool for establishing higher-confidence resolutions for particular types of lexico-syntactic contexts.

Many of the mistakes detected during the abovementioned evaluation would be corrected by semantic analysis, which was not part of the evaluation setup reported in McShane and Babkin (2016a).[25] For example, in (5.44), Construction 4 incorrectly posited *tears of joy* as the sponsor for *they*:

(5.44) Tears of joy and grief poured from the two teams as they lined up for the medal ceremony. (Gigaword)

However, since tears of joy are inanimate and therefore cannot be the agent of lining up, semantic analysis (described in the next section) will reject *tears of joy* as the coreferent for *they* and opt for *the two teams* instead.

We expect that further knowledge engineering will yield more constructions that have better-than-baseline predictive power and, therefore, can better inform the LEIA's combined analysis of semantics and reference, to which we now turn.

5.2.3 Semantically Vetting Hypothesized Pronominal Coreferences

All of the coreference votes hypothesized so far have relied on surface (lexico-syntactic) features. At this point, the LEIA invokes semantics: it checks (a) whether the coreferences are semantically valid and (b) whether they can help to disambiguate other constituents in their clause. Consider example (5.45):

(5.45) Mike talked at length with the surgeon before he started the operation.

Anyone reading this sentence outside of context should conclude that *he* most likely refers to the surgeon who is about to start a medical procedure. This is not the coreference decision that was offered by CoreNLPCoref or by our own constructions. Both of these corefer *he* with *Mike*. So, we want the LEIA—like a person—to override the surface-feature-based expectation using semantic reasoning. To understand how the LEIA does this, we must zoom out to the level of complete, multistage text analysis.

During Basic Semantic Analysis the LEIA computes four candidate interpretations of (5.45): either Mike or the surgeon could be starting either SURGERY or a MILITARY-OPERATION. The ontology indicates that SURGERY and MILITARY-OPERATION expect different kinds of AGENTS, as shown below:

```
SURGERY
   AGENT   default         SURGEON
           sem             MEDICAL-PERSONNEL
           relaxable-to    HUMAN

MILITARY-OPERATION
   AGENT   default         MILITARY-PERSONNEL
           sem             HUMAN
```

Based on this knowledge, the highest-scoring analysis of *he started the operation* will involve either a surgeon performing surgery or a member of the military performing a military operation, since these are the *default* AGENTS of their respective events. However, the given context makes no mention of a member of the military, so the only highest-scoring analysis that is available involves a surgeon performing surgery. There are two next-highest-scoring analyses: either Mike or the surgeon could be performing a military operation. These receive high (but not maximally high) scores because, as HUMANS, they fulfill the basic semantic constraint (listed on the *sem* facet) of the AGENT of MILITARY-OPERATION. The lowest-scoring option is that Mike is performing surgery, since it resorts to the *relaxable-to* facet for the AGENT slot: HUMANS can only in a pinch be the AGENTS of SURGERY (as when performing a lifesaving procedure on a battlefield). In short, the coreference relation that was preferred using surface heuristics was dispreferred using semantic ones, and semantic ones always win.

At least two questions might come to mind at this point. First, shouldn't reference resolution procedures for *the operation* have already determined which kind of operation it is? Ideally, yes, but (a) that decision, too, could be right, wrong, or as-yet undetermined (i.e., residually ambiguous), and (b) even without any previous context, people interpret the sentence as involving a surgeon performing surgery—and so, too, should LEIAs.

The second question is, What if something else entirely is going on, which might be indicated in the preceding context or might rely on information the interlocutor is expected to know? For example, Mike might be an anesthesiologist who starts the surgery by knocking out the patient before the surgeon cuts; or Mike might be a doctor who is planning to perform a small surgery but wants to check some details with a senior surgeon first; or Mike might be a general whose only trusted advisor happens to be the unit's surgeon, with whom he consults before launching any military operation. Language and the world allow for all these interpretations—and so does our modeling strategy. However, such reasoning does not happen *at this point in the process*. At this stage, the LEIA is using general linguistic and ontological knowledge to semantically vet coreference votes and, ideally, come up with the same interpretation that a person would outside of context. Later on, during Situational Reasoning, the agent again vets the posited coreference votes using all its situational awareness and world knowledge. This models a person having access to the whole context and the entire shared knowledge space.

5.2.4 Recap of Resolving Personal Pronouns during Basic Coreference Resolution

1. Some instances of pronouns are resolved by the externally developed CoreNLPCoref system: for example, Jack asked Suzy to marry him.
2. Some instances of pronouns are resolved more confidently by our internally developed lexico-syntactic constructions: But he said he believed she was right.
3. All posited coreference links are semantically vetted and, if necessary, overridden: Mike talked at length with the surgeon before he started the operation.

5.3 Pronominal Broad Referring Expressions

Broad referring expressions are called *broad* because they can corefer with either a noun phrase or a larger span of text.[26] Broad RefExes are commonly realized by the pronouns *this*, *that*, and *it*, but certain full noun phrases—headed by words like *suggestion* and *proposal*—can have broad reference as well.[27] This section addresses only pronominal broad RefExes.

Practically no coreference systems attempt to treat broad RefExes.[28] Resolving them can be relatively simple, as in (5.46), or beyond the state of the art, as in (5.47). In fact, in (5.47) the referent for *this* is not even in the text; it must be understood as something like "this person's behavior and how it is affecting me."

(5.46) They don't trust us. That's good. (COCA)

(5.47) She picked up a fork, stared at the food for a moment, then shook her head in despair. Fear had taken away her appetite. This can't go on, she thought angrily. Whoever he is, I won't let him do this to me. (COCA)

Since LEIAs will not be able to resolve all instances of broad RefExes anytime soon, we are supplying them with a method to independently detect which ones they *can* resolve with reasonable confidence and resolve only those (recall the simpler-first modeling principles introduced in chapter 2). Any examples not treated at this stage will be reconsidered during Situational Reasoning, when more resources can be brought to bear.

The current model includes five methods for detecting and resolving treatable instances of pronominal broad RefExes. These methods are described in the subsections below. For evaluations and additional details, see McShane (2015) and McShane and Babkin (2016a).

5.3.1 Resolving Pronominal Broad RefExes Using Constructions

Section 5.2.2 showed how lexico-syntactic constructions can be used to resolve third-person pronouns. Such constructions are similarly useful for resolving pronominal broad RefExes. The constructions which we have tested to date are illustrated by the following set of examples, each one introduced by an informal presentation of the associated construction.[29]

(5.48) [Ask/Wonder why CLAUSE, it's/it is because …]
If you've wondered why so many 80- and 90-year-old women are named Alice, it's because the president's daughter was the inspiration for the most popular name for girls born in the early years of the century. (COCA)

(5.49) [Why AUX SMALL-CLAUSE? It's/it is because …]
"Why is he busy? It's because of the pressure that's being put on him," … (Gigaword)

(5.50) [If/When/In_each_case_where/Whenever/Anytime_when CLAUSE, it's/it is because …]
If Myanmar seems oddly quiet, it is because many are tired of struggle and just want to improve their lives. (Gigaword)

(5.51) [Not only AUX (NEG) this/it/NP_{subj} … , this/it_{subj} …]
"Not only did it show that the emperor was very much a human being, it was also a grim reminder of the defeat and subservience of their nation." (Gigaword)

(5.52) [This/It/NP_{subj} (AUX) not only …, this/it_{subj} …]
The module not only disables the starter, it shuts down the fuel injection so the car won't run … (COCA)

(5.53) [This/It/NP_{subj} has/had nothing to do with …. This/it_{subj} ….]
"This provision has nothing to do with welfare reform. It is simply a budget-saving measure," … (Gigaword)

(5.54) This/It/NP_{subj} is/was not about …. This/it_{subj} is about ….]
"This war is not about diplomacy," he added. "It is about gangsterism …" (Gigaword)

(5.55) [This is not (a/an/the) N …. It_{subj} is/it's …]
"This is not the lottery. This is this man's life, …" (Gigaword)

(5.56) [That's why … that's why …]
That's why we stayed in the game and that's why we won. (Gigaword)

These constructions reflect pragmatic generalizations:

- Asking *why* something happened is often followed by an indication of why: (5.48) and (5.49).
- *If ... then* constructions can be used to explain potential situations: (5.50).
- Saying that something *not only* does one thing often leads to saying what *else* it does: (5.51) and (5.52).
- Saying that something *does not* involve something often leads to saying what it *does* involve: (5.53)–(5.55).
- Repetition structures, which feature a high degree of parallelism, often contain coreferential expressions: (5.56).

Some broad RefExes have propositional sponsors, whereas others have NP sponsors. Although finding NP sponsors might seem simpler, it actually isn't since the LEIA does not know beforehand whether the sponsor it is seeking is an NP or a proposition.

Vetting these constructions on a corpus showed that some of them must be specified more precisely in order to exclude false positives.[30] For example, although (5.57) matches the construction described in (5.52), the sponsor for *it* is not *the Defender,* as would be predicted by that construction.

(5.57) And for the first time in history, the Defender not only wants to introduce its own new rule for the class of boat to be raced, but also to keep this rule secret. It will be disclosed to challengers at a much later stage, putting all challengers at a huge disadvantage. (Gigaword)

There are at least three reasons why this context should not be interpreted as matching the cited construction; each one involves disrupting the simplicity and parallelism that give the construction predictive power: the voice is different in the two parts—active in the first, passive in the second; there is a subordinate clause intervening between the two parts; and there are additional NPs in the first part that are also candidate sponsors. Clearly, for this and all of the other constructions, additional knowledge engineering is needed to determine rule-out conditions—that is, ways in which the context can become too complex for the constructions to predict coreference relations.

5.3.2 Resolving Pronominal Broad RefExes in Syntactically Simple Contexts

As we said, the sponsor for a broad RefEx can be an NP, a span of text representing one or more propositions, or a meaning that is not explicitly presented at all. In this section, we focus on detecting contexts in which (a) the broad RefEx refers to a span of text and (b) the boundaries of that span of text can be automatically determined with high confidence due to the syntactic simplicity of the context.

Consider examples (5.58) and (5.59), in which *that* can be resolved confidently based on the fact that the previous clause in each case is syntactically simple and its left-hand edge represents a natural boundary for the most local context—that is, it is a sentence break.

(5.58) They live far from their homes. That makes them stronger than if they formed a real community. (Gigaword)

(5.59) "Strong Serbia is not to the liking of some powers abroad, and that's why they are trying to break it up with the help of the domestic traitors," he said. (Gigaword)

By contrast, in (5.60) the sentence preceding *it* contains two conjuncts (i.e., two clauses joined by *and*). This offers two options as the sponsor of *it*: either the entire sentence or only the most recent clause.[31]

(5.60) "Police will go pass some prostitutes on the corner and harass some kids having a disagreement. It's because we're young." (COCA)

Often, as in this case, such contexts involve benign ambiguity: it doesn't make much difference to subsequent reasoning which resolution option is chosen.

In order to exploit the notion of *syntactic simplicity* in a computer system, we have to formally define it. We do so using an approach we developed when working on verb phrase ellipsis (McShane & Babkin, 2016b). The gist is that simple syntactic structures have only one main verb; this excludes sentences containing clausal conjunction, relative clauses, subordinate clauses, *if* ... *then* structures, and so on.[32]

In order to test the hypotheses that the system could (a) automatically identify syntactically simple contexts and (b) automatically predict the sponsor for broad RefExes in them, we focused on the following set of constructions, in which the italicized verbs could appear in any inflectional form:

- [simple clause] + despite this/that
- [simple clause] + because of this/that
- [simple clause] + this/that *is* why/because
- [simple clause] + this/that *means*, *leads to*, *causes*, *suggests*, *creates*, *makes*

The reason for targeting these constructions is that the broad RefExes in them tend to have a clause-level (rather than an NP) sponsor. In other words, this example extraction method returned a lot of examples relevant to the study and few false positives.

Our experimentation showed that this coreference resolution strategy worked well for the examples it treated, but it did not treat very many examples because of the strict constraints on what a *simple* clause could look like. In order to expand the system's coverage, we experimented with relaxing the definition of *simple* (see McShane & Babkin, 2016a, for details). The results were mixed. For example, if simple clauses are permitted to be scoped over by modalities, then many more contexts are covered, but real-world reasoning is needed to determine if the modal meaning should be excluded from the resolution, as in (5.61a), or included in it, as in (5.61b).

(5.61) a. "I believe Jenny will swim faster than she ever has in Barcelona, and that means she has a good chance of bringing home five medals, though the color is still to be determined" … (COCA)

b. I believe Jenny will swim faster than she ever has in Barcelona, and that is why I bet big money on her.

Another way to expand the coverage of examples is to apply automatic sentence trimming to complex sentences. When trimming works correctly, it can turn nonsimple clauses into functionally simple ones. However, trimming is error-prone, often removing the wrong bits. We found this simplification method reliable only with respect to trimming away speaker attributions, as indicated by the strikethrough in (5.62).[33]

(5.62) "Energy efficiency is really the name of the game in terms of what we can do now,~~" she said, adding that she was disappointed that Bush did not adopt a more proactive stance on global warming, despite urging on the part of Blair. "~~That's why today I'm calling on the president to show real leadership," she said, adding it was unacceptable to adopt a stance that other nations blamed for high greenhouse gas emissions, such as China and India, take steps first. (Gigaword)

We have just begun to investigate how to best relax the definition of *simple* in order to achieve higher recall without losing too much precision. This appears to be a promising area for additional knowledge engineering.

5.3.3 Resolving Pronominal Broad RefExes Indicating Things That Must Stop

People want bad things to stop. So, given utterances like *This must stop!*, *This is unacceptable!*, and *This is awful!*, one expects *this* to refer to something bad. The corresponding hypothesis we explored was as follows: If we compiled a list of *bad* events/states, and if the context immediately preceding a statement like *This is bad!* included something on that list, then that thing should be the sponsor for the broad RefEx.

Testing this hypothesis required a list of bad events, which in the current jargon are called *negative sentiment terms*. Although we found an automatically compiled list of this type (Liu et al., 2005), it included many words that were either not events or not necessarily negative, such as *gibe* and *flirt*. So, to support our experimentation, we manually compiled our own list of over 400 negative sentiment terms, using introspection combined with manual inspection of both WordNet and Liu et al.'s list.[34] Then we tested our hypothesis against this list using the constructions *It must stop* (5.63) and *This is unacceptable* (5.64), along with various synonymous and near-synonymous variations.[35]

(5.63) "This war is in no way acceptable to us. It must stop immediately" ... (Gigaword)

(5.64) "... 1,200 people were detained and packed in here, in building 19–6. This is unacceptable in a member country of the Council of Europe" ... (Gigaword)

We carried out this experiment as a stand-alone task, outside of the LEIA's semantic analysis process, so it oriented around word strings, not the ontological concepts used in TMRs.

Our corpus analysis suggested that these constructions are useful for predicting the sponsors for broad RefExes. Still, certain enhancements were needed to avoid false positives.

Enhancement 1. Negative sentiment terms should guide coreference resolution only if more confident strategies have failed. For example, lexico-syntactic parallelism has stronger predictive power than a negative sentiment term. So the sponsor for *it* in (5.65) should be identified using construction 5 from section 5.2.2, which predicts that NPs that are sequential subjects of identical verbs will be coreferential.

(5.65) "This incident is unacceptable to the national authority and to the Palestinian people and free world. It is unacceptable at all levels." (Gigaword)

Enhancement 2. Syntactic analysis is needed to avoid false positives, both when matching a construction and when matching the events in our list. For example, *It must stop to refuel* does not match our construction since the clause includes the complement *to refuel* (meaning that *it* must refer to a vehicle, not an event). Also, the system must identify the part of speech of a potential sponsor. Thus, *rebels* can serve as a sponsor for a broad RefEx only when used as a present-tense verb (*Every time I tell Charlie to be quiet in class he rebels. This must stop!*), not as a plural noun.

Enhancement 3. Broad RefExes may refer to multiple entities viewed as a set. Analysis of such contexts requires dynamic list concatenation. For example, in (5.66), all the underlined negative events must be concatenated into the sponsor for *it*.

(5.66) "The stories we are hearing of the harassment of political opponents, detentions without trial, torture and the denial of medical attention are reminiscent of our experiences at the hands of apartheid police. It must stop now" ... (Gigaword)

In addition to investigating contexts that contained a readily identifiable bad event, we also investigated contexts that lacked such an event and found this generalization useful: *Any event that must be stopped must currently be going on.* There are at least three linguistic clues that an event is in progress: (a) a verb in the progressive aspect, (b) an adverbial expressing duration, and (c) a verb expressing an increase or a decrease in a property value (e.g., *grow ever louder*). For instance, in (5.67), the progressive aspect (*has been playing*) and the time adverbial (*for two hours straight*) suggest that *playing his recorder* is the sponsor of the broad RefEx, despite the fact that recorder playing can be quite nice if done well.

(5.67) That kid has been playing his recorder for two hours straight. This has to stop!

Of course, having multiple sources of evidence pointing to the same sponsor should increase the agent's confidence in its resolution decision.

The observation that bad things should stop is only one of many domain-independent generalizations that can guide the search for a broad RefEx's sponsor. Another obvious one involves the use of positive sentiment terms in the same way (*This must continue! This is fabulous/amazing!*). To reiterate a strategy from our discussion of the *theory-model-system* triad, we find it theoretically and methodologically preferable to use domain-independent generalizations and processing methods as much as possible. This strategy reduces LEIAs' cognitive load by allowing them to avoid using maximally deep and sophisticated reasoning for each problem they must solve.

5.3.4 Resolving Pronominal Broad RefExes Using the Meaning of Predicate Nominals

When a broad RefEx is used as the subject of a predicate nominal construction (i.e., *Broad-RefEx is/was NP*), it would seem that the meaning of the predicate nominal should indicate the semantic class of the sponsor. In some cases, this works well: in (5.68) *that* refers to a *year*, and 1971 is a year; in (5.69) *that* refers to a *place*, and *the prison* is a place.

(5.68) Back to 1971 for a moment. That was the year Texas Stadium opened, at a cost of $35 million. (COCA)

(5.69) The prison became for me the symbol of Soviet system. That was the place where there was an encounter between the last remnants of the freedom of Russia, between the last people who kept the survivors of freedom alive, and the leaders of the system which could be stable only if it controls the brains of all 200 million people. (COCA)

However, not all examples are so straightforward, as we found when we reviewed an inventory of automatically extracted examples of this kind. We identified the following five cases.

Case 1. When the predicate nominal is a proper name, there is almost never a textual sponsor: *This is World News Tonight with Peter Jennings.* (COCA)

Case 2. When the NP's meaning is vague, it does not usefully constrain the search for a sponsor. This was, unfortunately, the most common outcome in our investigation of the pattern *This/That [be] the [N]* using the online version of the COCA corpus. The most common vague (and, therefore, not useful) nouns were *way, problem, thing, reason, difference, case, point, subject, reason, reality, goal, theory, message, conclusion*. By contrast, the most common *useful* head nouns were *car, church, city, country, day, guy, location, man, person, place, plane, road, school, street, time, town, year, woman*.

Case 3. The optimal contexts contain exactly one candidate sponsor that either (a) matches the head noun of the NP, (b) is a synonym of that NP, or (c) is a hyponym or hypernym of that NP. For example, in (5.69), *prison* is a hyponym of *place*.

Case 4. In some cases, the context contains exactly one candidate sponsor whose identification should be possible given appropriate preprocessing. For example, proper name recognition is needed for *I love America. It is the place where I was born*; and date recognition is needed for example (5.68).

Case 5. Some contexts are tricky in some way—not necessarily too difficult to be automatically resolved using knowledge-based methods, but requiring some type of additional reasoning and/or resulting in some degree of uncertainty. Consider the examples below along with their analyses.

(5.70) [The sponsor is not grammatically identical to what is needed: *English* vs. *England*.]
But English soccer has a reputation it still can't shake off, no matter how hard it tries. This is the country that exported soccer violence back in the 1970s and '80s. (Gigaword)

(5.71) [The country *Czarist Russia* must be skipped over when working back through the text to find the sponsor.]
"My grandparents came to this country crammed into tight ship quarters from Czarist Russia because they believed this was the country where their votes would be counted" … (Gigaword)

(5.72) ['That' is whichever of the implied countries—the Czech Republic, Russia, Finland, or Sweden—has the most engaged big-time players.]
"Every year it is the same cast of characters, the Czechs, Russians, Finns and Swedes," Hitchcock said. "But it depends on the big-time players and if the big-time players are engaged then that is the country that wins." (Gigaword)

(5.73) [*The road* refers to a sequence of events that could end in another Great Depression.]
|Couric: If this doesn't pass, do you think there's a risk of another Great Depression? Palin: Unfortunately, that is the road that America may find itself on. (COCA)

An important detail is that, even if the system correctly points to the sponsor in these contexts, this does not fully resolve the meaning of the construction. For example, if *that* is resolved to 1971 in (5.58), the agent still has to combine the semantic interpretation of *1971 was the year* with the semantic interpretation of *Texas stadium opened.* We facilitate the agent's doing this by creating lexical senses for associated constructions. For example, the construction *NP$_{YEAR}$ [be] the year (when/that) EVENT* will generate the meaning representation EVENT (TIME YEAR).

5.3.5 Resolving Pronominal Broad RefExes Using Selectional Constraints

Events require particular kinds of objects to fill their case roles. For example, the concept CELEBRATE is most commonly used with a HUMAN as the AGENT and a HOLIDAY as the THEME. This information is recorded in the agent's ontology as follows:

CELEBRATE		
AGENT	default	HUMAN
	sem	SOCIAL-ORGANIZATION
THEME	default	HOLIDAY

Moving from ontology to language, we expect the verb *celebrate* to be used in clauses like *They celebrated Thanksgiving together* and *Roberto celebrated his birthday.* Expectations like these can help to automatically resolve broad RefExes in some contexts. The study reported in McShane (2015) explored this potential in contexts like the following:

(5.74) Enron says the deal looks favorable because it was negotiated in 1992. (Gigaword)

(5.75) The water was drinkable because it boiled for several minutes. (Gigaword)

We hypothesized:

If a broadRefEx (e.g., *it*) fills a semantically highly constrained case role slot (e.g., the THEME of NEGOTIATE or BOIL)
And if a typical filler of that slot is available in the immediately preceding context (e.g., BUSINESS-DEAL or WATER),
Then that typical filler is the sponsor for the broadRefEx.

Our experimental setup did not invoke semantic analysis. It was carried out using a simpler methodology involving lists of verbs and keywords that typically filled their case role slots.[36] We found that, although the intuition was correct and useful, four enhancements were needed to improve the precision of this reference resolution strategy. We first provide a set of relevant corpus examples and then use them to illustrate the four enhancements.

(5.76) Residents said they were running out of food in a city that had its electricity cut two days ago. Some wounded Iraqis bled to death, and a family was buried under the ruins of their house after it was bombed by a U.S. jet, Saadi said. (Gigaword)

(5.77) A holiday honoring Vid, the ancient Slavic god of healing, has become one of the most fateful days on the Serb calendar. Now known as St. Vitus Day, it is celebrated June 28. (Gigaword)

(5.78) In the worst atrocity, some 5,000 men, women and children were slaughtered in the border town of Halabja in March 1988 when it was bombed and shelled with cyanide gas. (Gigaword)

(5.79) Although Imayev was talking about fighting and encirclements, Kuraly, a village of about 5,000 people, could not have appeared more peaceful. Like many Chechen villages, it was bombed by Russian airplanes during the fighting that started in December. (Gigaword)

(5.80) The 20-year-old started playing cricket at the Soweto Cricket Club soon after it was built 10 years ago. (Gigaword)

Enhancement 1. Syntactic analysis must be used to avoid false positive keyword analyses. For example, if keywords are part of a nominal compound, they must be the final (head) element of that compound. Thus, if a context contains "the restaurant garage.… It was bombed," then *garage*, not *restaurant*, is a candidate sponsor for *it*.

Enhancement 2. Candidate sponsors must be ranked according to recency, with the most recent being favored—even though recency is not a fully reliable heuristic. For example, although in (5.76) both a *city* and a *house* can be bombed, the sponsor for *it* is the more proximate *house*.

Enhancement 3. Chains of coreference must be identified: for example, in (5.77) *a holiday honoring Vid* and *St. Vitus Day* are in a coreference chain, so pointing to either of them is a correct resolution of *it*.

Enhancement 4. Certain preprocessing results must be included in the sponsor-selection heuristics: restrictive postmodification (5.78), appositives (5.79), and proper nouns with meaningful headwords, as in (5.80), where *Soweto Cricket Club* is an entity of the type SOCIAL-CLUB.

Outstanding problems involve the usual suspects, such as vagueness—in (5.81), is *it* the library or the palace?—and indirect referring expressions, such as metonymy (5.82).

(5.81) After the meeting, Kinkel and Mubarak inaugurated a public library in a renovated palace overlooking the Nile. It was built with a German grant of 5.5 million marks (dlrs 3.9 million). (Gigaword)

(5.82) The stolen van Gogh, he said, has special value because it was painted in the last six weeks of the artist's life. (Gigaword)

To sum up this subsection: When a verb's argument is realized as a pronoun, selectional constraints on this argument can guide the search for the pronoun's sponsor. There are multiple ways to operationalize this knowledge about the kinds of arguments required by different verbs. For the corpus analysis reported here, word lists were used. By contrast, during full semantic analysis by LEIAs, the combination of lexicon and ontology provides this knowledge.

5.3.6 Recap of Resolving Pronominal Broad RefExes

1. Pronominal broad RefExes occurring in listed constructions are resolved: "If you've wondered why so many 80- and 90-year-old women are named Alice, it's because …." (COCA)
2. Pronominal broad RefExes in syntactically simple (or pruned to become simple) contexts are resolved: "They live far from their homes. That makes them stronger than if they formed a real community." (Gigaword)
3. Pronominal broad RefExes referring to undesirable things are resolved: "'This war is in no way acceptable to us. It must stop immediately ….'" (Gigaword)

4. Pronominal broad RefExes described by predicate nominals are resolved: "Back to 1971 for a moment. That was the year Texas Stadium opened …." (COCA)
5. Pronominal broad RefExes filling narrow selectional constraints are resolved: "Enron says the deal looks favorable because it was negotiated in 1992." (COCA)

All the strategies used to resolve pronominal broad RefExes have three noteworthy features: (a) they do not require domain-specific knowledge or reasoning, so they can be applied to texts in the open domain; (b) they were developed and tested with only a small knowledge-engineering effort but still yielded quite useful results; and (c) they employ readily computable heuristics, which minimizes the cognitive load for LEIAs; this both simulates human functioning and promises to make LEIAs more efficient.

5.4 Definite Descriptions

Definite descriptions (NP-Defs, noun phrases with *the*) are treated at multiple stages of processing an input. Here we review what has already been done with them (section 5.4.1), present what is new at this stage (section 5.4.2), and describe what remains to be done at later stages (section 5.4.3).

5.4.1 Definite Description Processing So Far: A Refresher

The following steps occur during Pre-Semantic Analysis:

- CoreNLPCoref posts coreference votes for some instances of NP-Defs.
- The CoreNLP preprocessor identifies proper names with *the*, suggesting that they do not require a textual sponsor: for example, *the CIA.*

During Basic Semantic Analysis, these processes happen:

- The LEIA identifies nonreferring instances of NP-Defs, such as those used in idioms (*He kicked the bucket*). Since these words/phrases do not generate TMR frames, they are not subject to coreference procedures. If CoreNLPCoref has posited coreference votes for such words/phrases, they are ignored.
- For all TMR frames generated by an NP-Def, the LEIA creates a COREF slot filled with the call to the meaning procedure *resolve-NP-Def.* This happens because the meaning-procedures zone of the lexical sense of *the* contains this function call, which is copied into the basic TMR. For example, the TMR for the input *The horse is eating lazily* is as follows:

```
HORSE-1
    AGENT-OF        INGEST-1
    COREF           resolve-NP-Def
    lex-sense       horse-n1
    lex-sense-sub   the-art1
```

INGEST-1	
AGENT	HORSE-1
URGENCY	0
TIME	*find-anchor-time*
lex-sense	*eat-v1*
lex-sense-sub	*lazily-adv1*
lex-sense-sub	*be-aux1*

This TMR says that two lexical senses were invoked to create the HORSE-1 frame: *horse-n1* (the main one) and *the-art1* (a subservient one). It also indicates that two meaning procedures remain to be run—*find-anchor-time* (i.e., determine the time of speech, to account for the present tense) and *resolve-NP-Def* (i.e., find the sponsor for HORSE-1, since the presence of the article *the* indicates that it might need one).

In short, the basic TMRs for inputs that include *the* reflect three types of information: which instances of NP-Defs weren't referring expressions to begin with (there will be no TMR frames for them); which instances of NP-Defs are proper nouns that do not require, but might still have, a textual sponsor; and which instances of NP-Defs require additional reference-oriented processing. "Additional processing" does not mean that there is always a sponsor; instead, it means that either a sponsor must be found or the agent must understand why one is not needed.

5.4.2 Definite Description Processing at This Stage

This stage includes six functions for processing NP-Defs, described below.

5.4.2.1 Rejecting coreference links with property value conflicts If CoreNLPCoref identified a sponsor for the NP-Def, the LEIA checks whether that sponsor is plausible on the basis of any property values mentioned in the text. For example, *the blue car* and *the red car* cannot corefer, nor can *the Swedish diplomat* and *the Hungarian diplomat*. The reason these pairs cannot corefer is because the descriptions use the same property (COLOR or HAS-NATIONALITY) with different values (blue/red, Sweden/Hungary). By contrast, *the red car* can corefer with *the expensive car* because there are no conflicting property values: one description talks about COLOR whereas the other talks about COST.[37] That is, a red car can be expensive—no problem.

Note that, at this stage, only properties presented in the text itself are considered, not properties that are expected to be known as part of general world knowledge. For example, BMWs are expensive, high-quality cars, so it is unlikely that one would be referred to as *the cheap car* or *the poorly made car*. Later on, during Situational Reasoning, the agent will, yet again, check whether posited coreference links make sense, but at that point it will consult its knowledge of both the world and the situation.

5.4.2.2 Running reference-resolution meaning procedures listed in lexical senses Certain definite descriptions require special reasoning to be resolved: for example, the sponsor for *the couple* must be two individuals, and the sponsor for *the trio* must be three. A special resolution function is recorded in the meaning-procedures zone of each such multiword lexical sense. The calls to these procedures are copied into the TMR during Basic Semantic Analysis and are run at this time.

5.4.2.3 Establishing that a sponsor is not needed As described in the introduction to this chapter, in some cases, NP-Defs do not require a sponsor. These cases include the following:

- *Universally known NP-Defs*: This covers such entities as *the earth*, *the sun*, and *the solar system*. We keep an ever growing list of such entities.[38] In the lexicon, each such noun includes at least two senses: one that requires *the* and refers to the universally known meaning, and another that does not require *the* and refers to a more general, related meaning.[39] Both of these analyses are generated during Basic Semantic Analysis. If CoreNLPCoref has suggested a sponsor for the given NP-Def, and if that sponsor has not been invalidated based on property values, then this analysis takes precedence over the candidate analysis "universally known and without a sponsor." Of course, a chain of coreference might employ the *universally known* interpretation—but that will be established by the *first* instance of NP-Def in a text, not the subsequent ones.
- *NP-Defs with restrictive modification*: Restrictive modifiers provide essential, non-optional, information about a noun.[40] These modifiers make the meaning of the noun phrase concrete in the real world, so there is no need for a sponsor. Restrictive modifiers can be detected using templates (stored as multiword senses of *the*), such as the following:
    - the + N + PP: *the streets of my hometown*
    - the + proper-N + N: *the French army*
    - the NUMBER + TEMPORAL-UNIT + (since/that/when) + CLAUSE: *Not exactly statuesque at five feet, but she'd grown a good three inches in <u>the two years or so since he'd saved her from slavery</u>.* (COCA)
- *NP-Defs that are proper nouns*: These will already have been identified by the CoreNLP named-entity recognizer, but it is at this point that the agent fills the COREF slot of the TMR with the filler *no-sponsor-needed.* (Recall that, although CoreNLPCoref is run during Pre-Semantic Analysis, the agent only consults its results at this stage.)
- *Generic uses of NP-Defs*: These are used primarily in definitional statements such as Wikipedia's *<u>The lion</u> (Panthera leo) is a species in the family Felidae.*[41]

5.4.2.4 Identifying bridging references Bridging references connect two entities that are not *co*referential but are semantically related in particular ways. Mentioning the sponsor in the context virtually introduces semantically related entities, and they can be referred to using *the*. At this time, our model of bridging references covers bridging via object meronymy and event scripts.

Bridging via object meronymy. *Meronymy* is the has-as-part relation. Object meronymy means that one object is part of another object. Since we all know, for example, that windows can be parts of offices, we can say things like *I walked into her office and the window was open*. We use *the* with *window*—even though no window was mentioned before—because the potential for a window was effectively introduced into the discourse when *office* was mentioned. In the ontology, object meronymic relations are recorded using the HAS-OBJECT-AS-PART relation, for example,

OFFICE-ROOM
  HAS-OBJECT-AS-PART   WINDOW, DESK, CEILING, DOOR, ...

To detect object meronymy, the agent checks whether the ontological concept for the RefEx (here, WINDOW) is listed in the HAS-OBJECT-AS-PART slot of any of the ontological concepts used within the window of coreference (i.e., the immediately preceding context). The answer is yes for OFFICE-ROOM. As long as the ontology lists the needed meronymic relationship, this analysis is straightforward.

When object meronymy explains a given use of *the*, two things happen to the nascent TMR:

- The RefEx's frame is supplemented by the meronymic information—here, WINDOW-1 (PART-OF-OBJECT OFFICE-ROOM-1).[42]
- The COREF slot in the RefEx's frame, whose appearance was originally triggered by the lexical sense of *the*, is removed, showing that there is no coreference and none is needed.

Bridging via an event script. Scripts are complex events—that is, they are events that contain subevents. For example, AIR-TRAVEL-EVENT includes subevents such as a MOTION-EVENT that takes the flier to the airport, followed by AIRPORT-CHECK-IN, AIRPORT-SECURITY-CHECK, BOARD-AIRPLANE, AIRPLANE-TAKE-OFF, and so on.[43] The main event and all of its subevents have expected participants and props—in our example, PILOT, FLIGHT-CREW, AIRPLANE, AIRPLANE-TRAY-TABLE, and so on. The mention of these events virtually introduces all relevant participants into the discourse—and, therefore, they can be referred to using an NP-Def. For example, all the underlined instances of NP-Def in (5.83) are licensed by the mention of *flight*:

(5.83) I had an awful flight last week. It was bumpy and the pilot didn't explain why. The tray table was broken and kept falling on my knees. The flight attendant was in a bad mood and was snarky with everyone. And the landing gave me whiplash.

When script-based bridging explains a given use of *the*, it can lead to modifications of the nascent TMR, depending on the ontological relationship between the sponsor and the

RefEx. If they are directly linked in the script by a relation, then that relation is used. When *flight* (which is described in its TMR as FLY-PLANE-1) licenses *the pilot*, the frame for *pilot* is expanded to

PILOT-1
  AGENT-OF FLY-PLANE-1

If, by contrast, the two entities are linked by a longer ontological path, then we use the generic RELATION instead. For example, we said above that *flight* can license *the tray table*. However, the actual ontological path between them contains multiple steps:

AIRPLANE-TRAY-TABLE-1
  LOCATION AIRPLANE-1

AIRPLANE-1
  INSTRUMENT-OF FLY-PLANE-1

The reason for this simplification is that our main goal at this point is to figure out why *the* was used. We can attain that goal, and record the results, without the additional complication of computing an ontological path. Impressionistically, we as humans understand that a tray table is related to a flight; it is not necessary to flesh out *how* when computing the meaning of our example text. As with the case of object meronymy, when we have explained the use of *the* by script-based reasoning, we remove the COREF slot because we are no longer looking for a coreferential sponsor.

5.4.2.5 Creating sets as sponsors for plural definite descriptions Sometimes the sponsor for plural definite descriptions must be dynamically composed from constituents, usually NPs, located in different parts of the text. In some cases, this is not difficult, as when the preceding context contains exactly two entities of the needed semantic type—in (5.84), NATION, and in (5.85), YEAR.

(5.84) For instance, in a well-known 1985 incident, a Coast Guard icebreaker navigated through the Northwest Passage, which <u>the United States</u> claims is an international strait, without seeking Canadian permission. In response, <u>Canada</u> "granted permission" (despite the lack of a request to that effect) for the voyage and, although <u>the two countries</u> agreed to the presence of Canadian observers onboard, the United States still disputed the Canadian claim of sovereignty over the waters. (COCA)

(5.85) Compared to the <u>1995</u> season, the <u>1996</u> season was strikingly different. The fleet was about 10% smaller than in <u>1995</u> and spent 23% fewer days fishing. Fishing was concentrated in the eastern region, whereas in <u>1995</u> it was concentrated in the western region (72%) between Papua New Guinea and Federated States of Micronesia (Fig. 3). Statistics on the number of sets/trip and trips/vessel, on the other hand, were essentially identical for <u>the two years</u> (Table 1). (COCA)

However, many cases are more difficult. For example, a context that refers to many people involved in a lawsuit—lawyers, a judge, defendants, spectators, witnesses—can conclude with, *In the end, the men were acquitted.* Interpreting *the men* requires reasoning that is more sophisticated than creating a set of all the HUMANs mentioned in the preceding context.

Our current algorithm for creating sets is rather simple:

- Identify the ontological class (i.e., concept mapping) of the NP-Def.
- Find all instances of that type, or its ontologically close hyponyms or hypernyms, in the window of coreference.
- If there are exactly two matching instances, create a set from them and corefer that set with the NP-Def.
- If there are not exactly two matching instances, postpone this resolution until Situational Reasoning.

5.4.2.6 Identifying sponsors that are hypernyms or hyponyms of definite descriptions We can refer to entities in more generic or more specific ways. The more generic references are *hypernyms* of the more specific ones; the more specific references are *hyponyms* of the more generic ones. For example, in a given context, *lung cancer, the cancer, the disease*, and *the affliction* can all refer to the same thing—for example, (5.86)—just as *the studio apartment*, *the apartment*, and *the residence* can all refer to the same thing.

(5.86) More than 30 years after declaring war on <u>cancer</u>, <u>the disease</u> refuses to surrender. (COCA)

The CoreNLPCoref engine that was run earlier does not identify sponsors for most contexts that require reasoning about subclasses and superclasses. LEIAs, by contrast, attempt such reasoning. We will illustrate how using example (5.86).

During Basic Semantic Analysis the LEIA will have generated candidate TMRs that analyze *cancer* as both CANCER-DISEASE and CANCER-ZODIAC. The only analysis for *disease* will be DISEASE (our current lexicon has only one non-construction-oriented sense of this word). The LEIA will check the ontology to see if either the pair [CANCER-DISEASE, DISEASE] or the pair [CANCER-ZODIAC, DISEASE] are in the same line of inheritance (i.e., if they are related as hypernym and hyponym). Only the former pairing is in the same line of inheritance, which both explains the use of the NP-Def *the disease* and allows the LEIA to disambiguate the analysis of *cancer*—all at one time.

The more specific the entities are (i.e., the deeper they are in the ontology), the more reliable the coreference-oriented reasoning. CANCER-DISEASE and DISEASE are quite specific, so there is a good chance they are coreferential. By contrast, if one of the entities is referred to as *the thing*—which maps to OBJECT in the ontology—it could corefer with any object in the text. Of course, the agent still must resolve inputs that include *the thing* along with many other vague referring expressions. However, this cannot be done reliably without invoking more knowledge, which, for our agents, is done during Situational Reasoning.

5.4.3 Definite Description Processing Awaiting Situational Reasoning

By the end of this stage of processing, many instances of NP-Defs will have been resolved. Processing the remaining ones requires additional types of knowledge and reasoning that will become available during Situational Reasoning. These include the following:

- Interacting with the visual/physical environment. For example, if someone tells a robot, *Give me the hammer*, the robot needs to identify the intended hammer in the physical environment.
- Using features from long-term memory that are missing in the language context. For example, a speaker might refer to *the sisters* under the assumption that the listeners know which individuals in the context are sisters.
- Mindreading—that is, making inferences about the speaker's plans and goals, about shared knowledge, and so on. For example, if someone comes into work and the first thing he says is *The box finally arrived*, this assumes that both of the participants are aware that a box was expected.

5.4.4 Recap of Definite Description Processing at This Stage

The following types of NP-Def processing are carried out at this stage:

1. Reject coreference links with property value mismatches: for example, *the red car* and *the blue car* do not corefer.
2. Run the reference resolution functions recorded in the lexical senses for specific words and phrases: for example, *the couple*, *the trio*.
3. Establish when a sponsor is not needed—for example, for universally known entities (*the sun*) and NP-Defs with restrictive postmodification (*the streets of my hometown*).
4. Identify bridging sponsors. For example, in the sentence *I walked into her office and the window was open*, *office* is the bridging sponsor for *the window*. When this is recognized, the TMR frame for WINDOW-1 will be enhanced to include PART-OF-OBJECT OFFICE-ROOM-1.
5. Create a set as the sponsor for plural NP-Defs: *Violists can be jealous of violinists for getting better parts, but the musicians still have to cooperate.*
6. Identify sponsors that are hypernyms or hyponyms of an NP-Def: *Go to the bank, the building on the corner.*

5.5 Anaphoric Event Coreference

Anaphoric event coreferences (AECs) can be expressed using verb phrase ellipsis[44] (5.87) or overt anaphors, such as *do it/that/this/so* (5.88).[45]

(5.87) Four family members can vote, and they all intend to __. (COCA)

(5.88) We want to keep the plant open—but we can't do it by ourselves. (COCA)

Both elided and overt-anaphoric verb phrases are detected during Basic Semantic Analysis using special lexical senses. Elided VPs are detected using lexical senses of modal and auxiliary verbs that anticipate a missing complement. Overt-anaphoric VPs are detected using the multiword senses for *do it*, *do that*, *do this*, and *do so*. The sem-strucs of those senses generate an underspecified EVENT in the TMR, which is flagged as requiring coreference resolution. This is, by the way, the same processing flow as is used for personal pronouns like *he* and *she*.

Fully resolving AECs involves answering up to five semantic questions, which we illustrate using the VP ellipsis–containing sentence in (5.89).

(5.89) John washed his car yesterday but Jane didn't __.

1. What is the verbal/EVENT head of the sponsor? Wash/WASH.
2. Do the elided event and its sponsor have instance or type coreference? Type coreference: there are two different instances of the WASH event.
3. Do the internal arguments have instance or type coreference (i.e., the same or different real-world referents)?[46] Type coreference: there are two different instances of AUTOMOBILE.
4. Are the meanings of modifiers in the sponsor clause copied or not copied into the resolution? Copied: *yesterday* applies to both propositions.
5. Are modal meanings in the sponsor clause copied or not copied into the resolution? This example does not contain modal meanings, but a slight variation on it does: *John tried to wash his car yesterday but Jane didn't.* In this case, the modal meaning *tried to* would be copied into the resolution: *Jane didn't try to wash her car yesterday.*

The results of this reasoning are shown in table 5.2, where the TMR frames are arranged to highlight the parallelism across the clauses. Note specifically that even though the second clause does not include the strings *wash*, *her car*, or *yesterday*, all these meanings are reconstructed during ellipsis resolution. So the meaning representation reads as if the input were "John washed his car yesterday but Jane didn't *wash her car yesterday*."

The subsections below further explore questions 1–5 above.

Table 5.2
Ellipsis-resolved meaning representation for *John washed his car yesterday but Jane didn't*

John washed his car yesterday but				Jane didn't*			
WASH-1				*WASH-2*			
	AGENT	HUMAN-1			*AGENT*	*HUMAN-2*	
	THEME	AUTOMOBILE-1			*THEME*	*AUTOMOBILE-2*	
	TIME	(*combine-time* -1 day)			*TIME*	*(combine-time -1 day)*	
					SCOPE-OF	*MODALITY-1*	
HUMAN-1				HUMAN-2			
	HAS-PERSONAL-NAME		'John'		HAS-PERSONAL-NAME		'Jane'
	GENDER		male		GENDER		female
	AGENT-OF		WASH-1		AGENT-OF		*WASH-2*
AUTOMOBILE-1				*AUTOMOBILE-2*			
	RELATION	HUMAN-1	; "his"		*RELATION*	*HUMAN-2*	
	THEME-OF	WASH-1			*THEME-OF*	*WASH-2*	
CONTRAST-1			; "but"	MODALITY-1			
	DOMAIN	WASH-1			TYPE	epistemic	
	RANGE	WASH-2			VALUE	0	; negation
					SCOPE	*WASH-2*	

* The elements in italics are reconstructed through VP ellipsis resolution.

5.5.1 What Is the Verbal/EVENT Head of the Sponsor?

Identifying the verbal/EVENT head of the sponsor is the starting point for all the other aspects of AEC resolution. Identification can range from very simple to very difficult, as illustrated by the contrast between (5.90) and (5.91). Note that in this section, since we are focusing on identifying the head of the sponsor clause, we underline only the head in the examples.

(5.90) I'm 51 now, and if I'm going to <u>do</u> it I'd better <u>do it</u> now. (COCA)

(5.91) "You have to guard students and weed out the bad kids and get them in an alternative setting or get out of school entirely," Roach said. "And there are federal laws with special education that prevent you from <u>doing that</u>." (COCA)

Finding the sponsor head for the second *do it* in (5.90) is easy because the context has the following combination of coreference-predicting property values (recall section 2.6 on simpler-first modeling):

- The coreferential RefExes are identical (*do it* ... *do it*), which reflects parallelism and simplicity.

- The coreferential RefExes are in an *if* ... *then* construction, which reflects prefabrication.
- The two parts of the construction contain one simple proposition each, which reflects simplicity.
- This expression, or close paraphrases of it (*If* someone$_i$ *is going to do it,* pronoun$_i$ *(had) better do it now/soon/fast*), should be familiar to every native speaker of English, which reflects prefabrication and ontological typicality.

Note that the first instance of *do it*, which serves as the sponsor for the latter instance, must itself be resolved as well, but that is a separate task that might be simple or complex.

By contrast, (5.91) is complex—one needs to apply human-level reasoning to understand what is disallowed. We, in fact, are not sure what the intended resolution is. It could be any of the propositions labeled A, B, C, and D in (5.92).

(5.92) You have to [$_A$ guard students and [$_B$ weed out the bad kids and [$_C$ get them in an alternative setting or [$_D$ get out of school entirely]]]].

As with all of our microtheories, we first focused on identifying the subset of cases that could be treated using heuristics that could be applied to any text, without requiring that the agent have specialized domain knowledge (like the federal laws about special education) or reasoning capabilities. These easier cases are automatically detected by the agent and resolved at this stage. All residual examples are postponed until Situational Reasoning.

We implemented two different versions of the sponsor-head-identification algorithm (reported in McShane & Babkin, 2016b; McShane & Beale, 2020) that relied on similar theoretical principles—simplicity, parallelism, and prefabrication. They were operationalized into quite different models and associated systems, the latter having a more developed theoretical substrate. The basic insight of both is that in some constructions, syntax alone can predict the sponsor for AECs. This harks back to the cornerstone observation of generative grammar: that grammaticality judgments can be made even for nonsensical sentences. So, even for a gibberish sentence like (5.93), we can easily reconstruct the elliptical gap as *gwaffed the gappulon.*

(5.93) The sloip <u>gwaffed the gappulon</u> and the loips did __ too.

The question is, how far can we push syntax-only sponsor-head selection before needing to resort to semantics? Or, stated differently, how can we formally capture the perceived simplicity of examples like (5.94)—(5.97), so that agents can confidently identify their sponsor heads, while postponing examples that require semantic reasoning?

(5.94) I want to <u>do</u> it and it would make me sad if I didn't __. (COCA)

(5.95) We needed to match Soviet technology for national defense purposes, and most Americans understood the dangerous consequences if we did not __. (COCA)

(5.96) In my opinion, none of these quarterbacks coming out should play in 1999, but you know some of them will __. (COCA)

(5.97) It's your money; do whatever you want to with it. If you want to make a big pile of it and burn it, you can do it. (COCA)

Note that although these examples are relatively simple—and were correctly treated by the model/system described in McShane and Beale (2020)—they are not trivially simple. For example, in resolving (5.94)–(5.96), the system needed to ignore the most proximate verbal candidates (*make/understood/know*) as well as exclude the sponsor-clause modals (*want/needed to/should*) from the resolution. And in resolving (5.97), the system needed to include a conjoined VP in the resolution, while excluding the sponsor-clause modal.

The goal of this research was to see how far we could push a lexico-syntactic approach before it broke, since focusing on only the absolutely simplest cases would offer too little coverage of examples to be very useful. Although both our 2016 and 2020 models focused on identifying the simpler cases, the models themselves are not simple. That is because not only did all of the component heuristics need to be formally defined in terms of the processors that could be used to compute them, but the limitations of those processors affected the extent to which the models could, at the current state of the art, be faithfully captured in an implementation.

Of course, syntax is not the only tool LEIAs have available at this stage for AEC coreference resolution. They can also apply selectional constraints unilaterally to attempt to understand what the underspecified EVENT could, in principle, mean and, therefore, which candidate sponsor is most fitting. For example, the sponsor head for the elided VP in (5.98) could be either *reading* or *munching*. But since dogs can't read, *munching* is the clear-cut semantically informed choice.

(5.98) Carol was reading a book and munching on corn chips. Her dog was __ too.

5.5.2 Is There Instance or Type Coreference between the Events?

If the elided event precisely corefers with its sponsor, there is *instance coreference* (5.99a), whereas if the reference is to a different *instance* of the same *type* of event, there is *type coreference* (5.99b).

(5.99) a. Jim tried to open the bottle but couldn't __.
b. Jim couldn't open the bottle but Jerry could __.

Note that even though the same bottle is being opened, as long as there are different agents, there must be different event instances. That is, event-instance coreference requires that all property values unify. However, either of the clauses can include additional modifiers, as in (5.100).

(5.100) Jim tried to open the bottle but couldn't __ without a bottle opener.

Our current algorithm for making the type-versus-instance coreference decision is as follows:

If the clauses have non-coreferential subjects; *This was determined earlier*
Then there is type coreference
Else
  If the clauses have coreferential subjects
  Then check all other property values
    If none have conflicting values
    Then this is instance coreference
    Else (i.e., if some property values conflict) it is type coreference.

5.5.3 Is There Instance or Type Coreference between Objects in the VPs?

If the sponsor clause includes objects (a direct object, an indirect object, or the object of a preposition), their reconstructions in the AEC clause might precisely corefer with those in the sponsor clause, which is instance coreference (5.101), or there might be different instances of the same type of object, which is type coreference (5.102). Starting with this set of examples, the entire sponsor is underlined.

(5.101) [Same fence, same alley]
I jumped the fence into the alley and Sally did __ too.

(5.102) [Sally walks her dog]
I walk my dog every morning and Sally does __ too.

Determining the instance-versus-type coreference of internal arguments (direct and indirect objects) is tricky and can require reasoning about how things generally work in the world. Our current algorithm is as follows:

Use instance coreference for internal arguments if either of these is true:

- The event coreference is instance coreference: John made this pizza, he really did __.
- The event coreference is type coreference and the sponsor-clause internal argument is NP-Def: He jumped over the fence and I did __ too.

Use type coreference for internal arguments if any of these are true:

- The event coreference is type coreference and the sponsor-clause internal argument has *a/an/some*: He ate a sandwich and I did __ too.

- The event coreference is type coreference and the sponsor-clause internal argument has a possessive modifier (*his*, *her*, *Martin's*, and so on): John washed his car and I did __ too.
- The event coreference is type coreference and the sponsor-clause internal argument has no article or determiner: John watches birds and I do __ too.

5.5.4 Should Adjuncts in the Sponsor Clause Be Included in, or Excluded from, the Resolution?

If the AEC clause specifies a value for PROPERTY-X, and if the sponsor clause *also* has a value for PROPERTY-X, then the sponsor clause's value is not copied during AEC resolution. For example, in (5.103), the interpretation of *today* fills the TIME slot in the AEC clause TMR, blocking the copying of the interpretation of *yesterday*.

(5.103) He arrived by train yesterday and she did __ today.

All other property values—here, *by train*—are copied over during AEC resolution, and, if applicable, the instance-versus-type coreference rules described above are invoked. For example, in (5.104), the meaning of *by the pool* fills the LOCATION slot and uses the same instance of SWIMMING-POOL because *the pool* is NP-Def.

(5.104) Grandma was drinking wine by the pool and Grandpa was __ too.

5.5.5 Should Modal and Other Scopers Be Included in, or Excluded from, the Resolution?

The algorithm for identifying AEC sponsor heads relies heavily on modality as a heuristic (see McShane & Beale, 2020, for details). For example, the pairing of the modals *might* and *might not* in (5.105) suggests not only that the sponsor head is *help* but also that the modal *might* should be excluded from the resolution (we don't want to end up with *But it also might not might help the cod*).

(5.105) It might help the cod. But it also might not __. (COCA)

Apart from modalities, other meanings can scope over propositions and be either included in, or excluded from, AEC resolutions:

(5.106) a. Whereas he vowed to tell the truth, she actually did it.
b. Whereas he vowed to tell the truth, she didn't __.

(5.107) a. He said he would come but he didn't __.
b. He said he would come but she didn't __.

The best way to study how scopers work is to collect corpus examples of all the kinds and combinations of scopers in the sponsor clause and the ellipsis clause, respectively. As a first step toward this, we created a list of nearly one hundred modal/aspectual correlations and searched for examples of them in the Gigaword corpus (Graff & Cieri, 2003). The search patterns grouped together inflectional forms of words and synonyms such as the following:

doesn't V … doesn't try to __

can start $V_{\text{PROGRESSIVE}}$ … won't __

As it turned out, many of the correlations were not attested at all. Others had many hits, but their behavior was entirely predictable a priori, leading to no new insights. For example, given a positive-negative pair of modal verbs (e.g., *could/couldn't*), the sponsor-clause modal should be excluded from the resolution. But some of the study results did suggest the need for specific rules, as illustrated by the following examples:

(5.108) a. They didn't want to go so they didn't __.
b. They didn't have to go so they didn't __.

(5.109) a. They didn't want to go and we didn't __ either.
b. They didn't have to go and we didn't __ either.
c. They didn't try to go and we didn't __ either.

In all these examples, the modal verb *didn't* occurs in both the sponsor clause and the ellipsis clause, and *didn't* in the first clause scopes over another modal (*want to, have to, or try to*) as well as the head verb (*go*). The key difference between the two sets of examples is that in (5.108), the clauses' subjects are coreferential, whereas in (5.109), they are not. When the subjects are coreferential, the additional modal (*want to, have to*) is excluded from the ellipsis resolution; by contrast, when the subjects are different, the additional modal is included in the resolution. From the human perspective, we can say that these resolution choices simply make sense. But from the system's point of view, these decisions need to be recorded as resolution rules.

At the time of writing, a more comprehensive microtheory of scoper treatment is under development. It needs to address not only modal and aspectual verbs but also any other words and multiword expressions that can take a verbal complement.

5.5.6 Recap of Anaphoric Event Coreference

To identify the sponsor for an elided or anaphoric event, the following questions must be answered:

1. What is the verbal/EVENT head of the sponsor?
2. Is there instance or type coreference between the events?
3. Is there instance or type coreference between objects in the VPs?
4. Should adjuncts in the sponsor clause be included in, or excluded from, the resolution?
5. Should modal and other scopers be included in, or excluded from, the resolution?

5.6 Other Elided and Underspecified Events

During Basic Semantic Analysis (the previous stage of processing), the agent (a) detected various types of ellipsis, (b) provisionally resolved them using generic EVENTS as placeholders, and (c) put flags in the TMRs indicating that their meanings needed to be specified. Those flags are, in fact, calls to procedural semantic routines that require various types of heuristic evidence that become available at different times. Here we consider two types of verbal ellipsis whose meaning can often be computed at this stage.

Aspectuals + OBJECTS. When an aspectual verb (e.g., *start*, *finish*, *continue*) takes an OBJECT as its complement (e.g., *He started a book*), this is a clear sign that an EVENT has been elided: after all, one can only start, finish, or continue *doing something* with an object.

One method of establishing the default interpretation in such contexts is querying the ontology. The question for the agent is, Are there any EVENTs for which the default meaning of the subject is the AGENT and the default meaning of the object is the THEME? Recall that, in the LEIA's ontology, case role fillers are described using facets that represent three levels of constraints: the default constraint (*default*), the basic semantic constraint (*sem*), and the expected potential for extended usages (*relaxable-to*). If there is exactly one EVENT whose *default* case role fillers fulfill the search criteria, then that event is a strong candidate as a default interpretation. This would occur if, during the analysis of the sentences *She started a book* and *The author started a book*, the LEIA found the following in the ontology:

```
READ
    AGENT      default    HUMAN
    THEME      default    BOOK-DOCUMENT

WRITE
    AGENT      default    AUTHOR
    THEME      default    BOOK-DOCUMENT
```

However, what if the ontology contained different information? What if it listed HUMAN as the default agent of WRITE? In that case, both sentences would be analyzed, by default, as referring to either reading or writing.

Apart from the complexities of decision-making during ontology acquisition (there often is no single perfect answer), many other details need to be worked through to achieve a strongly predictive microtheory of *aspect* + OBJECT resolution. For example, the sentence itself might not present the most specific possible information. The *she* in *She started a book* might refer to a well-known author, and her social role may or may not be relevant in the given context (she might be starting to read a book on an airplane, just like a nonauthor would do). The full inventory of contextually relevant properties will become available during the later stage of Situational Reasoning.

Conditions of change. Events are caused by events. If a text says that an *object* causes an event, this is a sure sign that a related event has been elided. For example, *The onions made her cry* actually means something like *Molecules released when the onion was cut came in contact with the tissues of her eye, causing a chain of events that led to crying.* Or, for someone who hasn't just googled "Why do onions make people cry?" the answer is "Some event involving the onion made her cry." Fleshing out the actual event(s) in question can involve either ontological knowledge, as for the onion example, or situational knowledge. For example, if *The raccoon caused the accident*, we know that something that it did caused the accident, which could be anything from running out into traffic to lunging at a dog who then ran into the street. Understanding how a raccoon could cause an accident does not involve knowledge of language, but knowledge of the world. However, before jumping to the conclusion that the need for extensive world knowledge and reasoning represents a dire state of affairs, remember the following: At this point in processing, the LEIA has already detected that an event was elided, and now it is evaluating whether it can easily determine—using ontological knowledge—which one. If it cannot, it will wait until the stage of Situational Reasoning to decide whether it cares. In many cases, it will not. After all, saying *A raccoon caused the accident* might have the discourse function of indicating that it was not the fault of one of the drivers.

5.7 Coreferential Events Expressed by Verbs

Events can be expressed by verbs or by noun phrases. For example, *Someone sneezed loudly* and *There was a loud sneeze* both generate an instance of SNEEZE (LOUDNESS .8) in the TMR. Does the way the event is presented in the text affect how we approach its reference treatment? Yes.

As we have already seen, every noun phrase—whether it instantiates an OBJECT or an EVENT in the TMR—is evaluated with respect to its coreference needs. When *a loud sneeze* is evaluated, the use of the article *a* will block the search for a sponsor. If, by contrast, the input refers to *the loud sneeze*, then the agent will try to identify its sponsor.

Coreference resolution for verbs, however, is approached differently. By default, the agent does not search for a sponsor for each EVENT referred to by a verb. There are two reasons why. First, most verbs introduce new events into the context, which means that the default answer to "Does it have a sponsor?" is "No." Second, when speakers *do* need to establish coreferences among events, they have other ways of doing so apart from using a coreferential verb. We have already seen five of these, repeated below with examples.

(5.110) [A coreferential EVENT can be elided.]
I wanted to kick it but I didn't __. (COCA)

(5.111) [A coreferential EVENT can be expressed by a broad referring expression.]
He didn't even seem sorry and that made me mad. (COCA)

(5.112) [A coreferential event can be expressed using an overt verbal anaphor.]
The focus of debate now is whether Congress will limit the size of the companies or ask the new regulator to do so. (COCA)

(5.113) [A coreferential event can be presented using a full NP. The TMR for this example will contain two instances of RAIN-EVENT with a coreference link.]
Sunday night it rained again, and the rain turned into snow. (COCA)

(5.114) [An event that is a subevent of its sponsor can be presented as a full NP. The ontological description of TRAVEL-EVENT, which is instantiated by *head to*, includes all of the subevents indicated by the underlined NPs.]
And so, decades after Grandma pledged allegiance to the American flag, worn out by my own struggles in the promised land, I heed her ancestral call and head to St. Kitts. The flight is smooth, the landing perfect, the welcome gracious. (COCA)

Of course, one *can* express coreferential events using a sequence of verbs, so the agent has to be on the lookout for such cases. Fortunately, sequences of coreferential verbs often occur in particular types of constructions. Specifically, the clause containing the coreferential verb often occurs immediately after the sponsor clause and includes either an additional modifier (5.115) or the specification of a previously unspecified or underspecified argument (5.116).[47]

(5.115) Mary jogged yesterday, and she jogged so far!

(5.116) Max read all night. He was reading *War and Peace*.

If a clause does not match one of our recorded verbal-repetition constructions, the agent does not seek out a coreferent for its main verb *at this stage*. If there is a coreference relation—either with some event reported in the text or with an event previously remembered by the agent—then this needs to be established during Situational Reasoning, when full reference resolution (i.e., grounding to agent memory) is undertaken.

As a point of comparison, consider how event coreference has been approached in mainstream NLP, where the emphasis has been to support applications involving information extraction, such as template filling, populating databases, and topic tracking. As Lu and Ng explain in their 2018 survey paper, for event mentions to be coreferential they need to be of the same type and have compatible arguments, as in their example:

(5.117) Georges Cipriani {left}$_{ev1}$ a prison in Ensisheim in northern France on parole on Wednesday. He {departed}$_{ev2}$ the prison in a police vehicle bound for an open prison near Strasbourg. (p. 5479)

Lu and Ng explain that event coreferences are more difficult than entity coreferences because (a) establishing event coreferences relies on a larger number of noisy upstream results, including establishing entity coreference relations, and (b) the coreferents can be realized as many types of syntactic categories. All of the works cited by Lu and Ng involve machine learning (ML) since practically all recent work adopts this approach. However, some primarily ML-oriented systems—such as the one reported in Lu and Ng (2016)—do incorporate some handcrafted rules as well.

Lu and Ng (2018) do not sugarcoat the difficulty of the event coreference task or the quite modest capabilities of state-of-the-art systems that they report in a tabular summary of published system evaluations. They explain that many event coreference evaluation scores are artificially high because event mentions are manually identified prior to the evaluation runs—a significant simplification since the identification task (known as *trigger detection*) is itself quite difficult. They summarize, "Both event coreference and trigger detection are far from being solved" (p. 5479). They note that the best results have been achieved for English and that "event coreference models cannot be applied to the vast majority of the world's low-resource languages for which event coreference–annotated data is not readily available" (p. 5484).

Whereas Lu and Ng (2018) focus on system building, Hovy et al. (2013) address a related issue: corpus annotation in service of ML. Their goal is to address some of the more difficult aspects of coreference that have been avoided by past annotation efforts and "to build a corpus containing event coreference links that is annotated with high enough inter-annotator agreement to be useful for machine learning." They establish three levels of event identity: *full identity*; *partial identity*, divided into (a) *membership*, which links multiple instances of the same type of event, and (b) *subevent*, which links events from the same script; and *no identity*. The novel aspect of the work is the second category, which they describe using interesting examples:

> In our work, we formally recognize partial event overlap, calling it *partial event identity*, which permits different degrees and types of event coreference. This approach simplifies the coreference problem and highlights various inter-event relationships that facilitates grouping events into 'families' that support further analysis and combination with other NLP system components. (p. 22)

The latter, as we interpret it, means that the annotations offer a coarse-grained linking of elements of input that require more detailed analysis using methods that are outside the purview of the reported work.

5.8 Further Exploration

1. Read, or at least browse through, "MUC-7 Coreference Task Definition" (Version 3.0, July 1997) by L. Hirschman and N. Chinchor, https://www-nlpir.nist.gov/related_projects/muc/proceedings/co_task.html. Note how complex the instructions are—how many phenomena have to be explicitly ruled in and ruled out.

2. Explore the state of the art in coreference engines using the Stanford CoreNLP interface at corenlp.run. Be sure to select the coreference annotator by clicking on the Annotations field and selecting "coreference" from the pull-down menu. Experiment with easy inputs and more difficult ones, as described in this chapter.

3. Use the online version of the COCA corpus (https://www.english-corpora.org/coca/) to explore anaphoric event coreference. The most reliable way to identify examples (and avoid false positives) is to search for relevant strings before a "hard" punctuation mark—for example, a period, semicolon, question mark, or exclamation point. Some of the many possible search strings are these (note the spacing conventions required by the interface):

could n't .	would n't .	should n't .	wo 'nt .		
might .	can .	did .	will .	may .	must .
do it .	do that .	did this .	does it .		
has to .	wants to .	tries to .	promised to .		

a. When do these search strings return false positive results (i.e., examples that don't show anaphoric event coreference)? How can they be avoided?

b. What is the sponsor in each example? Can you point to a text string that precisely reflects the sponsor? If not, is the sponsor extralinguistic, is the reference vague, or is something else going on?

c. If there is a textual sponsor, is there type or instance coreference of the verbs? Type or instance coreference of the internal arguments? Does the sponsor clause include modal or other scoping verbs? If so, are they included in or excluded from the resolution?

d. Is each example easy or difficult for a system to automatically resolve? Why?

4. Looking just at the table of contents at the beginning of the book, try to reconstruct what was discussed in each section of chapter 5 and recall or invent examples of each phenomenon.

6

Extended Semantic Analysis

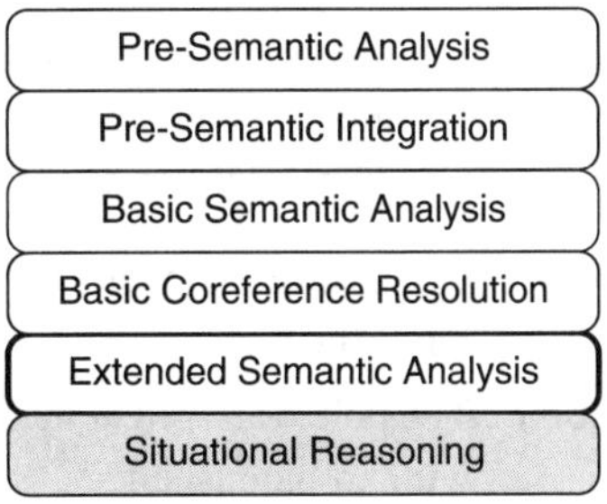

During Extended Semantic Analysis the LEIA looks beyond the local dependency structure (i.e., the main event in a clause and its arguments) in an attempt to resolve outstanding ambiguities, incongruities, and underspecifications that were identified during Basic Semantic Analysis. Like all processing so far, Extended Semantic Analysis uses methods that are applicable to texts in all domains. It does not involve Situational Reasoning, which will be invoked, if needed, later. Extended Semantic Analysis is triggered in the following situations:

1. Multiple TMR candidates received a high score because Basic Semantic Analysis could not resolve some ambiguities (section 6.1).
2. All TMR candidates received a low score because Basic Semantic Analysis encountered incongruities (section 6.2).
3. Data in the basic TMR—namely, calls to procedural semantic routines—indicate that more analysis of a specific kind is needed. Most often, an underspecified concept requires further specification (section 6.3).[1]
4. The TMR is a nonpropositional fragment that must be incorporated into the larger context (section 6.4).

Extended Semantic Analysis, like Basic Coreference Resolution, addresses difficult linguistic phenomena. For some of them, a complete solution will be beyond the state of the art for quite a while. But this is not necessarily detrimental to the agent's overall functioning.[2] Consider a real-life example: You cross paths with a colleague walking across campus, have a quick chat, and she wraps it up by saying, "Sorry, I've got to run to a dean thing." *Dean thing* is a nominal compound that leaves the semantic relation between the nouns unspecified. The *thing* in question could be about a dean, organized by a dean, required by a dean, or for deans only. Do you care which? Probably not. The speaker's point is that she has a good reason to cut the conversation short. Underspecification is a useful design feature of language, and it makes no sense to build agents who will not stop until they have tried long and hard to concretize every vague utterance.

6.1 Addressing Residual Ambiguities

During this stage, the LEIA's main knowledge source for resolving residual ambiguity (i.e., choosing from among multiple high-scoring candidate interpretations) is the ontology. The agent attempts to understand the context by looking for ontological connections between candidate interpretations of words. Consider the following minimal pair of examples:

(6.1) The police arrived at the port before dawn. They arrested the pirates with no bloodshed.

(6.2) The police arrived at the secret computer lab before dawn. They arrested the pirates with no bloodshed.

What comes to mind as the meaning of *pirates* in each case? Most likely, seafaring bandits for the first, and intellectual property thieves for the second. This is because a port suggests maritime activity, whereas a computer lab suggests intellectual activity. The reason why the agent cannot recognize the preferred reading of *pirates* during Basic Semantic Analysis is that the deciding clue—the location of the event—is in a different dependency structure. That is, when the *arrest* sentences are processed in isolation, both readings of *pirate* are equally possible because they both refer to types of HUMAN, and all HUMANS can be arrested. To disambiguate, the agent needs to extend its search space to the preceding sentence.[3]

Five types of ontological knowledge have proven useful for disambiguating such inputs. All of these heuristics involve relations between OBJECTS since it is OBJECT-to-OBJECT relations that were not covered by the dependency-based (largely OBJECT-to-EVENT) disambiguation of Basic Semantic Analysis. The heuristics are applied in the order in which they are presented.

6.1.1 The Objects Are Linked by a Primitive Property

OBJECTS in the LEIA's ontology are described by dozens of properties. Some of these, such as LOCATION and HAS-OBJECT-AS-PART, link OBJECTS to other OBJECTS, asserting their close ontological affinity. The following pair of examples shows how this knowledge is useful for disambiguation.

(6.3) "What a nice big stall!" "Well, that's a very big horse!"

(6.4) The horse was being examined because of a broken tooth.

When a LEIA encounters *horse* in an input, it must determine whether it refers to an animal, a sawhorse, or a piece of gymnastic equipment. When it considers the animal-oriented analysis HORSE, all of the concepts shown in the ontology excerpt below (as well as many others) are understood to be potential participants in the context.

HORSE		
AGENT-OF	sem	TROT, CANTER, GALLOP, BUCK-EVENT, ...
COLOR	sem	white, black, gray, bay, chestnut, buckskin,[4] ...
LOCATION	sem	BARN, **ANIMAL-STALL**, RIDING-ARENA, ...
HAS-OBJECT-AS-PART	sem	HOOF, MANE, TAIL, HEAD, LEG, **TOOTH**, ...

So, when analyzing (6.3), the LEIA recognizes that both HORSE and ANIMAL-STALL are in the candidate space; and when analyzing (6.4), it recognizes that both HORSE and TOOTH are in the candidate space. Finding these correlations helps to disambiguate *both* of the words in each context simultaneously—after all, *stall* can also mean a booth for selling goods, and *tooth* can also refer to a tool part.

6.1.2 The Objects Are Case Role Fillers of the Same Event

Another way to detect close correlations between OBJECTs is through a mediating EVENT. That is, the LEIA might be able to find an EVENT for which some interpretations of the OBJECTs in question fill its case role slots. Returning to our seafaring bandit example (6.1), the ontology contains a WATER-TRAVEL-EVENT for which PIRATE-AT-SEA is a typical filler of the AGENT case role, and PORT is a typical filler of both the SOURCE and DESTINATION case roles, as shown below.

WATER-TRAVEL-EVENT		
AGENT	default	SAILOR, **PIRATE-AT-SEA**
	sem	HUMAN
	relaxable-to	ANIMAL
THEME	sem	WATER-VESSEL
LOCATION	sem	BODY-OF-WATER
SOURCE	default	**PORT**
	sem	GEOGRAPHIC-ENTITY
DESTINATION	default	**PORT**
	sem	GEOGRAPHIC-ENTITY

Finding these fillers, the LEIA concludes that WATER-TRAVEL-EVENT is the ontological context of the utterance and selects PIRATE-AT-SEA (not INTELLECTUAL-PROPERTY-THIEF) as the analysis of *pirate*, and PORT (not PORT-WINE) as the analysis of *port*. The success of this search strategy depends on the coverage of the ontology at any given time—that is, it is essential that the ontology *have* an event that associates seafaring bandits and ports using its case roles.

6.1.3 The Objects Are Linked by an Ontologically Decomposable Property

The properties discussed in section 6.1.1, LOCATION and HAS-OBJECT-AS-PART, are primitives in the ontology. However, it is convenient—both for knowledge acquisition and for the agent's reasoning over the knowledge—to record some information using properties that are shorthand for more complex ontological representations. We call these *ontologically decomposable* properties because rules for their expansion must be specified in knowledge structures appended to the ontology. Consider the excerpt from the ontological description of INGEST that was introduced in chapter 2:

INGEST		
AGENT	sem	ANIMAL
	relaxable-to	SOCIAL-OBJECT
THEME	sem	FOOD, BEVERAGE, INGESTIBLE-MEDICATION
	relaxable-to	ANIMAL, PLANT
	not	HUMAN

Using the simple slot-filler formalism of the nonscript portion of the ontology, it is not possible to record who eats what—that horses eat grass, hay, oats, and carrots, whereas koalas eat only eucalyptus leaves.[5] That is, the portion of the knowledge structures below indicated in square brackets cannot be easily accommodated using the knowledge representation strategy adopted for the broad-coverage (nonscript) portion of the ontology.

HORSE
  AGENT-OF INGEST [THEME GRASS, HAY, OAT-FEED, CARROT]

KOALA
  AGENT-OF INGEST [THEME EUCALYPTUS-LEAF]

The reasons why property values cannot, themselves, be further specified by nested property values are both historical and practical. Historically speaking, the ontology was acquired decades ago, in service of particular goals (mostly disambiguating language inputs) and supported by a particular acquisition/viewing interface. Practically speaking, there are reasons to uphold this constraint. Namely, it simplifies not only the human-oriented work of knowledge acquisition, management, and visualization but also an agent's reasoning over the knowledge. Much more could be said about this decision within the bigger picture of knowledge representation and automatic reasoning, but we leave that to another time. The point here is that it is possible both to uphold the decision to allow only simple slot fillers (along with all of its benefits) and to provide the agent with more detailed knowledge. In fact, there are at least two ways to record such knowledge: as ontological scripts (described in sections 2.3.1 and 2.8.2) and using decomposable properties. We consider these in turn.

If an agent needs extensive specialist knowledge about certain animals—for example, to generate a computer simulation of their behavior or reason about it—then full ontological scripts must be recorded. For example, one could acquire an INGESTING-BY-KOALAS script, which would not only assert that the AGENT of this event is KOALA and that its THEME is EUCALYPTUS-LEAF but also describe many more details about this process: how the koala gathers the leaves, how long it chews them, how many it eats per day, and so on. In short, an ontology could contain many descendants of INGEST (INGESTING-BY-KOALAS, INGESTING-BY-HORSES, INGESTING-BY-WHALES) that provide extensive information about what and how different kinds of animals eat. However, unless all of these new events are going to provide much more information than simply what each animal eats, it is inefficient to create concepts for all of them.

A more streamlined solution for recording *who eats what* is to create an ontologically decomposable property like TYPICALLY-EATS that directly links an animal to what it eats. This allows knowledge acquirers to record in the ontology information like the following:

```
HORSE
   TYPICALLY-EATS      default      GRASS, HAY, OAT-FEED, CARROT, APPLE

KOALA
   TYPICALLY-EATS      default      EUCALYPTUS-LEAF
```

This is a shorthand for

```
ANIMAL
   AGENT-OF   sem       INGEST-#1

INGEST-#1
   THEME      default   INGESTIBLE
```

in which the notation -#1 indicates coreference between *ontological instances* of the concept (i.e., it is a method of indicating coreference among knowledge structures in static knowledge resources). The shorthand TYPICALLY-EATS is connected to its expansion using a rule encoded in the analysis algorithm.

Consider how such object associations can help in disambiguation. Given the input *The cow was eating grass*, the knowledge COW (TYPICALLY-EATS GRASS) allows the agent to simultaneously disambiguate *cow* as the animal COW (not a derogatory reference to a woman) and *grass* as the lawn material GRASS (not marijuana).

6.1.4 The Objects Are Clustered Using a Vague Property

The approach just described requires knowledge acquirers to introduce specific, decomposable properties, provide rules for their semantic expansion, and record the associated sets of concepts. The resulting knowledge is very useful but takes time to acquire.[6] A faster and cheaper type of knowledge acquisition is to have a vague RELATED-TO property[7] that can hold a large set of associated concepts. For example:

- Objects related to the outdoor space of someone's house include GRASS, WEED, FENCE, TREE, LAWNMOWER, SWIMMING-POOL, DRIVEWAY, GARDEN-HOSE, BUSH, PICNIC-TABLE, LAWN-CHAIR, JUNGLE-GYM.
- Objects related to a kitchen include UTENSIL, BLENDER, PANTRY, COUNTERTOP, SAUCEPAN, DISHWASHER, KITCHEN-TOWEL.

This shorthand not only is a useful knowledge-engineering strategy but also reflects the concept-association behavior of people. For example, a person asked to name ten things associated with a horse might well include among them *saddle*. The association is recalled without the person actually going through a semantic expansion like "a saddle is the thing a person sits on when riding a horse."

To emphasize, we are talking about inventories of related *concepts*, not ambiguous *words*. So, although various word-based resources—for example, the results of statistical word clustering or the results of human word-association experiments—can be useful to help detect such associations, the knowledge must be (manually) encoded as concepts in order to become part of an agent's ontology and unambiguously inform its reasoning.

6.1.5 The Objects Are Linked by a Short Ontological Path That Is Computed Dynamically

If the above approaches fail to disambiguate an input, the agent can try to establish what the utterance is about by searching for the shortest ontological path between all candidate interpretations of the OBJECTs in the local context. The problem with this last-ditch strategy, however, is that it is difficult to achieve high-quality shortest-path calculations in an ontology. The main reason for this is that shortest-path calculations depend on the effective assignment of traversal costs for different kinds of properties. For example, traversing an IS-A link will have a lower cost than traversing a HAS-OBJECT-AS-PART link because concepts linked by IS-A (e.g., DOG and CANINE) are more similar than concepts linked by HAS-OBJECT-AS-PART (e.g., DOG and EAR). The key to using this strategy is to apply it only if it results in a short, very low-cost path between concepts. If the path is not very low-cost, then this is not a reliable heuristic.[8]

To recap, so far we have seen five ontology-search strategies that an agent can use to resolve residual ambiguity. All of them rely on identifying closely related ontological OBJECTs in the immediately surrounding context. Optimizing the definition of the *immediately surrounding* context is as difficult as automatically determining the window of coreference for coreference resolution.

6.1.6 Reasoning by Analogy Using the TMR Repository

Another source of disambiguating heuristics is the TMR repository, which is a knowledge resource that records the agent's memories of past language-to-meaning mappings. Remembered TMRs can serve as a point of comparison for reasoning by analogy.[9] Reasoning by analogy is a big topic that we will touch on only to the extent needed for the goal at hand: lexical disambiguation.

A difficult problem in lexical disambiguation is the frequency with which a sentence can potentially have both a literal and a metaphorical reading. For example, *strike back* can mean "to hit physically" or "to retaliate nonphysically (e.g., verbally)." If the LEIA has previously encountered the expression *strike back*, the remembered meaning representations can serve as a vote for the associated analysis. However, slick as this approach might sound—and psychologically plausible as well—it is anything but straightforward to implement. There are at least three complications.

Complication 1. In different domains, different disambiguation decisions will be correct. For example, if a LEIA analyzes many texts about boxing and then turns its attention to texts about office interactions, it should not interpret every spat that involves confrontational language as an instance of physical assault simply because it has many TMRs about people punching each other's lights out. It follows that reasoning by analogy requires a nontrivial prerequisite: marking each remembered TMR with a domain in which it is applicable. This makes the applicability of this method rather problematic.

Complication 2. If the remembered TMRs are to be useful targets of reasoning by analogy, then they must not only belong to the same domain as the TMR being disambiguated but also be correct. But generating correct TMRs for every single input is beyond the state of the art. This means that a TMR repository is likely to contain a combination of correct and not-completely-correct TMRs. The most reliable way to ensure the quality of the repository would be to have people check and correct all the TMRs. However, this is realistic only for small repositories. One automatic method of assessing TMR quality is using the agent's own confidence estimates in its interpretations, which are computed and stored as a matter of course. However, for reasons explained in section 9.3, those estimates are not always reliable. Another method of automatically assessing the quality of TMRs relies on heuristics that must be computed outside the NLU module. Namely, if an input requires some action by the agent, and if the agent responds appropriately to it, then there is a good chance that the agent correctly understood it. Of course, automatically determining that the agent's action was appropriate requires task-level reasoning beyond what we detail in this book. Moreover, since not every input gives rise to an observable action by the agent, this is far from an all-purpose solution to evaluating the quality of the TMRs in the TMR repository.

Complication 3. The TMR repository might not contain any analyses relevant to the given input. This raises the questions, "Should different agents share their TMR repositories?" and "How can we best utilize similarity measures to exploit close but not exact matches?" As regards the latter, whereas "I'm going to kill him" does not usually refer to murder (though it can), "I'm going to smack him upside the head" may or may not involve physical violence. So the ontological similarity between HIT and KILL does not necessarily support reasoning by analogy in this instance.

If the domain-independent methods of resolving residual ambiguity described in the preceding subsections do not cover a particular input, then domain-specific methods (described in chapter 7) must be brought to bear. The reason we do not start with the latter is that committing to a specific domain largely erases the word-sense ambiguity problem to begin with. In fact, avoiding word-sense ambiguity by developing narrow-domain applications is a widely practiced strategy for developing agent systems. Although this strategy can work quite well for narrow domains, it will not advance the state of the art in making agents perform at the level of their human counterparts. After all, even when people are engaging in a narrowly defined task, they *will* engage in off-topic conversation—that is just part of being human.

6.1.7 Recap of Methods to Address Residual Ambiguity

- Prefer interpretations of OBJECTs that are linked by a primitive property in the ontology. For example, to analyze *The horse was being examined because of a broken tooth*, use the ontological knowledge HORSE (HAS-OBJECT-AS-PART TOOTH).
- Prefer interpretations of OBJECTs that fill case role slots of the same EVENT in the ontology. For example, to analyze *The police arrived at the port before dawn. They arrested the pirates with no bloodshed*, use the ontological knowledge WATER-TRAVEL-EVENT (AGENT PIRATE-AT-SEA) (DESTINATION PORT).
- Prefer interpretations of OBJECTs that are linked by an ontologically decomposable property. For example, to analyze *The horse wants some grass*, use the ontological knowledge HORSE (TYPICALLY-EATS GRASS), which expands, via a recorded reasoning rule, to HORSE (AGENT-OF INGEST (THEME GRASS)).
- Prefer interpretations of OBJECTs that are linked by the vague ontological property RELATED-TO. For example, to analyze *I need to tack up the horse; where's the bridle?*, use the ontological knowledge HORSE (RELATED-TO BRIDLE).
- Prefer interpretations of OBJECTs that are linked by a short ontological path. For example, to analyze *I need to tack up the horse; where's the bridle?*—assuming that the needed RELATED-TO information was *not* recorded—use the path HORSE (THEME-OF TACK-UP-HORSE (INSTRUMENT BRIDLE)).
- Use reasoning by analogy against the TMR repository. For example, if every past analysis of *strike back* generated the nonphysical interpretation, that is a vote in favor of the nonphysical interpretation for the new input—assuming that none of the complications discussed above confound the process.

6.2 Addressing Incongruities

Incongruity describes the situation when no analysis of an input aligns with the expectations recorded in the LEIA's knowledge bases. The subsections below describe four sources of incongruities—metonymy, preposition swapping, idiomatic creativity, and indirect modification—and the methods LEIAs use to resolve them.

6.2.1 Metonymy

In a metonymy, one entity stands for another. For example, in (6.5), *the spiky hair* refers to a particular person with spiky hair.

(6.5) The spiky hair just smiled at me.

Speakers of each language know which metonymic associations can exist between a named entity and what it stands for. (By contrast, metaphors can establish novel relations between entities.)

Metonymy leads to a sortal incongruity during Basic Semantic Analysis. This means that an event head and its dependents fail to combine in a way that aligns with ontological expectations. In (6.5), the problem is that HAIR is not a valid AGENT of SMILE-EVENT. Speakers of English readily understand the indirect reference because we know that people can be referred to metonymically by their physical features, clothing, or items closely associated with them.

Just as people are aware of typical metonymical relationships, so, too, must be LEIAs. To maintain an inventory of canonical metonymical replacements, our model introduces a dedicated knowledge resource, the LEIA's Metonymic Mapping Repository.[10] A subset of its content is illustrated by (6.6)—(6.10).

(6.6) [Producer for product]
Then your father bought an Audi with a stick shift. (COCA)

(6.7) [Social group for its representative(s)]
And for her heroic efforts, the ASPCA awarded her a gold medal. (COCA)

(6.8) [Container for the substance in it]
... A large pot boiled lid-rattlingly on the stove. (COCA)

(6.9) [Clothing for the person wearing it]
I want to dance with the big belt buckle.

(6.10) [Artist for a work of art]
In addition to the Rembrandts, there are five Vermeers, nearly a dozen Frans Halses, and the list goes on ... (COCA)

Carrying out such replacements is a simple and high-confidence method for dealing with the most typical metonymies.

If an input containing a potential metonymy does not match a recorded construction, then the agent attempts to determine how the kind of entity named in the text is related to the kind of entity expected by the ontology. For example, (6.11) says that either SPECTACLES or a set of DRINKING-GLASSes (the two meanings of *glasses* in the lexicon) are the AGENT of BORROW. But the ontology says that only HUMANS can BORROW things.

(6.11) The big glasses borrowed my bike.

So the agent must determine whether either SPECTACLES or a set of DRINKING-GLASSes might be standing in for a HUMAN—and, if so, based on what relation(s). It does so using the same kind of ontological search described in section 6.1.5 (Onyshkevych, 1997). Stated briefly, the agent computes the weighted distance between HUMAN and both of these concepts. The cumulative score for each reading is a function of the length of the path and of the cost of traversing each particular relation link. The shortest path turns out to be between HUMAN and SPECTACLES, so the metonymy can be resolved as *HUMAN RELATION (SPECTACLES (SIZE .8))*.

6.2.2 Preposition Swapping

Prepositions are a common source of performance errors by native and nonnative speakers alike.[11] In each of the examples below, the first preposition choice is canonical and the second is not—but it was attested in examples in the COCA corpus.

- translate language-X into [to] language-Y
- abide by [with] X
- be absolved of [from] X

Considering that English is the current worldwide *lingua franca*, with many speakers having nonnative fluency, it is a high priority for LEIAs to accommodate this type of close-but-not-perfect input. For example, it is not uncommon for subtitles of foreign films to be of high quality overall but to show the occasional odd preposition. The question is, How to detect instances of preposition swapping?

The first thing to say is the obvious: Not any preposition can be swapped for any other. So the preposition-swapping algorithm must be tightly constrained. According to our current algorithm, in order for the agent to hypothesize preposition swapping, all of the following must hold:

1. The lexicon must contain a fixed expression (i.e., an idiom or construction) that matches the input lexically and syntactically except for the preposition choice. So we are not talking about free combinations of prepositions and their complements.
2. All of the semantic constraints for that fixed expression must be met. In the example *translate X into [to] Y*, X must be a language or text and Y must be a language. These constraints are specified in the lexical sense for the construction *translate X into Y*. (Note that there is a different sense for *translate to* which means "result in," as in *Saving an extra $50 a month translates to $600 a year*.)
3. The preposition pair belongs to a list of preposition pairs that we have determined to be, or hypothesize to be, subject to swapping. These pairs either contain prepositions with similar meanings (*in/into, into/to, from/out of, by/with*) or contain at least one preposition that is extremely semantically underspecified, such as *of*.

A natural question is, If the LEIA successfully processes an input using this preposition-swapping repair method, should the attested preposition be recorded in the lexicon and treated ever after as canonical? The answer is no for three reasons. First, when LEIAs *generate* language, they should not generate less-preferred versions. Second, this recovery procedure works on the fly, so there is no reason to record the less-preferred version. Third, resorting to a recovery procedure models the additional cognitive load of processing unexpected input, which will result in a penalty to the overall confidence score of the TMR. This means that successful analyses that do *not* require recovering from unexpected input will be preferred, as they should be.

In addition, the LEIA's history of language analyses (recorded in the TMR repository) could be consulted when evaluating the likelihood of, and scoring penalty for, a preposition-swapping analysis. For example, if the TMR repository contains multiple examples of a particular preposition swap, the LEIA could reduce the penalty for that swap to a fraction of the norm. After all, maybe diachronic language change over the next couple of decades will result in *translate to* becoming a perfectly natural alternative to *translate into* for language-oriented contexts.

6.2.3 Idiomatic Creativity

The creative use of idioms[12] may or may not trigger extended processing. Let us begin with a case in which extended processing will not be triggered because no incongruity will be detected. Imagine that you are in your backyard entertaining a guest, and two deer sidle up, stomping on your freshly seeded grass. You clap your hands and make some noise, but they ignore you—they are fearless, suburban deer. So you go inside, grab your trumpet (lucky thing, you play the trumpet), and burst into a fanfare, at which time the deer bound out of sight into the woods. Your guest says, *Wow, you've killed two deer with one trumpet!* You laugh, but your companion LEIA won't get the joke—at least not at this stage of analysis. After all, everything lines up semantically. The event KILL requires an ANIMAL as the AGENT (you as a HUMAN fit), it requires a nonhuman ANIMAL as the THEME (the DEER fit), and it allows for a physical object to be the INSTRUMENT (the TRUMPET fits, even though it is not the preferred instrument of killing, which is WEAPON). Even if this utterance were taken out of context, any human would know it must have an indirect meaning: people just don't use trumpets to kill deer. Some day, LEIAs will have to have this depth of world knowledge as well. For now, no incongruity will be flagged for this example.

Not so, however, for the example, *You've killed two scratches with one rug!*, which might be said when a single throw rug works to cover two gouges in a wooden floor. This *will* lead to an incongruity because there is no meaning of *kill* that expects its THEME to be SCRATCH-MAR.

A similar split obtains between the examples *Don't put all your eggs in one boat* versus *Don't put all your eggs in one portfolio of statewide munis*—both of which were attested in the COCA corpus. In the first case, there will be no incongruity since eggs can be put into a boat. (We explain later how the idiomatic usage can be detected in a different way.) But in the second, there *will* be an incongruity because *portfolio of statewide munis* is an ABSTRACT-OBJECT, not a PHYSICAL-OBJECT; therefore, it is not a suitable DESTINATION for the physical event TRANSFER-POSITION. These pairs of examples illustrate how selectional constraints can flag an incongruity and suggest that the input might include idiomatic creativity.

If the input might be a play on an idiom, the agent must first identify the lexical sense that records the canonical form of that idiom. Although some global notion of fuzzy matching could be invoked, this is risky since *close but not quite* typically means that the input simply doesn't match the idiom. For example, *kick the pail* does not mean *die*, even though *pail* is a synonym of *bucket.*

There are two stages to processing creative idiom usages, detecting them and semantically analyzing them, which we consider in turn.

6.2.3.1 Detecting creative idiom use We prepare agents to detect creative idiom use in two ways: (1) by writing lexical senses that anticipate particular kinds of variations on particular idioms and (2) by implementing lexicon-wide rules that cover generic types of idiomatic creativity. We consider these in turn.

Writing lexical senses that anticipate particular kinds of variations on particular idioms. Many individual idioms allow for variations that people know or can easily imagine. The most reliable way to prepare agents to detect and analyze such variations is to record them in the lexicon. Table 6.1 illustrates such anticipatory lexical acquisition using results from an informal corpus study of idiom variation in the COCA corpus. Column 1 presents canonical forms of idioms, which will be recorded as one lexical sense, and column 2 presents variable-inclusive constructions, recorded as another lexical sense.[13] For example, the lexicon contains two senses of the verb *drop* to cover the data in row 1: one for the fixed form "at the drop of a hat" and the other for the variable-inclusive form "at the drop of a [N+]." Column 3 presents corpus-attested variations on the idioms, whose full examples are presented as (6.12)–(6.29).

Table 6.1
Canonical and variable-inclusive forms of idioms recorded as different lexical senses

Canonical form	Variable-inclusive form	Attested variations in cited examples
at the drop of a hat	at the drop of a [N+]	sixteenth note, hot dog, pin, pacifier, beaker, hare, backbeat
put all [one's] eggs in one basket	put all [one's] eggs in one [N+]	portfolio of statewide munis, blender, boat
	put all [one's] eggs in the [N+]	the stock market basket
	put all [one's] [N+] in one basket	
NEG judge a book by its cover	NEG judge a [N+] by its cover	bike, star
	NEG judge a book by its [N+]	title
[get information, be told] straight from the horse's mouth	[get information, be told] straight from the [N+]'s mouth	moose's, looney's
[get, be given] a dose of [one's] own medicine	[get, be given] a dose of [one's] own [N+]	frank talk
kill two birds with one stone	kill [NUM] [N+] with one [N+]	two topics/column

Note: [N+] indicates a head noun plus potential modifiers. [NUM] indicates a number.

(6.12) Each singer could turn the emotional temperature up or down *at the drop of a sixteenth note.* (COCA)

(6.13) Iris said that Judy Garland could cry *at the drop of a hot dog.* (COCA)

(6.14) I could break into *sobs at the drop of a pin.* (COCA)

(6.15) You take 100 pictures *at the drop of a pacifier.* (COCA)

(6.16) As trained scientists are wont to do *at the drop of a beaker*, he postulated a plausible theory. (COCA)

(6.17) I would run from New York and Columbia, like a hound *at the drop of a hare.* (COCA)

(6.18) These 12 tracks boast a startlingly powerful sound, shifting *at the drop of a backbeat* from a whispered seduction to a raging fury. (COCA)

(6.19) You would be foolish to *put all your eggs in one portfolio of statewide munis.* (COCA)

(6.20) Well, if you're running a startup that sells to other startups, you might be *putting all your eggs in one blender.* (COCA)

(6.21) I don't *put all my eggs in one boat.* (COCA)

(6.22) Diversifying, spreading your wealth around, not *putting all your eggs in the stock market basket*, is going to pay off. (COCA)

(6.23) By now we all know better than to *judge a bike by its cover*, but the Time RXR provides a deceptively smooth ride, especially as its angular, aggressive look screams race-stiff. (COCA)

(6.24) Don't *judge a star by its cover.* One of Kepler's seismic discoveries is the Sun-like star Kepler-37, which lies about 220 light-years from Earth in the constellation Lyra. (COCA)

(6.25) Sometimes you're not supposed to *judge a book by its title*, but in these types of books, there's an awful lot to the title. (COCA)

(6.26) Last week, I talked to both of them to get the story of moosebread, or "moose food," as they call it, *straight from the moose's mouth.* (COCA)

(6.27) We're going to hear it *straight from the loony's mouth.* (COCA)

(6.28) Mr. Vajpayee said India was not prepared to do that, and the president *got a dose of his own frank talk* as he sat at that state dinner last night and India's president took issue with the U.S. leader[']s description of the Indian subcontinent. (COCA)

(6.29) Put the two together and you *kill two topics with one column.* (COCA)

Let us work through the first example in the table. The canonical form of the idiom is shown in column 1: *at the drop of a hat.* Column 2 indicates that *at the drop of [any head noun with optional modifiers]* is an acceptable play on the idiom. However, note that the string *at the drop of* is immutable. Of course, like any aspect of knowledge acquisition, the decision about how to best formulate the idiom-extension template is best informed by a combination of intuition and corpus evidence.

In the second example, three different extended templates are all considered possible. They allow for different elements to be variable, but not all at the same time.

Earlier we said that the idiomaticity of *put all one's eggs in one boat* could not be detected on the basis of semantic incongruity because there is no incongruity—one can put eggs in a boat. So, how will the system know to consider an idiomatic interpretation? As long as we list the sense *put all [one's] eggs in one [N+]* in the lexicon, the Basic Semantic Analyzer will generate the idiomatic interpretation alongside the literal one; choosing between them will be undertaken during Situational Reasoning.

Note that a core aspect of acquiring idioms is listing all of their known variations, not just the one that pops to mind first. For example, although *let the cat out of the bag* is the most canonical form of this idiom, it often occurs as *the cat is out of the bag.* Since this variant is so common, we hypothesize that it is part of people's lexical stock and is most appropriately recorded as a separate lexical sense, leaving the more creative flourishes for dynamic analysis.

As concerns the agent's confidence in detecting idioms, the fixed senses offer the most confidence, whereas the variable-inclusive senses are more open to false positives. Our next method of detecting idiom play—using lexicon-wide rules—has broader coverage but is also more open to false positives.

Implementing lexicon-wide rules that detect generic types of idiomatic creativity. So far, we have experimented with three such rules.

Rule 1. Allow for two fixed NPs in the idiom to be swapped, as in (6.30).

(6.30) The man has a bad accent, he tells McClane it was raining "dogs and cats" instead of cats and dogs, and he refers to the elevator as "the lift." (COCA)

NP swapping might be the result of misspeaking, misremembering, or trying for comedic effect.[14]

Rule 2. If there are six or more fixed words in an idiom, allow for any one of them to be replaced, as in (6.31) and (6.32).

(6.31) [A play on the six-word idiom 'wake up and smell the coffee']
If you're an idler, wake up and smell the bushes burn. (COCA)

(6.32) [A play on the six-word idiom 'be a match made in heaven']
By contrast, the group feels its blend of dance sounds and political lyrics is a musical marriage made in heaven. (COCA)

The fixed-word threshold of six attempts to balance the desire to detect as much idiom play as possible against inviting too many false positives.

Rule 3. Allow for modifiers. Example (6.32) illustrates this type of wordplay: *a* musical *marriage made in heaven.*

6.2.3.2 Semantically analyzing creative idiom use Once creative idiom use has been detected, it must be semantically analyzed. The three steps of semantic analysis described below apply to all instances of idiomatic creativity, whether the idiom play is explicitly accounted for by a lexical sense with variables or is detected using lexicon-wide rules. There are slight differences in processing depending on the detection method, but we will mention them only in passing since they are too fine-grained to be of general interest. If the creative idiom use matches a lexical sense that specifically anticipates it, then (a) the procedural semantic routine comprising the three analysis steps below is recorded in the meaning-procedures zone of that sense; (b) that procedural semantic routine can be tweaked, if needed, to accommodate that particular instance of idiomatic creativity; and (c) the confidence in the resulting analysis is higher than when lexicon-wide rules are applied.

We will describe the three steps of analysis using example (6.15): *You take 100 pictures at the drop of a pacifier.*

Step 1. Generate the TMR using the recorded meaning of the basic form of the idiom. Our example plays on the recorded idiom *at the drop of a hat*, which means *very quickly.* So our example means that the person takes one hundred photographs very quickly (SPEED .9).

```
TAKE-PHOTOGRAPH-1
    AGENT           HUMAN-1[15]
    THEME           SET-1
    SPEED           .9

SET-1
    MEMBER-TYPE     PHOTOGRAPH
    CARDINALITY     100
```

Step 2. Explicitly record in the TMR the SPEECH-ACT that is implicitly associated with any utterance. Understanding this step requires a bit of background. The meaning of every utterance is, theoretically speaking, the THEME of a SPEECH-ACT. The AGENT of that SPEECH-ACT is the speaker (or writer) and the BENEFICIARY of that SPEECH-ACT is the hearer (or reader). In general, we do not have the agent generate a SPEECH-ACT frame for every declarative statement because it is cumbersome; however, the SPEECH-ACT does implicitly exist. In the case of plays on idioms, making the SPEECH-ACT explicit is just what is needed to offer a template for recording property values that would otherwise have no other way to attach to the TMR.

```
SPEECH-ACT-1
    THEME           TAKE-PHOTOGRAPH-1

TAKE-PHOTOGRAPH-1
    AGENT           HUMAN-1
    THEME           SET-1
    SPEED           .9

SET-1
    MEMBER-TYPE     PHOTOGRAPH
    CARDINALITY     100
```

Step 3. Add two properties to the SPEECH-ACT: a RELATION whose value is the meaning of the creatively altered constituent, and the feature-value pair 'WORDPLAY yes'. For our example, this says that the utterance reflects some form of wordplay (which might, but need not, involve humor) involving a baby's pacifier. Putting all the pieces of analysis together yields the following final TMR.

```
SPEECH-ACT-1
    THEME           TAKE-PHOTOGRAPH-1
    RELATION        PACIFIER-FOR-BABY-1
    WORDPLAY        yes
TAKE-PHOTOGRAPH-1
    AGENT           HUMAN-1
    THEME           SET-1
    SPEED           .9
SET-1
    MEMBER-TYPE     PHOTOGRAPH
    CARDINALITY     100
```

Although we have implemented the above algorithm, we haven't rigorously tested it on a corpus because variations on idioms—although entertaining and not unimportant for agent systems—are just not all that common, at least in available corpora (as observed by Langlotz, 2006, pp. 290–291, as well). However, work on idiomatic variation has broader implications. Idioms are just one type of construction, and constructions of all kinds are open to variation. So our approach to handling idiomatic variability by a combination of listing variable-inclusive senses and implementing lexicon-wide rules applies to the variability of nonidiomatic constructions as well.

6.2.4 Indirect Modification Computed Dynamically

As explained in section 4.1.6, most cases of indirect modification—for example, *responsible decision-making, rural poverty*—are best handled by lexical senses that anticipate, and then make explicit, the implied meaning. However, lexicon acquisition takes time and resources. This means that it is entirely possible that the lexicon will contain some sense(s) of a modifier but not every needed sense.

For example, say the lexicon contains only one sense of *responsible*, which expects the modified noun to refer to a HUMAN, as in *responsible adult* or *responsible dog owner.* If the LEIA encounters the input *responsible decision-making*, there will be an incongruity since RESPONSIBILITY-ATTRIBUTE can only apply to HUMANs. This will result in a low-scoring TMR that will serve as a flag for additional processing.

The good news about this state of affairs is that many instances of indirect modification share a similarity: they omit the reference to the agent of the action. So, a *vicious experiment* is an experiment whose agent(s) are vicious; an *honorable process* is a process whose agent(s) are honorable; and a *friendly experience* is an experience whose participant(s) are friendly. (We intentionally do not pursue a depth of semantic analysis that would distinguish between a person *behaving* viciously and a person having the general attribute of *being* vicious. It is early to pursue that grain size of description throughout a broad-coverage system.) In all such examples, the type of entity that was elided can be reconstructed with the help of knowledge recorded in the ontology.

- *Vicious experiment*: Although the attribute HOSTILITY (which is used to represent the meaning of *vicious*) can apply to any ANIMAL, the AGENT of EXPERIMENTATION must be HUMAN, so the elided entity in *vicious experiment* must be HUMAN.
- *Honorable process*: Although the AGENT of carrying out a PROCESS can be any ANIMAL, the attribute MORALITY (which is used to represent the meaning of *honorable*) applies exclusively to HUMANS, so the elided entity in *honorable process* must be HUMAN.
- *Friendly experience*: Since the property FRIENDLINESS can apply to any ANIMAL, and the AGENT of a LIVING-EVENT (used to represent the meaning of *experience*) can also be any ANIMAL, the elided entity in *friendly experience* is understood as ANIMAL.

Let us look in yet more detail at this ellipsis-reconstruction rule using the example *bloodthirsty chase.*

adj.$_{ModifiesAnimal}$ *EVENT* (e.g., bloodthirsty chase)

```
EVENT
   AGENT     ANIMAL-#1        ; an ontological instance of ANIMAL

ANIMAL-#1
   PROPERTY-IN-QUESTION    value-in-question
```

Rendered in plain English, this rule says, "If an adjective is supposed to modify an ANIMAL but it is being used to modify an EVENT, then introduce an ANIMAL into the meaning representation, make it the AGENT of that EVENT, and apply the modifier's meaning to it."

The TMR for *bloodthirsty chase* will convey that there is a CHASE event whose AGENT is an unspecified ANIMAL who wants to KILL ('wanting to kill' is the analysis of *bloodthirsty*). Although this representation reliably resolves the initially detected incongruity, it leaves a certain aspect of meaning—who is chasing whom—underspecified. This information may or may not be available in the context, as shown by the juxtaposition between (6.33) and (6.34).

(6.33) Lions regularly engage in bloodthirsty chases.

(6.34) The lion and rabbit were engaged in a bloodthirsty chase.

Strictly language-oriented reasoning can correctly analyze (6.33) but not (6.34). The first works nicely because the sense 'X engages in Y' maps the meaning of X to the AGENT slot of the EVENT indicated by Y. In essence, it interprets 'X engages in EVENT-Y' as 'X

EVENT-Ys' (here, *Lions chase*). When our *bloodthirsty* conversion rule is applied, the correct TMR interpretation will be generated. *Whom* lions chase is not indicated in the context. If knowledge about whom lions typically chase is available in the ontology, the agent will look for it only if prompted to do so by some application-specific goal.

The problem with (6.34) is that world knowledge is needed to understand that the lion and the rabbit are not collaborating as coagents chasing somebody else. There is no linguistic clue suggesting that the set should be split up. Specific knowledge engineering in the domain of predation would be needed to enable this level of analysis.

So far, "insert an agent" is the only indirect-modification rule for which we have compelling evidence. But literature offers creative singletons, such as Ross McDonald's gem, *She rummaged in the purse and counted five reluctant tens onto the table.* Although one might assume that incongruous modifiers should always be applied to the nearest available referring expression, we think it premature to jump to that conclusion.

6.2.5 Recap of Treatable Types of Incongruities

- Metonymy: *The spiky hair* (i.e., the person with the spiky hair) *just smiled at me.*
- Preposition swapping: *He was absolved from* (rather than *of*) *responsibility.*
- Idiomatic creativity: *Don't put all your eggs in the stock market basket* (instead of *in one basket*).
- Indirect modification: *The lion engaged in a bloodthirsty chase* (the lion, not the chase, was bloodthirsty).

6.3 Addressing Underspecification

Underspecification is detected when the basic TMR includes a call to a procedural semantic routine that has not yet been run.[16] This section considers three sources of underspecification: nominal compounds that were not covered by lexical senses, missing values in events of change, and underspecified comparisons.

6.3.1 Nominal Compounds Not Covered by Lexical Senses

Section 4.5 described two classes of NN compounds that are fully treated during Basic Semantic Analysis thanks to lexical senses that anticipate them:

- Fixed, frequent compounds that are recorded as head entries: for example, *attorney general, drug trial, gas pedal.*
- Compounds containing one element that is fixed (and therefore serves to anchor the compound in the lexicon) and one element that is semantically constrained. For example, one sense of the noun *fishing* expects an NN structure composed of any type of FISH followed by the word *fishing*; it can analyze inputs such as *trout fishing* and *salmon fishing.*

If an NN compound does not belong to either of the above classes, then during Basic Semantic Analysis all combinations of meanings of N1 and N2 are linked by the most generic relation, called RELATION. These candidate interpretations are evaluated and scored both within the clause structure (during Basic Semantic Analysis) and with respect to coreference (during Basic Coreference Resolution). Deeper analysis of the candidate interpretations is provided by the four strategies described below.

Strategy 1. *Using ontological constructions.* Some combinations of concepts have a prototypical relationship. For example, TEMPORAL-UNIT + EVENT means that the event occurs at the given time, so *Tuesday flight* is analyzed as FLY-EVENT (TIME TUESDAY). Similar analyses apply to *morning meeting*, *weekend getaway*, and so on. Since both components of such constructions are concepts, the constructions cannot be anchored in the lexicon. Instead, they reside in a dedicated knowledge resource, the Ontological Construction Repository, that is consulted at this stage of processing.

Ontological constructions can be further categorized into those showing *unconnected* constraints and those showing *connected* constraints. The latter category indicates that the candidate meanings of one of the nouns must be tested as a property filler of the candidate meanings of the other noun.

Examples of Unconnected Constraints

If N_1 is TEMPORAL-UNIT and N_2 is EVENT, then the interpretation is N_2 (TIME N_1).
Tuesday flight: FLY-EVENT (TIME TUESDAY)

If N_1 is ANIMAL-DISEASE or ANIMAL-SYMPTOM and N_2 is HUMAN (not MEDICAL-ROLE), then the interpretation is N_2 (EXPERIENCER-OF N_1).
polio sufferer: HUMAN (EXPERIENCER-OF POLIO)

If N_1 is SOCIAL-ROLE and N_2 is SOCIAL-ROLE, then the interpretation is HUMAN (HAS-SOCIAL ROLE $N_{1,}$ N_2).
physician neighbor: HUMAN (HAS-SOCIAL-ROLE PHYSICIAN, NEIGHBOR)

If N_1 is FOODSTUFF and N_2 is PREPARED-FOOD, then the interpretation is N_2 (CONTAINS N_1).
papaya salad: SALAD (CONTAINS PAPAYA-FRUIT)

Examples of Connected Constraints

If N_1 is EVENT and N_2 is ANIMAL, and if N_2 is a *default* or *sem* AGENT of N_1, then N_2 (AGENT-OF N_1).
cleaning lady: HUMAN (GENDER female) (AGENT-OF CLEAN-EVENT)

If N_1 is EVENT and N_2 is an ontologically recorded *sem* or *default* INSTRUMENT of N_1, then N_1 (INSTRUMENT N_2).

cooking pot: COOK (INSTRUMENT POT-FOR-FOOD)

If N_1 is OBJECT and N_2 is a filler of the HAS-OBJECT-AS-PART slot of N_1, then N_2 (PART-OF-OBJECT N_1).

oven door: DOOR (PART-OF OBJECT OVEN)

If N_1 is EVENT and N_2 is EVENT, and if N_2 is a filler of the HAS-EVENT-AS-PART slot of N_1, then N_2 (PART-OF-EVENT N_1).

ballet intermission: INTERMISSION (PART-OF-EVENT BALLET)

If N_2 is EVENT and N_1 is a *default* or *sem* THEME of N_2, then N_2 (THEME N_1).

photo exhibition: EXHIBIT (THEME PHOTOGRAPH)

If N_2 is described in the lexicon as HUMAN (AGENT-OF EVENT-X) (e.g., *teacher* is HUMAN (AGENT-OF TEACH)) and N_1 is a *default* or *sem* THEME of X (e.g., PHYSICS is a *sem* filler for the THEME of TEACH), then the NN analysis is HUMAN (AGENT-OF X (THEME N_1)).

physics teacher: HUMAN (AGENT-OF TEACH (THEME PHYSICS))

home inspector: HUMAN (AGENT-OF INSPECT (THEME PRIVATE-RESIDENCE))

stock holder: HUMAN (AGENT-OF OWN (THEME STOCK-FINANCIAL))

If N_1 is PHYSICAL-OBJECT and N_2 is PHYSICAL-OBJECT and N_1 is a *default* or *sem* filler of the MADE-OF slot of N_2, then N_2 (MADE-OF N_1).

denim skirt: SKIRT (MADE-OF DENIM)

If N_2 is PROPERTY and N_1 is a legal filler of the DOMAIN of N_2, then N_2 (DOMAIN N_1).

ceiling height: HEIGHT (DOMAIN CEILING)

These constructions not only offer high-confidence analyses of the semantic relation inferred by the NN but also can help to disambiguate the component nouns. For example, although *papaya* can mean PAPAYA-FRUIT or PAPAYA-TREE, in *papaya salad* it can be disambiguated to PAPAYA-FRUIT in order to match the associated construction above.

It is important to emphasize that these rules seek only high-confidence ontological relations, defined using the *default* and *sem* facets of the ontology. If a compound is semantically idiosyncratic enough that it would fit only the *relaxable-to* facet of recorded ontological constraints, then it is not handled at this point in analysis. For example, although the LEIA would be able to analyze the clausal input *He teaches hooliganism* using the *relaxable-to* facet of the THEME of TEACH (which permits anything to be taught), it would not analyze the corresponding NN compound, *hooliganism teacher*, using the NN rule that covers *science teacher* or *math teacher* because the NN analysis rule is more constrained. It requires N1 to satisfy the *default* or *sem* THEME of an EVENT—in this case, TEACH. So, when analyzing *hooliganism teacher*, the agent will leave the originally posited generic RELATION between the analyses of N1 (*hooliganism*) and N2 (*teacher*).

One might ask why we record any NN constructions in the lexicon since they could all, in principle, be recorded as more generic ontological constructions. For example, rather than record the construction "FISH fishing" in the lexical sense fishing-n1, we could record the construction "FISH FISH-EVENT" in the Ontological Construction Repository. In the latter scenario, when the system encountered the word *fishing*, it would recognize it as FISH-EVENT, resulting in the same analysis. The reason for the split has primarily to do with (a) convenience for acquirers and (b) the desire to post analyses at the earliest processing stage possible. If a given word, like *fishing*, is often used in compounds, and if it has no synonyms (or only a few synonyms that can readily be listed in its *synonyms* field of the lexical sense), then it is simpler and faster for the acquirer to record the information in the lexicon under *fishing* rather than switch to the Ontological Construction Repository and seek concept-level generalizations. Moreover, when the compound is recorded in the lexicon, it will be analyzed early, during Basic Semantic Analysis.

Strategy 2. *Recognizing NN paraphrases of N+PP constructions.* In some cases, a nominal compound is a paraphrase of an N+PP construction that is already recorded in the lexicon.[17] For example, one lexical sense of the noun *chain* expects an optional PP headed by *of,* and it expects the object of the preposition to mean a STORE or RESTAURANT. This covers inputs like *the chain of McDonald's restaurants*, whose TMR will be as follows:

```
SET-1
    MEMBER-TYPE    RESTAURANT-1
    CARDINALITY    >1
RESTAURANT-1
    HAS-NAME       'McDonald's'
```

Recording the meanings of typical N+PP constructions in the lexicon is done as a matter of course, since it assists with the difficult challenge of disambiguating prepositions. A pre-runtime lexicon sweep translates these N+PP constructions into corresponding NN constructions. Continuing with our example, this automatic conversion generates the NN construction STORE/RESTAURANT+*chain*, which covers the input *the McDonald's restaurant chain*, generating the same TMR as shown above.

The system computes these NN constructions prior to runtime, rather than storing them permanently in the lexicon, so that the constructions match the inventory of N+PP lexical senses, even if the form, scope, or inventory of the latter changes. If, for example, a new societal trend developed by which churches and schools could be organized into chains—giving rise to turns of phrase like *chain of churches* and *chain of elementary schools*—then knowledge acquirers would need to expand the lexical sense for *chain*+PP by allowing the object of the preposition to mean not only STORE and RESTAURANT but also CHURCH and SCHOOL.

Strategy 3. *Detecting a property-based relationship in the ontology.* In some cases, the meanings of the nouns in an NN compound are directly linked by some ontological property. For example, *hospital procedure* is lexically ambiguous since *procedure* can mean either HUMAN-EVENT (i.e., any event carried out by a person that involves particular subevents in a particular order) or MEDICAL-PROCEDURE. But since the ontology contains the description MEDICAL-PROCEDURE (LOCATION HOSPITAL), this analysis simultaneously disambiguates *procedure* and selects the correct relation between the concepts.

Strategy 4. *Identifying a short (but not direct) property-based path in the ontology.* In other cases, the meanings of the nouns in the NN are ontologically connected, but along a path that involves multiple properties. This is true, for example, of *hospital physician.* The interpretation with the shortest ontological path is PHYSICIAN (AGENT-OF MEDICAL-PROCEDURE (LOCATION HOSPITAL)). But there are actually a lot of ways in which HOSPITAL and PHYSICIAN could be linked using an ontological search. For example, since a hospital is a PLACE and a physician is a HUMAN, and since HUMANs go to PLACEs, then the physician could be the AGENT of a MOTION-EVENT whose DESTINATION is HOSPITAL. Similarly, since a hospital is a PHYSICAL-OBJECT, and since a PHYSICIAN is a HUMAN, and since any HUMAN can DRAW practically any PHYSICAL-OBJECT, then the PHYSICIAN could be the AGENT of a DRAW event whose THEME is HOSPITAL. The list of such analyses could go on and on. But the point is this: the use of an essentially elliptical structure like an NN compound requires that the speaker give the listener a fighting chance of figuring out what's going on. Using the compound *hospital physician* to mean a hospital that a physician is sketching is simply not plausible. That lack of plausibility is nicely captured by ontological distance metrics. The ontological path that goes from PHYSICIAN all the way up to HUMAN and from HOSPITAL all the way up to PHYSICAL-OBJECT is much longer than the path of our preferred reading.

Now, one could argue that PHYSICIAN (LOCATION HOSPITAL) is not the most semantically precise analysis possible, which is true. If we wanted a better analysis, we could create a construction that expected LOCATION followed by WORK-ROLE (which is a subclass of SOCIAL-ROLE), which would output meaning representations like the one below:

```
HUMAN
    HAS-WORK-ROLE    ^N2
    AGENT-OF         WORK-ACTIVITY-1

WORK-ACTIVITY-1
    LOCATION         ^N1
```

This construction would precisely analyze inputs like *hospital physician*, *bakery chef*, and *college teacher* as people fulfilling the listed work roles at the listed places. The point is that the agent will only attempt unconstrained ontology-based reasoning if there is no recorded construction to provide a more precise analysis.

The four analysis strategies for NN compounds just described, in addition to the lexically based ones described as part of Basic Semantic Analysis, still do not exhaust the analysis space for NNs. If an NN has not yet been treated, then the generic RELATION posited during Basic Semantic Analysis will remain, and the agent's last chance to generate a more specific analysis will be during Situational Reasoning.

So far we have concentrated on the analysis of two-noun compounds, but the approach can be extended to treating larger compounds. A LEIA's first step in treating any compound containing three or more nouns occurs much earlier than this. During Pre-Semantic Integration, the LEIA reambiguates the syntactic parser's bracketing of the internal structure of compounds containing three or more nouns. Then, during the various stages of semantic analysis, it seeks out islands of highest confidence among pairs of nouns and finally combines those partial analyses.

Consider the compound *ceiling height estimation*. The candidate bracketings are [[ceiling height] estimation] and [ceiling [height estimation]]. The analysis of the first bracketing will receive a very high score using two rules introduced above.

[ceiling height]

Rule: If N_2 is PROPERTY and N_1 is a legal filler of the DOMAIN of N_2, then N_2 (DOMAIN N_1).

Here: HEIGHT is a PROPERTY and CEILING is a legal filler of it, so HEIGHT (DOMAIN CEILING).

[[ceiling height] estimation]

Rule: If N_2 is EVENT and N_1 is a *default* or *sem* THEME of N_2, then N_2 (THEME N_1).

Here: ESTIMATE is an EVENT and HEIGHT is a *sem* THEME of ESTIMATE, so ESTIMATE (THEME HEIGHT).

Putting these two analyses together, the TMR for *ceiling height estimation* is

```
ESTIMATE-1
   THEME      HEIGHT-1

HEIGHT-1
   DOMAIN     CEILING-1
```

By contrast, the analysis of the second bracketing analysis, [ceiling [height estimation]], will receive a much lower score because there is no high-confidence rule to combine CEILING with ESTIMATE (CEILING is not a *sem* or *default* filler of the THEME case role of the event ESTIMATE).

Although it would be unwise to underestimate the potential complexity of processing large compounds, it is reasonable to assume that multinoun compounds require an extension to the algorithm presented here rather than a fundamentally different approach.

An important question is, *How well do our NN analysis strategies work*? For that answer, see section 9.2.1.

6.3.2 Missing Values in Events of Change

Events of change are events that describe a change in the value of some property: for example, *speed up*, *lose weight*, *increase*. Descriptions of events of change often convey two property values from which a third can be calculated. It is likely that people actually do this calculation, at least if the information is important to them, so LEIAs should as well. Consider some examples from the *Wall Street Journal* (1987–1989; hereafter WSJ), all of which present two values from which a third can be calculated:

(6.35) In 1985, 3.9 million women were enrolled in four-year schools. Their number increased by 49,000 in 1986. (WSJ)

(6.36) Interco shot up 4 to 71¾ after a Delaware judge barred its poison pill defense against the Rales group's hostile $74 offer. (WSJ)

(6.37) An index of longterm Treasury bonds compiled by Shearson Lehman Brothers Inc. rose by 0.79 point to 1262.85. (WSJ)

At this stage of processing, the LEIA can carry out these calculations and save them to memory along with the stated information. The functions for calculating are listed as meaning procedures in the lexical senses for the words indicating the events of change, such as *increase*, *shoot up*, and *rise* from our examples. (See McShane, Nirenburg, & Beale, 2008, for a more in-depth treatment of events of change).

6.3.3 Ungrounded and Underspecified Comparisons

Here we present the microtheory of ungrounded and underspecified comparisons as a whole, even though different classes of comparisons are analyzed to different degrees across stages 3–6 of NLU (3: Basic Semantic Analysis, 4: Coreference Resolution, 5: Extended Semantic Analysis, and 6: Situational Reasoning).

The first thing to say is that this microtheory is at a less advanced stage of development than some of our other ones. Although corpus analysis has informed it, we have not yet rigorously vetted it against a corpus. Still, this microtheory reflects a nontrivial modeling effort and nicely illustrates the distribution of labor across the modules of semantic and pragmatic analysis. Specifically, it underscores (a) that different types of heuristics become available at different stages of processing and (b) that an agent can decide how deeply to pursue the intended meaning of an input. For example, if *My car is better than that one* is used as a boast, then it doesn't matter which particular properties the speaker has in mind. However, if the speaker is advising the interlocutor about the latter's upcoming car purchase, then the properties in question absolutely *do* matter. Does the car handle well in snow? Have heated seats? An above-average extended warranty? The point is that, to function like people, agents need to judge how deeply to analyze inputs on the basis of their current interests, tasks, and goals.

Our current microtheory of comparatives classifies them according to two parameters: how/where the compared entities are presented, and how precise the comparison is.[18] We first define the value sets for these properties without examples and then illustrate their combinations with examples.

Values for *how/where the compared entities are presented*

1. They are both included in a comparative construction that is recorded in the lexicon.
2. The comparison involves a single entity, in what we call an *inward-looking* comparison (in contrast to an *outward-looking* comparison, in which two different entities are compared). So far, the most frequently encountered inward-looking comparisons involve either the change of an entity's property value over time or a counterfactual.
3. The compared-with entity is either located elsewhere in the linguistic context (i.e., not in a construction) or is not available in the linguistic context at all. These eventualities are combined because the agent engages in the same search process either way. Notably, searching for the point of comparison invokes some of the same features as coreference resolution: semantic affinity (comparability), text distance (the point of comparison should not be too far away), and the understanding that the point of comparison might not be in the linguistic context at all.

Values for *how precise the comparison is*

1. Specific: A specific property is referred to, such as INTELLIGENCE or HEIGHT.
2. Vague: The comparison is expressed as a value of evaluative modality (e.g., *better, worse*) or as a simile (*Your smile is like a moonbeam*).
3. Vague with an explanation: Either type of vague comparison mentioned above can be followed by an explanation of what is meant. Semantically, the explanation can either identify the particular property value(s) in question (which is quite useful), or it can supplement the vague comparison with an equally vague explanation (e.g., the comparison can be followed by a metaphor: *Your smile is like a moonbeam: it lights up my heart*). In practical terms, the explanation can be easy to detect because it participates in a construction with the comparison, or it can be difficult to detect because, in principle, any text that follows a vague comparison may or may not explain it.

The permutations of these feature values result in the nine classes of comparatives shown in table 6.2. The table includes an indication of which modules can be invoked to analyze associated examples. We say "*can be* invoked" because the agent can, at any time in NLU, decide to forgo deeper analysis of an input. In operational terms, this means that it can choose not to launch a procedural semantic routine that is recorded in the nascent TMR.[19]

Table 6.2
Classes of comparative examples and when they are treated during NLU

Class	Target	Precision	Basic Semantic Analysis	Basic Coreference Resolution	Extended Semantic Analysis	Situational Reasoning
1	The compared entities are in a comparative construction.	Specific	✓			
2		Vague	✓			✓
3		Vague + explanation	✓	✓	✓	✓
4	The comparison involves a single entity (it is inward-looking).	Specific	✓			
5		Vague	✓			✓
6		Vague + explanation	✓	✓	✓	✓
7	The point of comparison is located elsewhere in the text or is not available in the language context.	Specific	✓	✓	✓	✓
8		Vague	✓	✓	✓	✓
9		Vague + explanation	✓	✓	✓	✓

We will now work through each of the nine classes of comparatives, providing further details and examples.

Class 1. *The compared entities are in a comparative construction, and the comparison is precise.* Examples of this type are fully analyzed during Basic Semantic Analysis thanks to constructions recorded in the lexicon. In some cases, these constructions include calls to procedural semantic routines to compose the meanings of the many variable elements on the fly (something that is quite common for constructions overall). Two of the many comparative constructions recorded in the lexicon are shown below. Both are recorded as senses of their only invariable word: *than*. In (6.38), the property referred to is INTELLIGENCE, whereas in (6.39), it is AESTHETIC-ATTRIBUTE.

(6.38) [Subj Verb Comparative *than* NP Auxiliary/Modal/Copula]
Animals are smarter than we are. (COCA)

(6.39) [Subj Verb Comparative *than* NP]
"You think she's prettier than Mama?" (COCA)

The fact that the basic, construction-based proposition in (6.39) is scoped over by both modality ('you think') and an interrogative does not require multiple constructions. Instead, these proposition-level enhancements are handled by general rules.

Class 2. *The compared entities are in a comparative construction, and the comparison is vague.*

(6.40) [Subj Verb Comparative *than* NP]
I don't sleep because my real life is better than my dreams. $_{\text{(COCA)}}$

(6.41) [Subj Verb Modifier *but* Subj Verb Comparative]
A southerly breeze is adequate but a west wind is better. $_{\text{(COCA)}}$

(6.42) [Subj *be like* NP]
Tell Jack Hanna his life is like a zoo and he'll say, "Thanks!" $_{\text{(COCA)}}$

The difference in the TMRs resulting from class 1 and class 2 is that the TMRs for class 2 examples include a call to a procedural semantic routine that can, if run, attempt to concretize the vagueness. In some cases, the procedural semantic routine is recorded in the lexical sense for the construction itself. For example, "NP *be like* NP" is always vague in that it does not specify which properties and values are implied when comparing two nominals. In other cases, the procedural semantic routine is attached to the lexical description of a vague comparative word used in the construction, such as *better* or *worse*. No matter the source of these procedural semantic routines, they basically say, "This meaning is vague. It may or may not be important/useful to concretize it. This determination cannot be made until Situational Reasoning, when the agent knows its task and goals. Therefore, carry the call to this procedural semantic routine in the TMR until that stage. At that stage, determine whether a more precise interpretation is needed. If it is, use all available heuristic evidence to try to compute it. If that fails, ask the interlocutor for help."

In most cases, vague expressions are meant to be vague and their interpretation can be left as such. In addition, in many cases even the speaker/writer might be hard-pressed to come up with the specific connotation. For example, comparing the zookeeper Jack Hanna's life to a zoo was likely as much a witticism as anything else. However, the very inclusion of the call to the procedural semantic routine in the TMR carries information: it asserts that the agent is aware that the utterance and its interpretation are vague.

Class 3. *The compared entities are in a comparative construction, and the comparison is vague with an explanation.* As described earlier, it can be tricky to determine whether the text that follows a vague comparison actually concretizes it. Currently, the only way an agent can determine this is if the explanation participates in a construction with the comparison, as in the following examples.

(6.43) [Subj_1 Verb *like* NP {, : ;—} $\text{Clause}_{\text{Subj1/Not-Comparative}}$]
A career is like a flower; it blooms and grows.

(6.44) [Subj Verb *like* NP {,—} Modifier(s)]
… When he's on the basketball court, he moves like a rabbit, all quick grace and long haunches. $_{\text{(COCA)}}$

According to the construction used for (6.43),

- The "Subj Verb *like* NP" clause must be followed by another clause, but that latter clause cannot also be of the form "Subj Verb *like* NP." (This excludes, e.g., *A career is like a flower; it is like a rose*).
- The clauses must be joined by a non-sentential punctuation mark. This requirement will result in some losses (an explanation could be presented as a new sentence), but we hypothesize that those losses are justified by the reduction in false positives.
- The subject of the second clause must be coreferential with the subject of the first. (This excludes, e.g., *A career is like a flower; I am happy about that.*)

The construction used for (6.44), for its part, requires that the agent be able to identify any nonclausal syntactic entitles that semantically serve as modifiers. As we see, this can be hard, since *all quick grace* and *long haunches* serve as modifiers although they do not have the most typical form of a modifier (adjective, adverb, or prepositional phrase). For all constructions in this class (i.e., vague with an explanation), the explanation is semantically attached to the comparison in the TMR using the property EXPLAINS-COMPARISON. This is specified, of course, in the sem-struc of the comparative construction recorded in the lexicon.

As corpora show, people explain their comparisons quite frequently, which makes recording these kinds of constructions worth the effort. The constructions posited above will clearly overreach, and more knowledge engineering is needed to identify the sweet spot between coverage and precision.

It is noteworthy that, even given an optimal inventory of constructions, the resulting analyses can be unenlightening because of the actual language input. For example, (6.43) uses metaphorical language to describe the vague comparison—not a whole lot of help for automatic reasoning. However, it is still useful for agents to recognize that the text attempted to explain the comparison.

Class 4. *The comparison involves a single entity (it is inward-looking), and the comparison is specific.* The key to processing inward-looking comparisons is being able to automatically detect that the comparison is, in fact, inward-looking—that is, that no external point of comparison need be sought. So far, we have identified three semantic clues for inward-looking comparisons: the use of noncausative CHANGE-EVENTS (6.45), causative CHANGE-EVENTS (6.46), and counterfactuals (6.47).

(6.45) [Subj *gets/grows* Comparative]

a. I noticed, all summer long, I was getting healthier. (COCA)

b. That patch was moving. And it was getting larger. (COCA)

c. Jack felt his grin get bigger. (COCA)

d. The sobs grow louder. (COCA)

e. The centipede grew bolder. (COCA)

(6.46) [Subj *gets/makes* Direct-Object Comparative]

a. You should make the sanctions tougher. (COCA)

b. Also, government subsidies to get industry greener are short term when industry prefers long term commitment. (COCA)

(6.47) [Subj *could (not) / could (not) have* Verb Comparative]

a. "The replacement process could have been easier too," Swift says. (COCA)

b. Surely his heart couldn't beat any faster. (COCA)

As a reminder, CHANGE-EVENTS are events that compare the value of a particular property in the event's PRECONDITION and EFFECT slots. They are realized in language using a very large inventory of words and phrases: *increase*, *decrease*, *lose confidence*, *speed up*, *grow taller*, and so on. The constructions noted in (6.45) and (6.46) involve CHANGE-EVENTS. Their semantic descriptions (which include a procedural semantic routine) allow the agent to generate TMRs like the following—using the example *I was getting healthier.*

```
CHANGE-EVENT-1
    PRECONDITION    HEALTH-ATTRIBUTE-1
    EFFECT          HEALTH-ATTRIBUTE-2
    TIME            <find-anchor-time
    PHASE           continue             ; a shorthand for the full ASPECT frame

HEALTH-ATTRIBUTE-1
    DOMAIN          HUMAN-1
    RANGE           <HEALTH-ATTRIBUTE-2.RANGE

HEALTH-ATTRIBUTE-2
    DOMAIN          HUMAN-1
    RANGE           >HEALTH-ATTRIBUTE-1.RANGE
```

This TMR says that the value of the person's HEALTH-ATTRIBUTE is lower in the PRECONDITION of the CHANGE-EVENT than in its EFFECT; and that is, in fact, what *get healthier* means.

As regards counterfactuals, like those in (6.47), they, too, can be treated by lexicalized constructions. However, since counterfactuals have not to date been a priority of our R&D, we will say nothing further about what the associated meaning representation should look like.

All such lexicalized constructions can be fully analyzed as part of Basic Semantic Analysis.

Class 5. *The comparison involves a single entity (it is inward-looking), and the comparison is vague.* We have already explained how inward-looking comparisons are treated, and we have already explained how vague comparison words "carry along" calls to procedural semantic routines in their TMRs, in case the agent decides to try to concretize the basic interpretation. Those two functionalities need only be combined to treat this class of comparatives, illustrated by the following examples.

(6.48) [An inward-looking, noncausative CHANGE-EVENT with a vague comparison]
After the experiment the waking dreams got worse. (COCA)

(6.49) [A counterfactual with a vague comparison]
It wasn't the best start for the day but it could have been worse. (COCA)

Class 6. *The comparison involves a single entity (it is inward-looking), and the comparison is vague with an explanation.* Like the class above, this one uses already explained functions. We have not come across any examples of this class, but they are easily invented, as the following modification to (6.48) shows.

(6.50) After the experiment the waking dreams got worse: they changed into nightmares.

As discussed earlier, explanations can be detected using various types of constructions. For (6.50), the construction is largely similar to one posited earlier in that it requires the first clause to be followed by non-sentential punctuation, and the subjects of the two clauses must be coreferential.

Vague counterfactuals, by contrast, require a different type of explanatory construction, since counterfactuals are often explained by more counterfactuals, as our invented expansions of (6.49) below show.

(6.51) It wasn't the best start for the day but it could have been worse: my car could have broken down<the train could have been late; my boss could have been in one of his nasty moods>.

Although it would require human-level knowledge and reasoning to understand why one's car breaking down<the train being late; one's boss being in a nasty mood>would make for a bad day, the agent does not need this to hypothesize that the continuation explains the vague counterfactual. What it needs is a construction that expects a vague inward-looking counterfactual to be followed by a non-sentential punctuation mark and then a precise counterfactual. Will this rule always identify only explanations? Probably not. But it serves as a foothold for further work on this microtheory.

Classes 7–9. *The point of comparison is located elsewhere in the text or is not available in the language context at all, and the comparison is either specific (class 7), vague (class 8), or vague with an explanation (class 9).* We group these classes together because this part of the microtheory is, at the time of writing, underdeveloped. Part of the work belongs to the stage of analysis we are focusing on here (Extended Semantic Analysis), part must wait for Situational Reasoning, and much depends on difficult aspects of coreference resolution having worked correctly during the last stage of processing. In short, a lot is required to make the associated examples work.

What differentiates this class from the others is that the point of comparison might be anywhere in the linguistic context or not available at all. The salient features that differentiate examples are as follows:

- How the compared entity is realized: as a full nominal (*this book is better*), a pronoun (*it is better*), or an elliptical expression (e.g., *the second* __ *is better*).
- How the point of comparison is realized: as a full nominal with the same head as what it is compared with, as a full nominal with a different head from what it is compared with, as a pronoun, as an elliptical expression, or not at all (i.e., it is absent from the linguistic context).
- If applicable, the distance between the linguistically overt compared entities: that is, the point of comparison can be the most proximate preceding nominal, the next one back, and so on.

Below are some examples illustrating different combinations of the abovementioned parameter values.

(6.52) Your force field is good but my teleporting is better. (COCA)

(6.53) Whatever your secret was, you have to agree, mine is better. (COCA)

(6.54) I often tell my clients that the state of mind they want when negotiating or navigating conflict is curiosity, not certainty. If you can manage to be curious when things get tough, that curiosity will be your best friend. Curiosity is better. It's the mode that opens us to discovery. (COCA)

(6.55) I like the sweet potato idea! Way tastier than store bought white potato chips. (COCA)

(6.56) Let's see if we can find what he was reaching for. Here. My reach is better. (COCA)

(6.57) I met a guy last night who brought 80 pounds of screenplays out here in his suitcase. But he didn't bring his skis. I think my gambit is better. (COCA)

(6.58) And you think this is easier?!

(6.59) 50% tastier!

In the last two examples, there is no linguistic point of comparison. We can imagine the first being said as two people struggle to carry a sofa up a skinny and winding stairway, having just argued about various strategies. The last example is typical of advertising: the comparison is so vague that there is nothing legally binding about it.

Having delved deep into this model of processing comparatives, let us now take a step back to the big picture in order to more fully motivate why we present this cross-modular microtheory as part of this module of Extended Semantic Analysis. During this stage of processing, if a LEIA considers it worthwhile to attempt to concretize underspecified comparisons, it can apply additional resolution functions. Those functions still rely exclusively on the agent's broad-coverage knowledge bases. We have identified three such types of reasoning that can be applied at this stage. Working out the full microtheory, however, remains on agenda.

1. *The LEIA can semantically reason about whether the assertion following a vague comparison explains it.* So far, we have prepared the agent to detect explanations for

vague comparisons using lexico-syntactic constructions. However, (a) those constructions might overreach, identifying a text segment as an explanation when it is actually not, and (b) they do not cover all eventualities. Ontologically grounded reasoning could weigh in on this determination. For example, the second sentence in (6.60) explains the vague, inward-looking comparison, but there is no text-level clue to point that out. One needs to know that road salt corrodes car finishes—information that is entirely reasonable to expect in an ontology with moderate coverage of car-related information.

(6.60) Come pring, my car looked a lot worse. Road salt is a bear.

2. *The LEIA can attempt to concretize vague comparisons based on ontological generalizations.* Vague comparisons often rely on people's knowledge of the salient aspects of different entities.

(6.61) Her eyes were like a sunrise. (beautiful; bright)

(6.62) She ran like a deer. (gracefully)

(6.63) He's like a regular giraffe! (very tall)

Vague comparisons are like ellipsis: when using one, the speaker has to give the hearer a fair chance of interpreting it correctly. This means relying on the expectations of a shared ontology, including the canonically distinguishing features of entities.

We can prepare agents to reason about saliency by manually indicating the most salient property values for each concept (which might, by the way, differ in some cases across cultures) and/or by having agents dynamically learn this information from text corpora. For example, a sentence like *She skipped barefoot across the stepping stones as graceful as a deer running ...* $_{\text{(COCA)}}$ suggests that a salient property of deer is their gracefulness.

This stage of Extended Semantic Analysis is the appropriate place for carrying out salience-based reasoning about vague comparisons because (a) this extra reasoning will not always be necessary (and, therefore, should not be a part of Basic Semantic Analysis) and (b) to the extent that the ontology indicates the salient properties of entities, this reasoning can be carried out for texts in any domain (i.e., it does not rely on the situational awareness that becomes available only later in the NLU process).

3. *The LEIA can attempt to identify the points of comparison for classes 7–9.* As explained above, this involves (a) leveraging previously established coreference relations, (b) reasoning about which entities in a context are semantically comparable, (c) factoring in the text distance between mentioned entities (since there might be multiple entities in the preceding context that must be considered as candidate targets of the comparison), and (d) leaving open the possibility that the point of comparison is not in the text at all.

6.3.4 Recap of Treatable Types of Underspecification

- Many nominal compounds that are not covered by the lexicon are covered by ontological constructions recorded in the Ontological Construction Repository: TEMPORAL-UNIT + EVENT, as in *night flight.*
- Missing values in events of change can be calculated and recorded: *Their earnings grew from $10,000 to $15,000* [change: + $5,000].
- Some types of underspecified comparisons can be made explicit: *John got stronger* (than he was before).

6.4 Incorporating Fragments into the Discourse Meaning

Since LEIAs understand inputs incrementally, they are routinely processing midsentence fragments. Those are not the kinds of fragments we are talking about here.[20] We are talking about fragments that remain nonpropositional when the end-of-sentence indicator is reached. A LEIA's basic approach to analyzing fragments is as follows:[21]

1. Generate whatever semantic interpretation is possible from the fragment itself.
2. Detect the as-yet unfilled needs in that semantic interpretation.
3. Attempt to fill those needs using all available heuristics.
4. Once those needs are filled, verify that the original semantic interpretation is valid. Otherwise, amend it.

This process is best illustrated using an example:

(6.64) "My knee was operated on. Twice." "When?" "In 2014."

The TMR for the first utterance, *My knee was operated on*, is

SURGERY-1
  THEME KNEE-1
  TIME *<find-anchor-time*

HUMAN-1
  HAS-OBJECT-AS-PART KNEE-1
  COREF *identify-speaker*

KNEE-1
  THEME-OF SURGERY-1
  PART-OF-OBJECT HUMAN-1

When the LEIA encounters the input *Twice*, which occurs as an independent sentence, it will find only one lexical sense, which describes this adverb as a verbal modifier that adds the feature ASPECT (ITERATION 2) to the EVENT it modifies. Since no EVENT is available in the local dependency structure, an EVENT instance is posited in the meaning representation without any associated text string.

ASPECT-1	
ITERATION	2
SCOPE	EVENT-1

EVENT-1	
SCOPE-OF	ASPECT-1
textstring	*none*

The feature-value pair *textstring none* triggers the search for a coreferential EVENT (in the same way as *find-anchor-time* triggers the search for the time of speech). The algorithm is currently quite simple: it identifies the main EVENT (i.e., the event to which any subordinate or relative clauses would attach) in the previous clause. The reason why this simple algorithm works pretty well is that understanding sentence fragments would impose too great a cognitive load on the listener if the intended link to the rest of the context were not easily recoverable.

In our context, the search for the most recent event instance will identify SURGERY-1 as a candidate, leading to the following TMR for *My knee was operated on. Twice.*

SURGERY-1	
THEME	KNEE-1
TIME	< *find-anchor-time*
COREF-OF	EVENT-1

HUMAN-1	
HAS-OBJECT-AS-PART	KNEE-1
COREF	*identify-speaker*

KNEE-1	
THEME-OF	SURGERY-1
PART-OF-OBJECT	HUMAN-1

ASPECT-1	
ITERATION	2
SCOPE	EVENT-1

EVENT-1	
SCOPE-OF	ASPECT-1
textstring	*none*
COREF	SURGERY-1

Although the LEIA does not need to pretty-print these results to effectively reason with them, it is easier for people to understand the TMR if we remove the coreference slots and replace EVENT-1 with SURGERY-1. This yields the following TMR:

SURGERY-1	
THEME	KNEE-1
TIME	*<find-anchor-time*
SCOPE-OF	ASPECT-1

HUMAN-1	
HAS-OBJECT-AS-PART	KNEE-1
COREF	*identify-speaker*

KNEE-1	
THEME-OF	SURGERY-1
PART-OF-OBJECT	HUMAN-1

ASPECT-1	
ITERATION	2
SCOPE	SURGERY-1

The next utterance is *When?*, another fragment. For each question word, the lexicon contains a sense that expects the word to be used as a fragmentary utterance. This reflects expectation-driven knowledge engineering—that is, preparing the system for what it actually will encounter, not only what grammar books say is the most typical.

For certain question words (e.g., *When? Where? How?*) the semantic representation (i.e., the sem-struc zone of the lexical sense) posits an EVENT that is flagged with coreference needs like we just saw for *twice*. For other question words (e.g., *Who? How many?*) the semantic representation posits an OBJECT that is similarly flagged for coreference resolution. Returning to our example, for the independent utterance *When?* the procedure seeks out the most recent main EVENT, just like our last meaning procedure did. Formally, the initial, sentence-level meaning representation for *When?* is

REQUEST-INFO-1	
THEME	EVENT-1.TIME

EVENT-1	
textstring	*none*
THEME-OF	REQUEST-INFO-1

When the coreference is resolved and the structure is pretty-printed, it looks like this.

REQUEST-INFO-1	
THEME	SURGERY-1.TIME

The final fragment in our example is *In 2014*. This is a bit more challenging because the preposition *in* is highly polysemous. One rule of thumb used by LEIAs when resolving polysemous words is to select the interpretation that matches the narrowest selectional constraints. In this case, the LEIA selects *in-prep10* (the tenth prepositional sense of *in*) because that sense asserts that the object of the preposition must refer to a MONTH, YEAR, DECADE, or CENTURY. The preprocessor has already provided the knowledge that 2014 is a date, which the LEIA translates (during CoreNLP-to-LEIA tag conversion) into the appropriate ontological subtree that holds all date-related concepts. Since this constraint is met, the LEIA can confidently disambiguate *in* as the property TIME applied to some EVENT. As before, the TMR for *In 2014* refers to an as-yet unresolved EVENT.

YEAR-1	
textpointer	2014
ABSOLUTE-TIME	(YEAR 2014)
TIME-OF	EVENT-1
EVENT-1	
textpointer	*none*
TIME	YEAR-1

When the event is contextually grounded—that is, when it is linked to the SURGERY in question—the meaning representation looks as follows:

YEAR-1	
textpointer	2014
ABSOLUTE-TIME	(YEAR 2014)
TIME-OF	SURGERY-1

Putting all these pieces together, we can see what the agent learns from the dialog, *"My knee was operated on. Twice." "When?" "In 2014."*

SURGERY-1	
THEME	KNEE-1
TIME	YEAR-1
SCOPE-OF	ASPECT-1
HUMAN-1	
HAS-OBJECT-AS-PART	KNEE-1
COREF	*identify-speaker*
KNEE-1	
THEME-OF	SURGERY-1
PART-OF-OBJECT	HUMAN-1
ASPECT-1	
ITERATION	2
SCOPE	SURGERY-1
YEAR-1	
ABSOLUTE-TIME	(YEAR 2014)

Of course, the entire dialog history (the series of TMRs) is also available to the agent, but the most important thing is what it stores to memory, which is the information shown above.

This example was useful in showing (a) how lexical senses can posit concepts that are not directly attested in the text and (b) how coreference resolution can be carried out with the help of associated meaning procedures. The results of the coreference ground the meaning of the fragment in the context.

There are many variations on this theme, which are treated as a matter of course using constructions in the lexicon that have associated procedural semantic routines.[22] For example, the lexical sense that covers questions of the form "Who Verb-Phrase?" instantiates a REQUEST-INFO frame whose THEME is the AGENT, EXPERIENCER, or BENEFICIARY of the given EVENT. The agent dynamically determines which case role is correct based on the meaning of the EVENT. So the TMRs for the following questions are

Who had surgery?

REQUEST-INFO-1
  THEME SURGERY-1.EXPERIENCER

SURGERY-1
  TIME *<find-anchor-time*

Who wrote that book?

REQUEST-INFO-2
  THEME WRITE-1.AGENT

WRITE-1
  THEME BOOK-1
  TIME *<find-anchor-time*

Who got a prize?

REQUEST-INFO-3
  THEME AWARD-EVENT-1.BENEFICIARY

AWARD-EVENT-1
  THEME AWARD-1
  TIME *<find-anchor-time*

The format of the slot filler—EVENT.CASE-ROLE—shows what is expected in the upcoming context. In the first example, the LEIA is waiting for an utterance whose meaning is compatible with an EXPERIENCER—that is, it must be an ANIMAL (which includes HUMANS). So if the next input is an ANIMAL, it will interpret it as the filler of that CASE-ROLE.

Note that the use of fragments is not limited to dialogs, since a speaker can answer his own question and the resulting analysis will be identical. Note also that the agent, during incremental processing, might initially get the wrong analysis, as would be the case if the dialog were *"Who got a prize?" "Antonio said that Mary did."* Initially, the agent might think that Antonio did. This is fine, and is exactly what a person would do if the speaker of the second utterance made a long pause, or coughed or laughed, after the first word.

When fragments are used outside of prototypical language strategies like these, their interpretation must be postponed until Situational Reasoning, when the agent can leverage its understanding of the domain script and the related plans and goals to guide the interpretation.

6.5 Further Exploration

1. Explore idiomatic creativity using the online search engine of the COCA corpus (https://www.english-corpora.org/coca/). Your searches need to be seeded by actual idioms, but the search strings can allow for various types of nonstandard usages. For example, the search string *kill _mc _nn with _mc _nn*, which covers the construction [*kill*+cardinal-number+any-noun+*with*+*cardinal-number*+any-noun], returns hits including the canonical *kill two birds with one stone* as well as *kill three flies with one stone, kill two birds with one workout*, and others.

2. Try to find examples of preposition swapping—and other performance errors—by watching foreign films and TV series with subtitles. We found the subtitles to the Finnish TV series *Easy Living* particularly interesting in this respect since they were of high quality overall with the occasional slip in preposition choice or use of an idiomatic expression.

3. Explore how numerical values in events of change are expressed using the search engine of the COCA corpus. Sample search strings include *increased by _mc*, which searches for [*increased by*+cardinal-number], *by _mc to _mc*, which searches for [*by*+cardinal-number+*to*+cardinal number], and countless more that use different verbs (e.g., *increase, decrease, go up, rise*) and different presentations of numbers (e.g., *by* cardinal-number *to* cardinal-number; *from* cardinal-number *% to* cardinal-number *%*). Consider the following questions about cognitive modeling:

- Do you think that you actually remember all the numbers from such contexts?
- If not, which ones do you remember and in which contexts?
- What should intelligent agents remember and not remember?
- Should they be more perfect than people in this respect (calculating and remembering everything), or should they be more humanlike?

4. Looking just at the table of contents at the beginning of the book, try to reconstruct what was discussed in each section of chapter 6 and recall or invent examples of each phenomenon.

7

Situational Reasoning

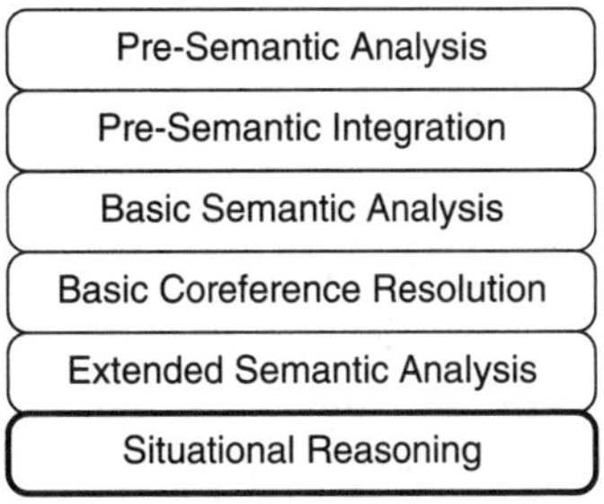

Up until this point, all of the LEIA's language analysis methods have been generic: they have leveraged the agent's lexicon and ontology and have not relied on specialized domain knowledge or the agent's understanding of what role it is playing in a particular real-world activity.

Considering that LEIAs are modeled after people, one might ask, "Don't people—and, therefore, shouldn't LEIAs—always know what context they are in? And, therefore, shouldn't this knowledge always be the starting point for language analysis rather than a late-stage supplement?" Although there is something to be said for this observation, the fact is that people actually can't always predict what their interlocutor's next utterance will be about. For example, while making dinner, you can be talking not only about food preparation but also about the need to buy a new lawn mower and a recent phone message that you forgot to mention. In fact, people shift topics so fast and frequently that questions like "Wait, what are we talking about?" are not unusual. The ubiquity of topic switching provides a theoretically motivated reason to begin the process of NLU with more generic methods—since the given utterance might be introducing a new topic—and then invoke context-specific ones if needed.

In addition to theoretical motivations, there are practical motivations for starting with generic methods of NLU and progressing to domain-specific ones as needed.

1. Ultimately, agents need to be able to operate at a human level in all contexts. Therefore, NLU capabilities should be developed in a maximally domain-neutral manner.
2. Many types of domain-specific reasoning are more complicated and computationally expensive than the generic NLU described so far, since they often involve capabilities like inferencing and mindreading.[1] Therefore, they should be used only if needed.
3. It can be difficult to automatically detect the current topic of conversation since objects and events from many domains can be mentioned in a single breath. Moreover, it is not the case that every mention of something shifts the topic of conversation to that domain. For example, saying, "Look, the neighbor kid just hopped over the fence for his baseball," does not mean that everyone should now be poised to think about strikes and home runs.

4. Although developers can, of course, assert that an agent will operate in a given domain throughout an application (and LEIAs can be manipulated to function that way as well), associated successes must be interpreted in the light of this substantial simplification. After all, committing to a particular domain largely circumvents the need for lexical disambiguation, which is one of the most difficult challenges of NLU.

Since Situational Reasoning is grounded in a particular domain and application, we might be tempted to delve directly into system descriptions. That will come soon enough, as applications are the topic of the next chapter. In this chapter we will (a) say a few more words about the OntoAgent cognitive architecture, all of whose components are available for Situational Reasoning, and (b) introduce our general model of integrating situational knowledge and reasoning into natural language understanding.

7.1 The OntoAgent Cognitive Architecture

As we have been demonstrating throughout, natural language understanding is a reasoning-heavy enterprise, and there is no clear line between reasoning *for* language understanding and reasoning *beyond* language understanding. We couldn't agree more with Ray Jackendoff's (2007) opinion that we cannot, as linguists, draw a tight circle around what some call *linguistic meaning* and expect all other aspects of meaning to be taken care of by someone else. He writes:

> If linguists don't do it [deal with the complexity of world knowledge and how language connects with perception], it isn't as if psychologists are going to step in and take care of it for us. At the moment, only linguists (and to some extent philosophers) have any grasp of the complexity of meaning; in all the other disciplines, meaning is reduced at best to a toy system, often lacking structure altogether. Naturally, it's daunting to take on a problem of this size. But the potential rewards are great: if anything in linguistics is the holy grail, the key to human nature, this is it. (p. 257)

It is only for practical purposes that this book concentrates primarily on the linguistic angles of NLU; doing otherwise would have required multiple volumes. However, as we introduced in chapter 1, full NLU is possible only by LEIAs that are modeled using a full cognitive architecture. The cognitive architecture that accommodates LEIAs is called OntoAgent. Its top-level structure is shown in figure 7.1.

This architecture is comprised of the following components:

- Two input-oriented components: perception and interpretation;
- The internal component covering attention and reasoning;
- Two output-oriented components: action specification and rendering; and
- A supporting service component: memory and knowledge management.

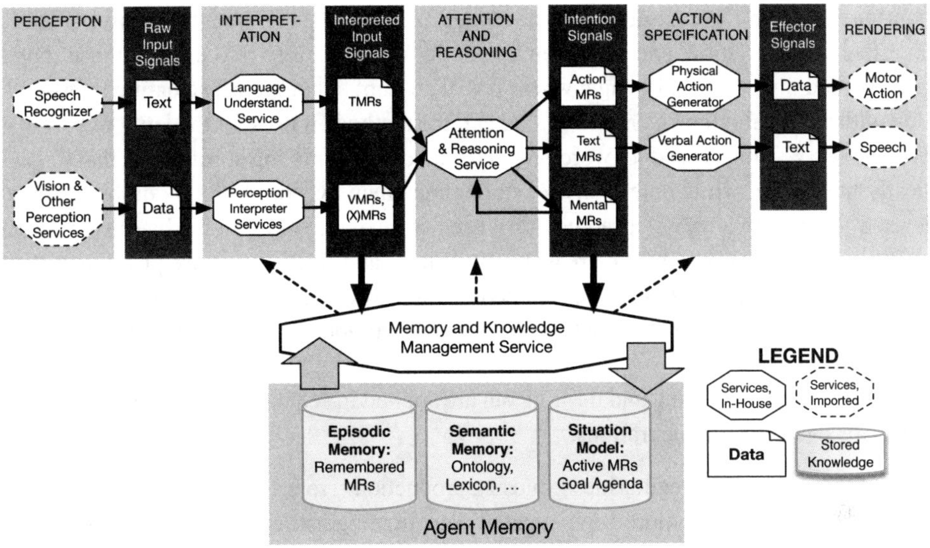

Figure 7.1
A more detailed view of the OntoAgent architecture than the one presented in figure 1.1.

The NLU capabilities described in this book are encapsulated in the Language Understanding Service. Just as it interprets textual inputs (which are, in some cases, transcribed by a speech recognizer), the Perception Interpreter Services must interpret other types of perceptual inputs—vision, nonlinguistic sound, haptic sensory inputs, and even interoception, which is the agent's perception of its own bodily signals. Whereas the NLU processes described so far relied on the agent's Semantic Memory (ontology and lexicon), the Situational Reasoning it now brings to bear can rely on its Situation Model and Episodic Memory as well.

This chapter continues to concentrate on our objective of describing the program of R&D that we call Linguistics for the Age of AI by discussing methods for treating the following phenomena using all the resources available to LEIAs operating in a situated context: (a) fractured syntax, (b) residual lexical ambiguity, (c) residual speech-act ambiguity, (d) underspecified known expressions, (e) underspecified unknown word analysis, (f) situational reference, and (g) residual hidden meanings.

7.2 Fractured Syntax

As described in section 3.2 and illustrated by figure 3.4, if syntactic mapping (i.e., the process of aligning elements of the parse with word senses in the lexicon) does not work out perfectly, the agent can choose to either (a) establish the best syntactic mapping it can and continue through the canonical middle stages of NLU (Basic Semantic Analysis, Basic Coreference Resolution, and Extended Semantic Analysis) or (b) circumvent those stages and proceed directly to this stage, where it will attempt to compute a semantic interpretation with minimal reliance on syntax. We call the methods it uses for the latter *fishing* and *fleshing out*. Fishing is used for wordy inputs: it involves extracting constituents that semantically fit together while potentially leaving others unanalyzed. Fleshing out, by contrast, is used for fragmentary, telegraphic utterances: it involves filling in the blanks given the minimal overt constituents. The reason why fishing and fleshing out are postponed until Situational Reasoning is that the agent needs some knowledge of what is going on to guide the analysis. In the absence of context, even people cannot make sense of highly fractured utterances.

The functions for fishing and fleshing out are applied in sequence. The fishing algorithm performs the following operations:

1. It strips syntactic irregularities—mostly, production errors such as repetitions and self-corrections—using more sophisticated detection algorithms than were invoked during Pre-Semantic Integration. For example, if two entities of the same syntactic category are joined by one of our listed self-correction indicators (e.g., *no*; *no, wait*; *no, better*), then strip the first entity along with the self-correction indicator: for example, *Give me* <u>*the wrench, no, better,*</u> *the screwdriver* → *Give me the screwdriver.*
2. It identifies NP chunks in the parse, since even for syntactically nonnormative inputs, NP chunking tends to be reliable enough to be useful.
3. It generates the most probable semantic analyses of the individual NP chunks on the basis of their constituents (e.g., the best semantic correlation between an adjective and its head noun) as well as domain- and context-oriented preferences. For example, if the agent is building a chair, then the preferred interpretation of *chair* will be CHAIR-FURNITURE not CHAIRPERSON. As we said earlier, the agent has to have some understanding of the domain/context in which it is operating in order for the fishing process to have a fair chance of working.
4. It uses the preferred NP analyses generated at step 3 as case role fillers for the events represented by verbs in the input. The disambiguation of the verbs is guided by both the meanings of the surrounding NPs and the most likely meaning of the verb in the given domain/context. Continuing with our chair-building example, the verb *give* has a preferred meaning of TRANSFER-POSSESSION, not DONATE or any of the other candidate analyses of that word.
5. It attempts to account for residual elements of input—such as modals, aspectuals, negation, and adverbials—by attaching the candidate analyses of residual elements to the just-generated TMR chunks, using, essentially, an *ordered bag of concepts* approach. For example, if the modal verb *should* occurs directly to the left of a verb whose

preferred analysis (from step 4) is TRANSFER-POSSESSION, then the meaning of *should*—that is, obligative MODALITY with a value of 1—will scope over that instance of TRANSFER-POSSESSION.

If fishing does not yield an actionable result, the agent can attempt fleshing out. In this process, the agent asks itself, "Could this (partial) meaning representation I just generated actually be a question I can answer? Or a command I can carry out? Or a piece of information I need to remember?" An example will best illustrate the idea.

Assume that a person who is collaboratively building a chair with a robot says, "Now the seat onto the base," which is an elliptical utterance that lacks a verb (semantically, an EVENT). Among the candidate interpretations of *the seat* and *the base* are CHAIR-SEAT and FURNITURE-BASE, which are concepts that figure prominently in the BUILD-CHAIR script. The use of these highly relevant concepts (a) triggers the agent's decision to go ahead and pursue fleshing out and (b) allows the agent to commit to these interpretations of *seat* and *base*.

Now the agent must analyze the other two contentful words in the input: *now* and *onto*. In the context of instruction giving, sentence-initial *now* can indicate the speech act REQUEST-ACTION, so the agent can assume that it is being asked to do something. *Onto*, for its part, means DESTINATION when applied to physical objects. So, on the basis of the co-occurrence of the above possibilities, as well as the ordering of words in the input, the agent can hypothesize that (a) it is being called to do some action, (b) the action involves as its THEME a specific instance of CHAIR-SEAT, and (c) the DESTINATION of that action is a particular instance of FURNITURE-BASE. This results in the initial TMR:

```
REQUEST-ACTION-1
    AGENT          HUMAN-1 ("speaker")
    THEME          EVENT-1
    BENEFICIARY    ROBOT-1

EVENT-1
    THEME          CHAIR-SEAT-1
    DESTINATION    FURNITURE-BASE-1
    COREF          seek-specification
```

The procedural semantic routine *seek-specification* is included in the EVENT-1 frame because the agent hypothesized that some event was elliptically referred to, and now it has to try to figure out which one. It hits pay dirt when it finds, in its BUILD-CHAIR script, the subevent

```
ATTACH
    THEME          CHAIR-SEAT
    DESTINATION    FURNITURE-BASE
```

It can then replace the unspecified EVENT from its initial TMR with ATTACH, leading to the following fleshed-out and actionable TMR.

REQUEST-ACTION-1
  AGENT HUMAN-1 ("speaker")
  THEME ATTACH-1
  BENEFICIARY ROBOT-1

ATTACH-1
  AGENT ROBOT-1
  THEME CHAIR-SEAT-1
  DESTINATION FURNITURE-BASE-1

The algorithms for fishing and fleshing out are actually much more complicated than what we just described. The reason we choose not to detail those algorithms here is that most of the reasoning is extralinguistic and depends centrally on what ontological knowledge is available for the given domain, what the agent's goals are, what its understanding of its interlocutor's goals are, what it is physically capable of doing, and so on. All of this is more appropriately presented within a comprehensive description of an agent application. If this sounds like a hint at our next book, it is.

7.3 Residual Lexical Ambiguity: Domain-Based Preferences

It often happens that Basic Semantic Analysis generates multiple candidate interpretations of an input. In some cases, Basic Coreference Resolution or Extended Semantic Analysis provides definitive evidence to select among them. But often, the agent reaches the stage of Situational Reasoning with multiple candidates still viable. This residual ambiguity can be treated in a straightforward manner by preferring analyses that use domain-relevant concepts. If the LEIA is building a chair, then the associated script will include many instances of CHAIR-FURNITURE and none of CHAIRPERSON. Accordingly, "I like chairs" (which allows for either interpretation on general principles) should be analyzed using CHAIR-FURNITURE. This approach might remind one of distributional semantics, with the crucial difference that, for our agents, meaning representations contain unambiguous concepts rather than ambiguous words.

7.4 Residual Speech Act Ambiguity

Recall that utterances that offer an indirect speech act reading always offer a direct speech act meaning as well. "I need a hammer" can mean *Give me one* or simply *I need one*—maybe I know that you aren't in a position to give me one or that we don't have one to begin with. During Basic Semantic Analysis the LEIA detected the availability of both interpretations for typical formulations of indirect speech acts, which are recorded as pairs of senses in the lexicon. It gave a scoring preference to the indirect reading, but the direct meaning remained available. Now it can use reasoning to make the final decision. If the indirect interpretation is something that the agent can actually respond to—if it is a question the

LEIA can answer or a request for action that the LEIA can fulfill—then that interpretation is selected. This would be the case, for example, if a chair-building LEIA were told, "I need a hammer" (a request: *Give me a hammer*) or "I wonder if we have any more nails" (which could be a question: *Do we have any more nails?* or a request: *Give me some nails*). By contrast, if the LEIA cannot fulfill the request ("I would love a sandwich right now") or does not know the answer to a question ("I wonder why they need so many chairs built today"), then it chooses the direct interpretation and does not attempt any action.

7.5 Underspecified Known Expressions

What does *good* mean? If a vague answer is enough—and it often is—then *good* indicates a positive evaluation of something. In fact, the independent statements, "Good," "Great," "Excellent," and others have lexical senses whose meaning is "The speaker is (highly) satisfied with the state of affairs." This is important information for a task-oriented LEIA because it implies that no reparative action is needed. By contrast, "What a mess" and "This looks awful" indicate that the speaker is unhappy and that the LEIA might consider doing something about it.

However, there are cases in which a vague expression actually carries a more specific meaning. If I ask someone who knows me well to recommend a *good restaurant*, then I expect him or her to take my location, budget, and preferences for food and décor into account. If someone recommends a *good student* for a graduate program, that student had better be sharp, well prepared, and diligent. And the features contributing to a *good résumé* will be very different for people researching résumé design principles than for bosses seeking new hires. Similarly, as we saw in section 6.3.3, underspecified comparisons—such as *My car is better than this one*—may or may not require that some particular property value(s) be understood.

For applications like recommendation systems, web search engines, and product ratings, notions of *good* and *bad* span populations (they do not focus on individuals) and tend to generalize across features (e.g., a restaurant might get an overall rating of 3 out of 5 even though the food is exceptional). By contrast, in agent applications, the features of individuals—including their preferences, character traits, and mental and physical states—can be key. For example, in the domain of clinical medicine, on which we have worked extensively (see chapter 8), the best treatment for a patient will depend on a range of factors that the LEIA knows about thanks to a combination of its model of clinical knowledge and the features it has learned about the individual during simulation runs (through, e.g., dialog and simulated events). Clearly, this is all highly specific to particular domains, situations, and applications.

7.6 Underspecified Unknown Word Analysis

Up to this point, unknown words have been treated as follows. During Pre-Semantic Integration they were provided with one or more candidate lexical senses that were syntactically specific but semantically underspecified. Then, during Basic Semantic Analysis, the semantic interpretation was narrowed to the extent possible using the unidirectional application of selectional constraints recorded in the ontology. For example, assuming that *kumquat* is an unknown word, given the input, *Kerry ate a kumquat*, the LEIA will understand it to be a FOOD because of its ontological knowledge about possible THEMES of INGEST. Now the question is, *Can knowledge of, and reasoning specific to, a particular domain narrow the interpretation still further?* In some cases, it can.

Let's assume that the agent is operating in a furniture-building domain and receives the utterance "Pass me the Phillips." Let's assume further that it does not have a lexical sense for *Phillips*, or even the full-form *Phillips-head screwdriver.* The agent can narrow down its interpretation to the set of objects that it is able to pass, under the assumption that its interlocutor is operating under sincerity conditions. If the agent understands which action the human is attempting to carry out—something that can be provided either by language ("I'm trying to screw in this screw") or through visual perception—it can narrow the interpretation still further.

7.7 Situational Reference

The objective of situational reference is to anchor all the referring expressions (RefExes) a LEIA encounters to the corresponding object and event instances in its situation model. At this stage of the process, reference resolution transcends the bounds of language and incorporates RefExes obtained as a result of the operation of perception modalities other than language (see figure 7.1). Resolving these extended coreferences is called *grounding* in agent systems. (For other work on grounding see, e.g., Pustejovsky et al., 2017.) Multichannel grounding is likely to require a nontrivial engineering effort because most agent systems will need to import at least some external perception services, whose results need to be interpreted (using the Perception Interpreter Services) and then translated into the same ontological metalanguage used for the agent's knowledge bases and reasoning functions. Stated plainly, developers of integrated agent systems cannot develop all functionalities in-house; they need to incorporate systems developed by experts specializing in all aspects of perception, action, and reasoning; and few R&D teams include all these types of specialists. After object and event instances have been interpreted and grounded, it is a different decision whether or not to store them to episodic memory.

At this stage of NLU, Situational Reasoning, three reference-related processes occur, all in service of the grounding just described: (a) the agent vets the correctness of previously identified sponsors for RefExes using the situational knowledge that is now available, (b) it identifies sponsors in the linguistic or real-world context for RefExes as yet lacking a sponsor, and (c) it anchors the meaning representations associated with all RefExes in the agent's memory. We consider these in turn.

7.7.1 Vetting Previously Identified Linguistic Sponsors for RefExes

Let us begin by recapping coreference processing to this point. Sponsors for many RefExes have been identified using methods that are largely lexico-syntactic. The only kind of semantic knowledge leveraged so far has involved CONCEPT-PROPERTY-FILLER triples recorded for the open-domain ontology—that is, not limited to particular domains for which the agent has been specially prepared. To repeat just two examples from Basic Coreference Resolution: (a) the property HAS-OBJECT-AS-PART provides heuristics for detecting bridging constructions (e.g., ROOM (HAS-OBJECT-AS-PART WINDOW)), and (b) the *default* fillers of case roles suggest preferences for pronoun resolution (e.g., the AGENT of SURGERY should best be a SURGEON).

Now, at this stage, the agent incorporates additional knowledge bases and reasoning to determine whether previously posted coreference decisions are correct. The process incorporates (a) the agent's knowledge/memories of contextually relevant object and event instances; (b) its ontological knowledge, recorded in scripts, of the typical events in the particular domain in which it is operating; and (c) its understanding of what, exactly, it and its human interlocutor are doing at the time of the utterance. The vetting process, as currently modeled, is organized as the following series of five checks.

Check 1. *Do stable properties unify?* We define *stable properties* as those whose values are not expected to change too often. For people, these include MARITAL-STATUS, HEIGHT, HAS-SPOUSE, HAS-PARENT, and so on. For physical objects, they include COLOR, MADE-OF, HAS-OBJECT-AS-PART, and so on. At this point in microtheory development, we are experimenting with *an* inventory of stable properties without assuming it to be the optimal one, and we are well aware of the changeability of practically any feature of any object or event given the right circumstances or the passing of a sufficient amount of time. What this check attempts to capture is the fact that a blue car is probably not coreferential with a red one, and a 6′3″ man is probably not coreferential with a 5′2″ one.

Formally, the agent must first identify which entities in its memory are worth comparing with the entity under analysis; then it must check the value of the relevant property to see if it aligns. The problem, of course, is that although a feature-value conflict can suggest a lack of coreference, lack of a conflict does not ensure coreference. Consider some examples:

(7.1) John doesn't like Rudolph because he's 6′2″ tall.

- CoreNLPCoref corefers *he* and *Rudolph*, which is probably what is intended, but there is no way for the LEIA to know that, since the engine's overall precision in resolving third-person coreference is not extremely high.
- The LEIA checks its memory for Rudolph's height. If his HEIGHT is 6′2″, then the LEIA confirms that the coreference link could be correct—but it need not be, since John could also be 6′2″. If Rudolph's height is known and is something other than

6′2″, then the agent rejects the coreference link. If it doesn't know Rudolph's height, or can't find any Rudolph in its memory, then this check abstains.

(7.2) Madeline would prefer not to barhop with Justine because she's married.

- CoreNLPCoref corefers *she* and *Madeline*, which may or may not be what is intended (this sentence equally allows for either interpretation of the coreference).
- The LEIA checks whether it knows Madeline's marital status. If Madeline is married, then it confirms that the coreference link could be correct, even though it is possible that both women are married. If Madeline is not married, then it rejects the coreference link. In all other cases, this check abstains.

To recap, feature checking cannot confidently assert that a coreference link is correct, but it can exclude some candidate coreference links when the entities' feature values do not unify.

Check 2. *Can known* SOCIAL-ROLES *guide sponsor preferences?* Prior knowledge of people's social roles can help to confirm or overturn previously posited coreference links. For example, given the following inputs and sufficient background knowledge about the individuals in question, the LEIA should be able to confirm their social roles.

(7.3) [The HUMAN referred to by *he* should have the SOCIAL-ROLE PRESIDENT.]
President George H. W. Bush offered "a kinder, gentler" politics. He lasted one term. Clinton called himself "a New Democrat." He got impeached. (COCA)

(7.4) [The HUMAN referred to by *he* should have the SOCIAL-ROLE SURGEON.]
Last week he operated on an infant flown in from Abu Dhabi. (COCA)

Another example in which this check can prove useful is our example from section 5.2.3: *Mike talked at length with the surgeon before he started the operation.* Upstream analysis suggested that *he* should corefer with *the surgeon* on the basis of ontological expectations—and that is true. But we also pointed out that Mike could be an anesthesiologist or a general practitioner preparing to carry out a minor surgery in his office. In the latter case, there is bona fide ambiguity without knowing more about the people involved in the context.

In considering how best to use social roles to guide coreference assignments, there is one important decision to be made: Should the social roles in question be constrained to those mentioned in the given discourse, or should any known social role found in the agent's memory about this individual be invoked? There is no simple answer. People typically have more than one social role. For example, someone whose profession is TEACHER can have any number of other social roles that are more contextually relevant, such as PARENT, COACH, HOMEOWNER, and so on.

Check 3. *Do the case role fillers of known* EVENTS *guide sponsor preferences?* This check will fire only if event coreference was already established—for example, due to a repetition structure. In that case, the agent will prefer that the case roles of coreferential events have parallel fillers (i.e., the same AGENTS, the same THEMES, and so on). Consider in this regard example (7.5).

(7.5) "Roy hit Dennis after Malcolm told off Lawrence." "Why did he hit him?"

- CoreNLPCoref corefers *Malcolm*, *he*, and *him*, which is incorrect.
- The LEIA, instead, establishes the coreference between the instances of *hit* and then lines up their case role fillers: Roy_1 hit_2 Dennis_3 / he_1 hit_2 him_3.

Check 4. *Can domain-specific ontological knowledge guide sponsor preferences?* Let us return to the chair-building domain. If *it* in the utterance *Hit it hard* could refer to either a NAIL or a CHAIR-BACK, and if the furniture-building scripts include many instances of hitting nails and none of hitting chair backs, then the preferred resolution of *it* will be NAIL. (Of course, one *can* need to hit a chair back, but both a human and an agent would be best advised to double-check the speaker's meaning before doing that.)

Yet another case in which domain-specific ontological knowledge can guide sponsor selection involves elided events that, until now, have remained underspecified. Consider the example *Help me*, which is recorded as a lexical sense that detects the elided event (help you *do what*?) but requires situational reasoning to resolve it. The LEIA needs to determine which event in the script its collaborator is pursuing and whether or not it (the LEIA) can assist with it. Of course, detecting its collaborator's current activity requires sensory inputs of a kind we have not yet discussed (see chapter 8), but the basic principle should be clear. Naturally, knowing which subevent of the script is currently being pursued narrows the search space and increases the agent's confidence in its interpretation.

Check 5. *Can some aspect of general knowledge about the world guide sponsor preferences?* This is the class of phenomena illustrated by the Winograd challenge problems (Levesque et al., 2012).[2] For example:

(7.6) a. <u>The trophy</u> doesn't fit into the brown suitcase because <u>it</u> is too large.
b. The trophy doesn't fit into <u>the brown suitcase</u> because <u>it</u> is too small.

(7.7) a. <u>Joan</u> made sure to thank Susan for all the help <u>she</u> had received.
b. Joan made sure to thank <u>Susan</u> for all the help <u>she</u> had given.

(7.8) a. <u>Paul</u> tried to call George on the phone, but <u>he</u> wasn't successful.
b. Paul tried to call <u>George</u> on the phone, but <u>he</u> wasn't available.

The knowledge necessary to support such reasoning can be recorded in the ontology in a straightforward manner: a precondition for A being INSIDE-OF B is that A is smaller than B; a typical sequence of events (a tiny script) is A HELPS B and then B THANKS A; when one tries to do something, one can either succeed or fail; in order for a

COMMUNICATION-EVENT (like calling) to be successful, the person contacted must be available. No doubt, a lot of such knowledge is needed to support human-level reasoning about all domains—something pursued, for example, in the Cyc ontology acquisition effort (Lenat, 1995). However, since for the foreseeable future LEIAs will have this depth of knowledge only for specialized domains, this kind of reasoning is assigned to the current module, all of whose functionalities require knowledge support beyond what is available for the open domain.

7.7.2 Identifying Sponsors for Remaining RefExes

Some of the RefExes that do not yet have a sponsor need to be directly grounded in the physical context.[3] This kind of reference resolution is, strictly speaking, outside the scope of this book. So here we will just briefly comment on how agents interpret nonlinguistic percepts in order to accomplish the physical grounding of objects and events.

As we explained with respect to figure 1.1 in chapter 1, no matter how a LEIA perceives a stimulus—via language, vision, haptics, or otherwise—it must interpret it and record that interpretation in the ontologically grounded metalanguage. The results are stored in knowledge structures that we call XMRs: meaning representations (MRs) of type X, with X being a variable. When the input is text, the XMR is realized as a TMR (a *text* meaning representation), whereas when the input is vision, the XMR is realized as a VMR (a *visual* meaning representation), and so on (see figure 7.1).

All XMRs have a set of generic properties as well as a set of properties specific to their source. The VMR below grounds the event of assembling along with all of the objects filling its case roles. Specifically, it expresses the situation in which the robot has seen its human collaborator assemble a chair leg using a bracket and a dowel. Although the formalism looks somewhat different from the pretty-printed TMRs we have been presenting throughout, it is actually entirely compatible.

```
@INPUTS.VMR.1={
    INSTANCE-OF           @ONT.VMR;
    REFERS-TO-GRAPH       "VMR#1";
    STATUS                "UNDERSTOOD";
    SIGNAL                "INPUT";
    TYPE                  "VISUAL";
    TIMESTAMP             1549721130;
    SOURCE                @SELF.ROBOT.1;
    ROOT                  @VMR#1.ASSEMBLE.1;
};
@VMR#1.ASSEMBLE.1={
    INSTANCE-OF           @ONT.ASSEMBLE;
    AGENT                 @ENV.HUMAN.1;
    THEME                 @ENV.ARTIFACT-LEG.1;
```

```
    INSTRUMENT          @ENV.BRACKET.1;
    INSTRUMENT          @ENV.DOWEL.1;
};
```

The process of interpreting the visual scene sufficiently to generate a VMR is very involved. And even when it is accomplished, this still does not yet fully ground an object or event. That is done when the agent incorporates the VMRs into memory, which is the process to which we now turn.

7.7.3 Anchoring the TMRs Associated with All RefExes in Memory

The last step of reference processing is anchoring the mentions of objects and events perceived in any way in the agent's memory. Memory management is a complex issue, involving decisions such as what to store in long-term memory versus what to discard as unimportant; when to forget previously learned information, if at all, depending on how closely LEIAs are intended to emulate people; and how to merge instances of given types of events (e.g., when several human collaborators teach a robot to perform a particular procedure in similar but not identical ways). These issues are far removed from NLU per se, and we will not discuss them further here.

To recap, during Situational Reasoning, various aspects of reference resolution are addressed: previously posited coreference links are semantically checked; some as-yet ungrounded referring expressions are grounded; and referring expressions are stored in memory if the agent decides to do so.

7.8 Residual Hidden Meanings

The question "Is there a deeper meaning?" is one that LEIAs need to consider but should not pursue too actively, as they could quickly drive their human partners crazy trying to do something about every utterance. Humans think aloud, complain, and engage in phatic exchanges without intending them to be acted on. Utterances often contain no hidden (underlying, implied) meanings, and even if they do, people often miss them. This is made clear by the frequency of such clarifications as "I was actually asking you to help me," "Are you being sarcastic?", "Does she *really* drink twenty cups of tea a day?", and "Come on, I was only joking."

As regards associated linguistic phenomena, we have made a start on modeling the detection of noncanonical indirect speech acts, sarcasm, and hyperbole, but we have not yet ventured into humor.

As-yet undetected indirect speech acts. As we saw in section 4.4, most indirect speech acts are conventionalized and can be detected using constructions recorded in the lexicon: for example, "We need to X," and "I can't do this by myself!" However, some indirect speech acts cannot be captured by lexical senses because there are no invariable words to anchor them in the lexicon. For example, an NP fragment in isolation (i.e., not in a paired discourse pattern, like a question followed by an answer) often means *Give me NP*, as long as the object in question can, in fact, be given to the speaker by the interlocutor. This last check is sufficient to exclude an indirect speech act reading for utterances like "Nuts!" in any context that does not involve either a machine shop (where *nuts* pair with *bolts*) or eating. So, if someone says, "Chair back!" to our chair-building LEIA, it can hypothesize that the user wants to be given a chair back and can see whether that is within its capabilities (it is). We discussed the treatment of bare NPs in section 4.3.4.

Another generalization is that expressing a negative state of affairs can be a request to improve it. Acting on this generalization, however, requires understanding which states of affairs are bad, which are good, and what LEIAs can do to repair the bad ones. For example, if a person tells a furniture-building LEIA, "This nail is too short," then he or she probably wants to be given not just any longer one but one that is the next size longer, if there is such an option. This generalization applies to any multivalued objects that a LEIA can give to its collaborator: one can have large and small hammers, long and short nails, heavyweight and lightweight clamps, and so on. By contrast, if the only chair back that is available is too heavy, then the human needs to figure out what to do about it. In short, much of the reasoning involved is both domain- and task-dependent, and we will approach compiling an inventory of appropriate reasoning rules in bottom-up fashion.

Sarcasm. Although detecting sarcasm might seem like an unnecessary flourish for LEIAs, it can actually have practical importance. As we discuss in chapter 8, mindreading is an important aspect of human communication. People (largely subconsciously) construct models of each other and make decisions based on those models. It actually matters whether "I love mowing the lawn!" means that I really do love it (and, therefore, don't get in the way of my fun) or that I don't love it (and if you don't do it next week I'll be mad). One way of preparing LEIAs to detect at least some realizations of sarcasm is to describe events and states in the ontology as typically desirable or undesirable—which is a default that can, of course, be overridden for a particular individual by concrete information stored to memory.

Hyperbole. People exaggerate all the time. (Get it?) *Grandma drinks twenty cups of tea a day. If you go one-half-mile-an-hour over the speed limit on that street, they'll give you a ticket.* In terms of formal meaning representations, hyperbole is best captured by converting the stated numbers into their respective abstract representations. For our examples, this correlates with *drinking a very large amount of tea* and *going very slightly over the speed limit.*

LEIAs detect exaggerations by comparing the stated value with expectations stored in the ontology—to the extent that the needed information is available. For example, if the ontology says that people are generally not more than seven feet tall, then saying that someone is twenty feet tall is surely an exaggeration. However, although our current ontology includes typical heights of people, it does not cover every type of knowledge, such as normal daily beverage consumption or the minimal speed infraction for getting a ticket. This need for significantly deep knowledge is why we postpone hyperbole detection until the stage of script-based reasoning. If our furniture-building robot is told, "We have to build this chair in two minutes flat!" but the LEIA's script says that the average building time is two hours, then it must interpret this as *very fast*. So, the basic TMR, generated during Basic Semantic Analysis, will include a duration of two minutes, but it can be modified at this stage to convert *two minutes* into the highest value on the abstract scale of SPEED.

As concerns detecting hyperbole, over time, the agent's TMR repository can be useful for this. Imagine that a LEIA encounters the example about Grandma drinking twenty cups of tea a day and has no way of knowing that it is an exaggeration. So the TMR states, literally, that Grandma drinks twenty cups of tea a day. However, a developer might review this TMR (which is always an option), recognize the hyperbole, and change the representation to an abstract indication of quantity—namely QUANTITY 1, which is the highest value on the abstract scale {0,1}. (The lack of a measuring unit is the clue that this is an abstract value.) The agent now has the combination of (a) the original input, (b) its mistaken analysis, and (c) the corrected analysis. This provides the prerequisites for it to reason that if Michael is said to drink twenty Cokes a day, this, too, is an exaggeration. Of course, there is nothing simple about language or the world: after all, it might be fine for a marathon runner training in a hot climate to drink twenty cups of water a day.

7.9 Learning by Reading

Learning by reading is an extension of the new-word learning the agent undertakes during Basic Semantic Analysis. There, the LEIA analyzes the meanings of new words using the semantic dependency structure and knowledge recorded in the ontology. This typically results in a coarse-grained analysis. For example, from the input *Jack is eating a kumquat*, the agent can learn only that *kumquat* is a FOOD—which is a good start, but only a start. To supplement this analysis, the agent can explore text corpora, identifying and processing sentences that contain information about kumquats. Typically, as with *kumquat*, the word has more than one sense—in this case, the word can refer to a tree or its fruit. So the agent needs to first cluster the sentences containing the word (using knowledge-based or statistical methods) into different senses and then attempt to learn the syntactic and semantic features of those senses. It can either link new word senses to the most applicable available concept in the ontology or posit a new concept in the most appropriate position in the ontological graph. What is learned can be used in various ways: it can improve a runtime analysis, it can serve as an intermediate result for semiautomatic knowledge acquisition (in order not to corrupt the quality of the knowledge bases), or it can directly modify the knowledge bases, provided that a necessary threshold of confidence is achieved.

Learning by reading has long been understood as a cornerstone of AI, since it will allow agents to convert large volumes of text into interpreted knowledge that is useful for reasoning. However, it is also among the most difficult problems, as evidenced by past experimentation both within our group (e.g., English & Nirenburg, 2007, 2010) and outside of it (e.g., Barker et al., 2007). At this point in the evolution of our program of NLU, we are preparing agents to learn by identifying and algorithmically accounting for eventualities. High-quality learning is, however, directly dependent on the size and quality of the knowledge bases (especially, lexicon and ontology) that are used to bootstrap the process. No doubt, more manual knowledge engineering is needed before we can expect agents to excel at supplementing those knowledge bases automatically.

8

Agent Applications: The Rationale for Deep, Integrated NLU

The last chapter introduced some of the ways in which NLU is fostered by its integration in a comprehensive agent environment. In fact, it would be impossible to fully appreciate the need for ontologically grounded language understanding without taking into consideration the full scope of interrelated functionalities that will be required by human-level intelligent agents. All these functionalities rely on the availability of high-quality, machine-tractable knowledge, and this reality dwarfs the oft-repeated cost-oriented argument against knowledge-based NLU: that building the knowledge is too expensive. The fact is that agents need the knowledge *anyway.*

The second, equally compelling rationale for developing integrated, knowledge-based systems is that they will enable agents to explain their decisions in human terms, whether they are tasked with teaching, collaborating, or giving advice in domains as critical as defense, medicine, and finance. In fact, *explainable AI* has recently been identified as an important area of research. However, given that almost all recent work in AI has been statistically oriented, the question most often asked has been to what extent statistical systems can *in principle* explain their results to the human users who will ultimately be held responsible for the decision-making.

This chapter describes application areas that have served as a substrate for our program of work in developing LEIAs. As with the language-oriented chapters, the description is primarily conceptual, since specific system implementation details become ever more obsolete with each passing day. The goal of the chapter is to contextualize NLU in overall LEIA modeling without the discussion snowballing into a fundamental treatment of every aspect of cognitive systems.

8.1 The Maryland Virtual Patient System

Maryland Virtual Patient (MVP) is a prototype agent system that provides simulation-based experience for clinicians in training. Specifically, it would allow medical trainees to develop clinical decision-making skills by managing a cohort of highly differentiated virtual patients in dynamic simulations, with the optional assistance of a virtual tutor.

The benefits of simulation-based training are well-known: it offers users the opportunity to gain extensive practical experience in a short time and without risk. For example, "The evaluation of SHERLOCK II showed that technicians learned more about electronics troubleshooting [for US Airforce aircraft] from using this system for 24 hr than from 4 years of informal learning in the field" (Evens & Michael, 2006, p. 375).

Development of MVP followed the demand-side approach to system building, by which a problem is externally identified and then solved using whatever methods can be brought to bear. This stands in contrast to the currently more popular supply-side approach, in which the choice of a method—these days, almost always machine learning using big data—is predetermined, and R&D objectives are shaped to suit.

The physician-educators who conceived of MVP set down the following requirements:

1. It must expose students to virtual patients that demonstrate sophisticated, realistic behaviors, thus allowing the students to suspend their disbelief and interact naturally with them.
2. It must allow for open-ended, trial-and-error investigation—that is, learning through self-discovery—with the virtual patient's anatomy and physiology realistically adjusting to both expected and unexpected interventions.
3. It must offer a large population of virtual patients suffering from each disease, with each patient displaying clinically relevant variations on the disease theme; these can involve the path or speed of disease progression, the profile and severity of symptoms, responses to treatments, and secondary diseases or disorders that affect treatment choices.
4. It must be built on models with the following characteristics:
    a. They must be explanatory. Explanatory models provide transparency to the medical community who must endorse the system. They also provide the foundation for tutoring, since they make clear both the *what* and the *why* of the simulation.
    b. They must integrate well-understood biomechanisms with clinical knowledge (population-level observations, statistical evidence) that bridges the gaps when causal explanations are not available.
    c. They must allow these nonexplanatory *clinical bridges* to be replaced by biomechanical causal chains if they are discovered, without perturbation to the rest of the model.
    d. They must be sufficient to support automatic function and realism, but they need not include every physiological mechanism known to medicine. That is, creating useful applications does not impose the impossible precondition of creating full-blown virtual humans.
5. It must cover diseases that are both chronic and acute and both well and poorly understood by the medical community.

6. It must allow students to have control of the clock—that is, to advance the simulation to the next phase of patient management at will, thus simulating the doctor's choices about when a patient is to come for a follow-up visit.
7. It must offer optional tutoring support that can be parameterized to suit student preferences.
8. It must allow virtual patients to make all kinds of decisions that real patients do, such as when to see the doctor, whether to agree to tests and interventions, and whether to comply with the treatment protocol.

The virtual patients in MVP are *double agents* in that they display both physiological and cognitive function, as shown by the high-level system architecture in figure 8.1.[1] Physiologically, they undergo both normal and pathological processes in response to internal and external stimuli, and they show realistic responses to both expected and unexpected interventions. Cognitively, they experience symptoms, have lifestyle preferences, can communicate with the human user in natural language, have memories of language interactions and simulated experiences, and can make decisions (based on their knowledge of the world, their physical, mental, and emotional states, and their current goals and plans). An optional tutoring agent provides advice and feedback during the simulation. The other medical personnel include the agents that carry out tests and procedures and report their results.

It is noteworthy that the MVP vision and modeling strategy not only fulfill the desiderata for virtual patient models detailed in the National Research Council's 2009 joint report (Stead & Lin, 2009), but they were developed before that report was published. A short excerpt illustrates the overlap:

> In the committee's vision of patient-centered cognitive support, the clinician interacts with models and abstractions of the patient that place the raw data in context and

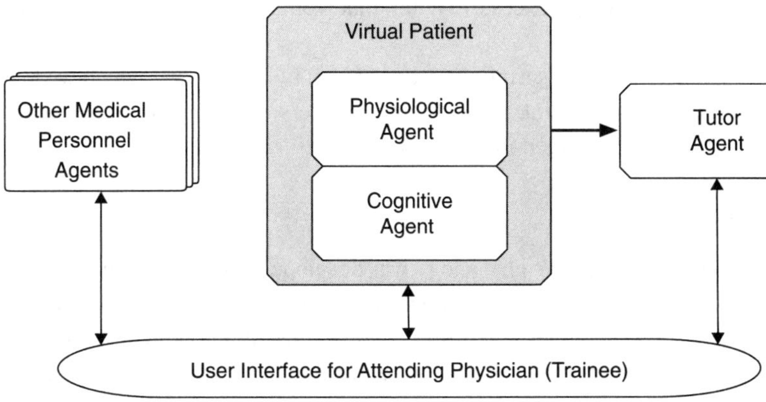

Figure 8.1
The Maryland Virtual Patient (MVP) architecture.

> synthesize them with medical knowledge in ways that make clinical sense for that patient.... These virtual patient models are the computational counterparts of the clinician's conceptual model of a patient. They depict and simulate a theory about interactions going on in the patient and enable patient-specific parameterization and multicomponent alerts. They build on submodels of biological and physiological systems and also of epidemiology that take into account, for example, the local prevalence of diseases. (p. 8)

MVP is a prototype system whose knowledge bases, software, and core theoretical and methodological foundations were developed from approximately 2005 to 2013 (e.g., McShane, Fantry, et al., 2007; McShane, Nirenburg, et al., 2007; McShane, Jarrell, et al., 2008; McShane, Nirenburg, & Jarrell, 2013; Nirenburg, McShane, & Beale, 2008a, 2008b, 2010a, 2010b). The system that was demonstrated throughout that period has not been maintained, but the knowledge bases, algorithms, methodology, and code remain available for reimplementation and enhancement. We refer to the system using the present tense to focus on the continued availability of the conceptual substrate and resources.

The obvious question is, *Why hasn't the work on MVP continued?* The reason is logistical: For a pedagogical system to be adopted by the medical community, large-scale evaluations—even of the prototype—are needed, and this is difficult to accomplish given the levels of funding typically available for research-oriented work. And without a formal evaluation of the prototype, it proved difficult to sustain sufficient funding to expand it into a deployed system. We still believe that MVP maps out an exciting and necessary path toward developing sophisticated, high-confidence, explanatory AI.

The description of MVP below includes the modeling of the virtual patient's physiology and cognition, a sample system run, the under-the-hood traces of system functioning, and a discussion of the extent to which such models can be automatically learned from the literature and extracted from domain experts. The descriptions attempt to convey the nature and scope of the work, without excessive detail that would be of interest only to experts in the medical domain.

Of course, ideally, system descriptions are preceded by demos—of which we had many during the period of development. In lieu of that, readers might find it useful to first skim through the system run described in section 8.1.4.

8.1.1 Modeling Physiology

The model of the virtual patient's physiology was developed in-house using the same ontology and metalanguage of knowledge representation as are used for NLU. Diseases are modeled as sequences of changes, over time, in the values of ontological properties representing aspects of human anatomy, physiology, and pathology. For each disease, some number of conceptual stages is established, and typical values (or ranges of values) for each property are associated with each stage. Values at the start or end of each stage are recorded explicitly, with values between stages being interpolated. Disease models include a combination of fixed and variable features. For example, although the number of stages for a given disease is fixed, the duration of each stage is variable. Similarly, although the values for some physiological properties undergo fixed changes across patients (to ensure that the disease manifests appropriately), the values for other physiological properties are variable within a specified range to allow for different instances of virtual patients to differ in clinically relevant ways.

Roughly speaking, diseases fall into two classes: those for which the key causal chains are well understood and can drive the simulation, and those for which the key causal chains are not known. The models for the latter types of diseases rely on clinical observations about what happens and when (but not why). Most disease models integrate both kinds of modeling strategies in different proportions.

To develop computational cognitive models that are sufficient to support realistic patient simulations in MVP, a knowledge engineer leads physician-informants through the process of distilling their extensive and tightly coupled physiological and clinical knowledge into the most relevant subset and expressing it in the most concrete terms. Not infrequently, specialists are also called on to hypothesize about the unknowable, such as the preclinical (i.e., presymptomatic) stage of a disease and the values of physiological properties between the times when tests are run to measure them. Such hypotheses are, by nature, imprecise. However, rather than permit this imprecision to grind agent building to a halt, we proceed in the same way as live clinicians do: by developing *a* model that is reasonable and useful, with no claims that it is the only model possible or that it precisely replicates human functioning.[2]

The selection of properties to be included in a disease model is guided by practical considerations. Properties are included if (a) they can be measured by tests, (b) they can be affected by medications or treatments, and/or (c) they are central to a physician's mental model of the disease. In addition to using directly measurable properties, we also include abstract properties that foster the creation of a compact, comprehensible model. For example, when the property PRECLINICAL-IRRITATION-PERCENTAGE is used in scripts describing esophageal diseases, it captures how irritated a person's esophagus is before the person starts to experience symptoms. Preclinical disease states are not measured because people do not go to the doctor before they have symptoms. However, physicians know that each disease process has a preclinical stage, which must be accounted for in an end-to-end, simulation-supporting model. Inventing useful, appropriate abstract properties reflects one of the creative aspects of computational modeling.[3]

Once an approach to modeling a disease has been devised and all requisite details have been elicited from the experts, the disease-related events and their participants are encoded in ontologically grounded scripts written in the metalanguage of the LEIA's ontology.[4] MVP includes both domain scripts and workflow scripts. Domain scripts describe basic physiology, disease progression, and responses to interventions, whereas workflow scripts model the way an expert physician would handle a case, thus enabling automatic tutoring.

8.1.2 An Example: The Disease Model for GERD

GERD—gastroesophageal reflux disease—is one of the most common diseases worldwide.[5] It is any symptomatic clinical condition that results from the reflux of stomach or duodenal contents into the esophagus. In laymen's terms, acidic contents backwash into the esophagus because the sphincter between the two—called the lower esophageal sphincter (LES)—is not functioning properly. The acidity irritates the esophagus, which is not designed to withstand such acid exposure.

What follows is a summary of the model for GERD. Even if you choose to skip over the details, do notice that the modeling involves an explanatory, interpretive analysis of physiological and pathological phenomena, reflecting the way physicians think about the disease. This is not merely a compilation of factoids from the medical literature, which would not be sufficient to create an end-to-end, simulation-supporting model.

The development of any model begins by selecting the properties that define it. That selection process is informed by the descriptions provided by domain experts. The description of GERD begins with its cause: one of two abnormalities of the LES. Either the LES has an abnormally low basal pressure (< 10 mmHg) or it is subject to an abnormally large number or duration of so-called *transient relaxations*. Both of these result in the sphincter being too relaxed too much of the time, which increases acid exposure to the lining of the esophagus. Clinically speaking, it does not matter which LES abnormality gives rise to excessive acid exposure; what matters is the amount of time per day this occurs. We record this feature as the property TOTAL-TIME-IN-ACID-REFLUX.

Although TOTAL-TIME-IN-ACID-REFLUX earns its place in the model as the variable that holds the results of the test called pH monitoring, it does not capture—for physicians or knowledge engineers—relative GERD severity. For that we introduced the abstract property GERD-LEVEL. The values for GERD-LEVEL correlate (not by accident) with LES pressure:

- If GERD is caused by a hypotensive (too loose) LES, then the GERD-LEVEL equals the LES pressure. So, a GERD-LEVEL of 5 indicates an LES pressure of 5 mmHg.
- If GERD is caused by excessive transient relaxations, then the GERD-LEVEL reflects the same amount of acid exposure as would have been caused by the given LES pressure. So a GERD-LEVEL of 5 indicates a duration of transient relaxations per day that would result in the same acid exposure as an LES pressure of 5 mmHg.

Key aspects of the model orient around GERD-LEVEL (rather than LES pressure, transient relaxations, or TOTAL-TIME-IN-ACID-REFLUX) because this is much easier to conceptualize for the humans building and vetting the model. For example, as shown in table 8.1, GERD-LEVEL is used to determine the pace of disease progression, with lower numbers (think "a looser LES") reflecting more acid exposure and faster disease progression. (The full list covers the integers 0–10.)

Table 8.1
Sample GERD levels and associated properties

GERD-LEVEL	TOTAL-TIME-IN-ACID-REFLUX **in hours per day**	**Stage duration in days**
10	*less than 1.2*	*a non-disease state*
8	1.92	160
5	3.12	110
3	4.08	60

The conceptual stages of GERD are listed below. Each stage is associated with certain physiological features, test findings, symptom profiles, and anticipated outcomes of medical interventions. All these allow for variability across patients.

1. Preclinical stage: Involves the nonsymptomatic inflammation of the esophagus. It is called *preclinical* because patients do not present to doctors when they have no symptoms.
2. Inflammation stage: Involves more severe inflammation of the esophagus. Symptoms begin.
3. Erosion stage: One or more erosions (areas of tissue destruction) occur in the esophageal lining. Symptoms increase.
4. Ulcer stage: One or more erosions have progressed to the depth of an ulcer. Symptoms increase even more.
5. Post-ulcer stage, which takes one of two paths:
    a. Barrett's metaplasia: A premalignant condition that progresses to cancer (an additional stage) in some patients.
    b. Peptic stricture: An abnormal narrowing of the esophagus due to changes in tissue caused by chronic overexposure to gastric acid. It does not lead to cancer.

Patients differ with respect to the end stage of GERD if it is left untreated. Some lucky individuals will never experience more than an inflamed esophagus; their disease process simply stops at stage 2. By contrast, other patients will end up with esophageal cancer. For those patients progressing to the late stage of the disease, there is a bifurcation in disease path—Barrett's metaplasia versus peptic stricture—for reasons that are unknown.

The ontological scripts that support each stage of simulation include the patient's basic physiological property changes, how the patient will respond to interventions if the user (i.e., a medical trainee) chooses to administer them, and the effects of the patient's lifestyle choices. Sparing the reader the code in which scripts are written, here is an example, in plain English, of how GERD progresses in a particular instance of a virtual patient who is predisposed to having erosion as the end stage of disease. In this example, the disease is left untreated throughout the entire simulation.

- During PRECLINICAL-GERD, the value of the property PRECLINICAL-IRRITATION-PERCENTAGE (an abstract property whose domain is MUCOSA-OF-ESOPHAGUS) increases from 0 to 100.[6]
- When the value of PRECLINICAL-IRRITATION-PERCENTAGE reaches 100, the script for PRECLINICAL-GERD is unasserted and the script for the INFLAMMATION-STAGE is asserted.
- During the INFLAMMATION-STAGE, the mucosal layer of the esophageal lining (recorded as the property MUCOSAL-DEPTH applied to the object ESOPHAGEAL-MUCOSA) is eroded, going from a depth of 1 mm to 0 mm over the duration of the stage.
- When MUCOSAL-DEPTH reaches 0 mm, the script for the INFLAMMATION-STAGE is unasserted, with the simultaneous assertion of the script for the EROSION-STAGE.
- At the start of the EROSION-STAGE, between one and three EROSION objects are created whose DEPTH increases from .0001 mm upon instantiation to .5 mm by the end of the stage, resulting in a decrease in SUBMUCOSAL-DEPTH (i.e., the thickness of the submucosal layer of tissue in the esophagus) from 3 mm to 2.5 mm.
- When SUBMUCOSAL-DEPTH has reached 2.5 mm., the EROSION-STAGE script remains in a holding pattern since the patient we are describing does not have a predisposition to ulcer.

Over the course of each stage, property values are interpolated using a linear function, though other functions could be used if they were found to produce more lifelike simulations. So, halfway through PRECLINICAL-GERD, the patient's PRECLINICAL-IRRITATION-PERCENTAGE will be 50, and three quarters of the way through that stage it will be 75.

The length of each stage depends on the patient's TOTAL-TIME-IN-ACID-REFLUX (see table 8.1). For example, a patient with a GERD-LEVEL of 8 will have a TOTAL-TIME-IN-ACID-REFLUX of 1.92 hours a day and each stage will last 160 days.

Some lifestyle habits, such as consuming caffeine, mints, and fatty foods, increase GERD-LEVEL manifestation in patients who are sensitive to those substances. In the model, if a patient is susceptible to GERD-influencing lifestyle habits and is engaging in those habits, then the effective GERD-LEVEL reduces by one. This results in an increase in acid exposure and a speeding up of each stage of the disease. If the patient is not actively engaging in the habit—for example, he or she might be following the doctor's advice to stop drinking caffeinated beverages—the GERD-LEVEL returns to its basic level. This is just one example of the utility of introducing the abstract property GERD-LEVEL into the model.

Each test that can be run is described in the ontology by the properties it measures, the clinically relevant ranges of values it can return, and expert interpretations of the results (see table 8.6 in section 8.1.5.2). When tests are launched on the patient at any time during the simulation, their results are obtained by the system accessing the relevant feature values from the patient's dynamically changing physiological profile.

We now turn to two aspects of physiological modeling that we incorporated into the model after its initial implementation: (a) accounting for why patients have different end stages of the disease and (b) modeling partial (rather than all-or-nothing) responses to medications. The fact that we could seamlessly incorporate these enhancements, without perturbation to the base model, is evidence of the inherent extensibility of the models developed using this methodology.

Enhancement 1. *Accounting for why patients have different end stages of GERD.* Although it is unknown why patients have different end stages of GERD if the disease is left untreated, physicians have hypothesized that genetic, environmental, physiological, and even emotional factors could play a role.[7] To capture some hypotheses that have both practical and pedagogical utility, we introduced three abstract properties into the model:

- MUCOSAL-RESISTANCE reflects the hypothesis that patients differ with respect to the degree to which the mucosal lining of the esophagus protects the esophageal tissue from acid exposure and fosters the healing of damaged tissue. A higher value on the abstract {0,1} scale of MUCOSAL-RESISTANCE is better for the patient.
- MODIFIED-TOTAL-TIME-IN-ACID-REFLUX combines MUCOSAL-RESISTANCE with the baseline TOTAL-TIME-IN-ACID-REFLUX to capture the hypothesis that a strong mucosal lining can functionally decrease the effect of acid exposure. For example, patients with an average MUCOSAL-RESISTANCE (a value of 1) will have the stage durations shown in table 8.1. Patients with an above-average MUCOSAL-RESISTANCE (a value of greater than 1) will have a lower MODIFIED-TOTAL-TIME-IN-ACID-REFLUX, whereas patients with a below-average MUCOSAL-RESISTANCE (a value of less than 1) will have a higher MODIFIED-TOTAL-TIME-IN-ACID-REFLUX. For example:
    - If a patient's TOTAL-TIME-IN-ACID-REFLUX is 3.12 hours, but the patient has a mucosal resistance of 1.2, we model that as a MODIFIED- TOTAL-TIME-IN-ACID-REFLUX of 2.5 hours (3.12 multiplied by .8), and the disease progresses correspondingly slower.
    - By contrast, if the patient's TOTAL-TIME-IN-ACID-REFLUX is 3.12 hours, but the patient has a MUCOSAL-RESISTANCE of .8, then the MODIFIED-TOTAL-TIME-IN-ACID-REFLUX is 3.75 hours (3.12 multiplied by 1.2), and disease progression is correspondingly faster.
- DISEASE-ADVANCING-MODIFIED-TOTAL-TIME-IN-ACID-REFLUX is the total time in acid reflux required for the disease to manifest at the given stage. This variable permits us to indicate the end stage of a patient's disease in a more explanatory way than by simply asserting it. That is, for each patient, we indicate how much acid exposure is necessary to make the disease progress into each stage, as shown in table 8.2. If the acid exposure is not sufficient to support disease progression into a given stage (as

Table 8.2
Computing, rather than asserting, why patients have different end stages of GERD. Column 2 indicates each patient's MODIFIED-TOTAL-TIME-IN-ACID-REFLUX per day. The cells in the remaining columns indicate the total time in acid reflux needed for GERD to advance in that stage. Cells with gray shading indicate that the disease will not advance to this stage unless the patient's MODIFIED-TOTAL-TIME-IN-ACID-REFLUX changes—which could occur, for example, if the patient took certain types of medications, changed its lifestyle habits, or had certain kinds of surgery.

Patient	MODIFIED TOTAL-TIME-IN-ACID-REFLUX	Preclinical	Inflammation	Erosion	Ulcer	Peptic stricture
John	1.92	1.92	1.92	2.3	2.5	3.12
Fred	2.8	1.92	1.92	2	2.7	3.12
Harry	4.08	1.92	1.92	3	3.5	4.0

shown by cells with gray shading), the patient's disease will be at its end stage. For example, John is a patient whose disease will not progress past the inflammation stage, even if left untreated, because his MODIFIED-TOTAL-TIME-IN-ACID-REFLUX is not high enough to support the erosion stage of GERD. By contrast, Fred's disease will advance into the ulcer stage, and Harry's disease will advance to peptic stricture.

Enhancement 2. *Modeling complete and partial responses to medication.* In order to capture the contrast between complete and partial responses to medications, medication effects are modeled as decreases in MODIFIED-TOTAL-TIME-IN-ACID-REFLUX, as shown in table 8.3. The table indicates the decrease in acid exposure caused by each medication for each patient, along with the resulting MODIFIED-TOTAL-TIME-IN-ACID-REFLUX. Explained in plain English:

- For each day that John takes an H2 blocker, his MODIFIED-TOTAL-TIME-IN-ACID-REFLUX will be 1.42, which is not a disease state. If he already has the disease, healing will occur. The other, more potent medication regimens will also be effective for him.
- For Fred, the H2 blocker is not sufficient to promote complete healing (it brings the MODIFIED-TOTAL-TIME-IN-ACID-REFLUX down to 2.5), but it would be sufficient to not permit his disease to progress to the ulcer stage. Or, if Fred were already in the ulcer stage, the ulcers would heal to the more benign level of erosions. If Fred took a PPI once or twice daily, his MODIFIED-TOTAL-TIME-IN-ACID-REFLUX would be < 1.92, meaning that his esophagus would heal completely over time.
- For Harry, the H2 blocker would barely help at all—he would still progress right through the stricture stage. Taking a PPI once a day would heal ulcers and block late stages of disease. Taking a PPI twice a day would heal the disease completely, unless Harry had already experienced a stricture: there is no nonoperative cure for a peptic

Table 8.3
Modeling complete and partial responses to medications. The reduction in MODIFIED-TOTAL-TIME-IN-ACID-REFLUX is listed first, followed by the resulting MODIFIED-TOTAL-TIME-IN-ACID-REFLUX in brackets.

Patient	MODIFIED-TOTAL-TIME-IN-ACID-REFLUX	H2 blocker	PPI once daily	PPI twice daily
John	1.92	−.5 [1.42]	−1.25 [.67]	−1.5 [.42]
Fred	2.8	−.3 [2.5]	−1 [1.8]	−2.25 [.55]
Harry	4.08	−.1 [3.98]	−.8 [3.28]	−2.2 [1.88]

stricture, a detail that we will not pursue at length here but which is covered in the model (the STRICTURE object generated by the simulation remains a part of the patient's anatomy).

To recap, these enhancements to the original GERD model permit each patient's end stage of disease progression to be calculated rather than asserted, and they permit medications to have varying degrees of efficacy.

One important point remains before we wrap up this overview of disease modeling. Any disease that has known physiological preconditions will arise any time those preconditions are met. For example, say a virtual patient is authored to have the disease achalasia, which is caused by a hypertensive LES (the opposite of GERD). And say a system user chooses to treat the achalasia using a surgical procedure that cuts the LES, changing it from *hyper*tensive to *hypo*tensive. Then the disease processes of GERD will automatically begin because the LES-oriented precondition has been met. There is no need for the person authoring the achalasia patient to say anything at all about GERD. This example illustrates why physiological models should be as causally grounded as possible, particularly as more and more interventions are added to the environment, making available all kinds of side effects outside those pertaining to the given disease.

8.1.3 Modeling Cognition

Virtual patients need many cognitive capabilities. Their language understanding capabilities have already been amply described. Their language generation involves two aspects: generating the content of what they will say, and generating its form. The content derives from reasoning and is encoded in ontologically grounded meaning representations. The form is constructed by templates, which proved sufficient for the prototype stage of this application but would need to be enhanced for a full-scale application system. Two other necessary cognitive capabilities of virtual patients are (a) learning new words and concepts through language interaction and (b) making decisions about action. We consider these in turn.

8.1.3.1 Learning new words and concepts through language interaction Learning is often a prerequisite to decision-making. After all, no patient—real or virtual—should agree to a medical procedure without knowing its nature and risks. Table 8.4 shows a brief dialog, which was demonstrated in the application system, between a virtual patient (P) and the human user playing the role of doctor (D). This dialog features the learning of ontology and lexicon through language interaction in preparation for the patient's decision-making about its medical treatment.

When the virtual patient processes each of the doctor's utterances, it automatically creates text meaning representations that it then uses for reasoning and learning. The text meaning representation for the first sentence is

```
ACHALASIA-1
    EXPERIENCER    HUMAN-1    ("the virtual patient")
```

Table 8.4
Learning lexicon and ontology through language interaction

Dialog	Ontological knowledge learned	Lexical knowledge learned
D: You have achalasia.	The concept ACHALASIA is learned and made a child of DISEASE.	The noun *achalasia* is learned and mapped to the concept ACHALASIA.[8]
P: Is it treatable? D: Yes.	The value for the property TREATABLE in the ontological frame for ACHALASIA is set to *yes*.	
D: I think you should have a Heller myotomy.	The concept HELLER-MYOTOMY is learned and made a child of MEDICAL-PROCEDURE. Its property TREATMENT-OPTION-FOR receives the filler ACHALASIA.	The noun *Heller myotomy* is learned and mapped to the concept HELLER-MYOTOMY.
P: What is that? D: It is a type of esophageal surgery.	The concept HELLER-MYOTOMY is moved in the ontology tree: it is made a child of SURGICAL-PROCEDURE. Also, the THEME of HELLER-MYOTOMY is specified as ESOPHAGUS.	
P: Are there any other options? D: Yes, you could have a pneumatic dilation instead, …	The concept PNEUMATIC-DILATION is learned and made a child of MEDICAL-PROCEDURE.	The noun *pneumatic dilation* is learned and mapped to the concept PNEUMATIC-DILATION.
D: … which is an endoscopic procedure.	PNEUMATIC-DILATION is moved from being a child of MEDICAL-PROCEDURE to being a child of ENDOSCOPY.	
P: Does it hurt? D: Not much.	The value of the property PAIN-LEVEL in PNEUMATIC-DILATION is set to .2 (on a scale of 0–1).	

The patient knows to make ACHALASIA a child of DISEASE in the ontology because the lexical sense it uses to process the input "You have X" asserts that X is a DISEASE. This sense is prioritized over other transitive meanings of the verb *have* because the discourse context is a doctor's appointment and the speaker is a doctor. A similar type of reasoning suggests that a Heller myotomy is some sort of MEDICAL-PROCEDURE. Our short dialog also shows two examples of belief revision: when the virtual patient learns more about the nature of the procedures HELLER-MYOTOMY and PNEUMATIC-DILATION, it selects more specific ontological parents for them, thereby permitting the inheritance of more specific property values.[9]

8.1.3.2 Making decisions about action Virtual patients carry out dynamic decision-making in a style that approximates human decision-making—at least to the degree that we can imagine how human decision-making works. For example, whenever a decision needs to be made, the virtual patient first determines whether it has sufficient information to make it—an assessment that is based on a combination of what it actually knows, what it believes to be necessary for making a good decision, and its personality traits. If it lacks some knowledge it needs to make a decision, it can posit the goal of obtaining this knowledge, which is a metacognitive behavior that leads to learning.

Formally speaking, a goal is an ontological instance of a property, whose domain and range are specified. Goals can appear on the agent's goal agenda in four ways:

- *Perception via interoception.* The moment the patient perceives a symptom, the symptom appears in its short-term memory. This triggers the addition of an instance of the goal BE-HEALTHY onto the agenda. We assume that achieving the highest possible value of BE-HEALTHY (1 on the abstract scale {0,1}) is a universal goal of all humans, and in cases in which it seems that a person is not fulfilling this goal, he or she is simply prioritizing another goal, such as EXPERIENCE-PLEASURE.
- *Perception via language.* Any user input that requires a response from the virtual patient (e.g., a direct or indirect question) puts the goal to respond to it on the agenda.
- *A precondition of an event inside a plan is unfulfilled.* For example, most patients will not agree to an intervention about which they know nothing. So, one of the events inside the plan of decision-making about an intervention is finding out values for whichever features of it are of interest to the individual.
- *The required period of time has passed since the last instances of the events BE-DIAGNOSED or BE-TREATED have been launched.* This models regular checkups and scheduled follow-up visits for virtual patients.

The goal BE-HEALTHY is put on the agenda when a virtual patient begins experiencing a symptom. It remains on the agenda and is reevaluated when (a) its intensity or frequency (depending on the symptom) reaches a certain level, (b) a new symptom arises, or (c) a certain amount of time has passed since the patient's last evaluation of its current state of health, given that the patient has an ongoing or recurring symptom or set of symptoms: that is, "I've had this mild symptom for too long. I should see a doctor."

When making decisions about its health care, the virtual patient considers the following types of features, which are used in the decision-making evaluation functions described below.

1. Its physiological state (particularly the intensity and frequency of symptoms), which is perceived via interoception and remembered in its memory. It is important to note that neither the patient nor the virtual tutor in the MVP system have omniscient knowledge of the patient's physiological state. The simulation system knows this, but the intelligent agents functioning as humans do not.
2. Certain character traits: TRUST, SUGGESTIBILITY, and COURAGE. The inventory can, of course, be expanded as needed.
3. Certain physiological traits: PHYSIOLOGICAL-RESISTANCE, PAIN-THRESHOLD, and the ABILITY-TO-TOLERATE-SYMPTOMS. These convey how intense or frequent symptoms have to be before the patient feels the need to do something about them.
4. Certain properties of tests and procedures: PAIN, UNPLEASANTNESS, RISK, and EFFECTIVENESS. PAIN and UNPLEASANTNESS are, together, considered typical side effects when viewed at the population level. The patient's personal individual experience of them is described below.
5. Two time-related properties: the FOLLOW-UP-DATE, that is, the time the doctor told the patient to come for a follow-up, and the CURRENT-TIME of the given interaction.

Most of these properties are scalar attributes whose values are measured on the abstract scale {0,1}.[10] All subjective features are selected for each individual virtual patient by the patient author. That is, at the same time as a patient author selects the physiological traits of the patient—such as the patient's response to treatments if they are administered—he or she selects certain traits specific to the cognitive agent, as well as the amount of relevant world knowledge that the patient has in its ontology. Two evaluation functions, written in a simple pseudocode, will suffice for illustration.

Evaluation function 1. *SEE-MD-OR-DO-NOTHING.* This function decides when a patient goes to see the doctor, both initially and for follow-up visits.

```
IF FOLLOW-UP-DATE is not set
   AND SYMPTOM-SEVERITY > ABILITY-TO-TOLERATE-SYMPTOMS
   THEN SEE-MD
; This triggers the first visit to the doctor.
```

ELSE IF FOLLOW-UP-DATE is not set
AND SYMPTOM-SEVERITY < ABILITY-TO-TOLERATE-SYMPTOMS
AND the SYMPTOM has persisted > 6 months
THEN SEE-MD
; *A tolerable symptom has been going on for too long.*

ELSE IF there was a previous visit
AND at the time of that visit SYMPTOM-SEVERITY <=.3
AND currently SYMPTOM-SEVERITY >.7
AND (SYMPTOM-SEVERITY – ABILITY-TO-TOLERATE-SYMPTOMS) > 0
THEN SEE-MD
ELSE DO-NOTHING
; *There was a big increase in symptom severity from low to high, exceeding the patient's ability to tolerate these symptoms. This triggered an unplanned visit to the doctor.*

ELSE IF there was a previous visit
AND at the time of that visit SYMPTOM-SEVERITY is between .3 and .7
AND currently SYMPTOM-SEVERITY >.9
AND [SYMPTOM-SEVERITY – ABILITY-TO-TOLERATE-SYMPTOMS] > 0
THEN SEE-MD
ELSE DO-NOTHING
; *There was a big increase in symptom severity from medium to very high, triggering an unplanned visit to the doctor.*

ELSE IF there was a previous visit
AND at the time of that visit SYMPTOM-SEVERITY >.7
AND currently SYMPTOM-SEVERITY >.9
THEN DO-NOTHING
; *Symptom severity was already high at the last visit—do not do an unplanned visit to the doctor because of it.*

ELSE IF the TIME reaches the FOLLOW-UP-TIME
THEN SEE-MD
; *Go to previously scheduled visits.*

ELSE DO-NOTHING

As should be clear, patients with a lower ability to tolerate symptoms will see the doctor sooner in the disease progression than patients with a higher ability to tolerate symptoms, given the same symptom level. Of course, one could incorporate any number of other character traits and lifestyle factors into this function, such as the patient's eagerness to be fussed over by doctors, the patient's availability to see a doctor around its work schedule, and so on. But even this inventory allows for considerable variability across patients—plenty, in fact, to support rigorous training of future physicians.

Evaluation function 2. *AGREE-TO-AN-INTERVENTION-OR-NOT.* Among the decisions a patient must make is whether or not to agree to a test or procedure suggested by the doctor, since many interventions carry some degree of pain, risks, side effects, or general unpleasantness. Some patients have such high levels of trust, suggestibility, and courage that they will agree to anything the doctor says without question. All other patients must decide whether they have sufficient information about the intervention to make a decision and, once they have enough information, they must decide whether they want to (a) accept the doctor's advice, (b) ask about other options, or (c) reject the doctor's advice. A simplified version of the algorithm for making this decision (which suffices for our purposes) is as follows:

IF a function of the patient's TRUST, SUGGESTIBILITY, and COURAGE is above a threshold OR the RISK associated with the intervention is below a threshold (as for a blood test)
  THEN the patient agrees to intervention right away.

ELSE [*] IF the patient feels it knows enough about the RISKS, SIDE-EFFECTS, and UNPLEASANTNESS of the intervention (as a result of evaluating the function DETERMINE-IF-ENOUGH-INFO-TO-EVALUATE)
  AND a call to the function EVALUATE-INTERVENTION establishes that the above risks are acceptable
  THEN the patient agrees to the intervention.

ELSE IF the patient feels it knows enough about the RISKS, SIDE-EFFECTS, and UNPLEASANTNESS of the intervention
  AND a call to the function EVALUATE-INTERVENTION establishes that the above risks are not acceptable
  THEN the patient asks about other options.
    IF there are other options
    THEN the physician proposes them and control is switched to [*].
    ELSE the patient refuses the intervention.

ELSE IF the patient does not feel it knows enough about the intervention (as a result of evaluating the function DETERMINE-IF-ENOUGH-INFO-TO-EVALUATE)
  THEN the patient asks for information about the specific properties that interest it, based on its character traits (e.g., a cowardly patient will ask about RISKS, SIDE-EFFECTS, and UNPLEASANTNESS, whereas a brave but sickly person might only ask about SIDE-EFFECTS).
  IF a call to the function EVALUATE-INTERVENTION establishes that the above RISKS are acceptable

THEN the patient agrees to the intervention.
ELSE the patient asks about other options
IF there are other options
THEN the physician proposes them and control is switched to [*].
ELSE the patient refuses the intervention.

This evaluation function makes use of two functions that we do not detail here, EVALUATE-INTERVENTION and DETERMINE-IF-ENOUGH-INFO-TO-EVALUATE (see Nirenburg et al., 2008a). These details are not needed as our point is to illustrate (a) the kinds of decisions virtual patients make, (b) their approach to knowledge-based decision-making, and (c) the kinds of dialog that must be supported to simulate the necessary interactions.

8.1.4 An Example System Run

To illustrate system operation, we present a sample interaction between a medical trainee named Claire and a virtual patient named Michael Wu. *Sample* is the key word here, as there are several substantially different paths, and countless trivially different paths, that this simulation could take based on what Claire chooses to do. She could intervene early or late with clinically appropriate or inappropriate interventions, or she could do nothing at all; she could ask Mr. Wu to come for frequent or infrequent follow-ups; she could order appropriate or inappropriate tests; and she could have the tutor set to intervene frequently, only in cases of imminent mistakes, or not at all. However, since Mr. Wu is a particular *instance* of a virtual patient, he has an inventory of property values that define him, which put some constraints on the available outcomes of the simulation. His physiological, pathological, psychological, and cognitive profile is established before the session begins, using the patient-creation interface described in section 8.2.5.1.

Psychological traits: trust [.2], suggestibility [.3], courage [.4]

Physiological traits: physiological resistance [.9], pain threshold [.2], ability to tolerate symptoms [.4]

Knowledge of medicine: minimal, meaning that the patient does not know the features of any interventions the user might propose

Disease(s) explicitly authored for this patient:[11] achalasia

Duration of each stage of the disease: preclinical [7 months], stage 1 [7 months], stage 2 [8 months], stage 3 [8 months], stage 4 [9 months]

Response to treatments if they are launched: BoTox [effective, wearing off over 12 months], pneumatic dilation [effective with regression], Heller myotomy [effective permanently]

Claire does not have direct access to any of this information and must learn everything about Mr. Wu through dialog, tests, and procedures.

When Claire launches the simulation, she must wait for Mr. Wu to present to the office. He makes this decision using the decision function in section 8.1.3.2. We use numbers in square brackets to indicate the key points of this simulation run.

[1] Mr. Wu presents with the chief complaint "difficulty swallowing." This is day 361 of the progression of his disease, which includes the preclinical stage and a portion of the first symptomatic stage. (Claire, of course, will not know this temporal information.) Mr. Wu has had symptoms for some time but until now the evaluation function SEE-DOCTOR-OR-DO-NOTHING has returned the answer DO-NOTHING.

[2] When Mr. Wu presents at the office, this initiates the first dialog with Claire. She types in unconstrained English text (note the extensive use of elliptical expressions). Mr. Wu (being a virtual patient) analyzes it into TMRs, makes a decision about how to answer, and generates a response. The first interaction runs as follows:

Claire : So, you have difficulty swallowing?

Mr. Wu : Yes.

Claire : Do you have difficulty swallowing solid food?

Mr. Wu : Yes.

Claire : Liquids?

Mr. Wu : No.

Claire : Do you have chest pain?

Mr. Wu : Yes, but it's mild.

Claire : Any heartburn?

Mr. Wu : No.

Claire : Do you ever regurgitate your food?

Mr. Wu : No.

Claire : How often do you have difficulty swallowing?

Mr. Wu : Less than once a week.

Claire : It's too early to take any action. Please come back in 9 months.

Mr. Wu : OK.

As we see, Claire decides to do nothing—an important kind of decision in clinical medicine, and one that is difficult to teach since a doctor's natural response to a patient asking for help is to do *something.*

[3] After nine months (on day 661 of the disease progression) Mr. Wu comes back for his follow-up. The cognitive simulation engine has regularly been running the evaluation function SEE-DOCTOR-OR-DO-NOTHING (since he is still symptomatic), but it has always returned DO-NOTHING—that is, do not schedule a new appointment before the scheduled

follow-up. Claire again asks Mr. Wu about his difficulty swallowing, chest pain, and regurgitation, using paraphrases of the original formulations (for variety and, in system demonstrations, to show that this is handled well by the NLU component). Mr. Wu responds that he has moderate chest pain, experiences regurgitation a few times a week, and has difficulty swallowing solids daily and liquids occasionally. Note that the progression of difficulty swallowing from solids to liquids is a key diagnostic point that the user should catch: this suggests a motility disorder rather than an obstructive disorder.

[4] Claire posits the hypothesis that Mr. Wu has a motility disorder and advises Mr. Wu to have a test called an EGD (esophagogastroduodenoscopy). Mr. Wu evaluates whether he will accept this advice using the function EVALUATE-INTERVENTION, described in section 8.1.3.2. Since he is concerned about the risks, he asks about them. When Claire assures him that they are extremely minimal, he agrees to the procedure.

[5] A lab technician agent virtually runs the test and delivers the results. This involves querying the physiological model underlying the simulation at the given point in time. A specialist agent returns the results with the interpretation: "Narrowing of LES with a pop upon entering the stomach. No tumor in the distal esophagus. Normal esophageal mucosa." These results include both positive results and pertinent negatives.

[6] Claire reviews the test results, decides that it is still too early to intervene, and schedules Mr. Wu for another follow-up in four months.

[7] When Mr. Wu presents in four months and Claire interviews him, the symptom that has changed the most is regurgitation, which Mr. Wu now experiences every day. Note that throughout the simulation the patient chart is automatically populated with responses to questions, results of tests, and so on, so Claire can compare Mr. Wu's current state with previous states at a glance.

[8] Claire suggests having another EGD and Mr. Wu agrees immediately, not bothering to launch the evaluation function for EGD again since he agreed to it the last time.

[9] Then Claire suggests having two more tests: a barium swallow and esophageal manometry. Mr. Wu asks about their risks (that remains his only concern about medical testing), is satisfied that they are sufficiently low, and agrees to the procedures. Lab technicians and specialist agents are involved in running the tests and reporting results, as described earlier. The barium test returns "Narrowing of the lower esophageal sphincter with a bird's beak," and the manometry returns "Incomplete relaxation of the LES, hypertensive LES, LES pressure: 53."

[10] Claire decides that these test results are sufficient to make the diagnosis of achalasia. She records this diagnosis in Mr. Wu's chart.

[11] Claire suggests that Mr. Wu have a Heller myotomy. He asks about the risks and pain involved. Claire responds that both are minimal. Mr. Wu agrees to have the procedure. Claire tells him to come back for a follow-up a month after the procedure.

[12] Mr. Wu has the procedure.

[13] Mr. Wu returns in a month, Claire asks questions about symptoms, and there are none. She tells Mr. Wu to return if any symptoms arise.

8.1.5 Visualizing Disease Models

If a cognitive modeling strategy and the applications it supports are to be accepted by researchers, educators, and domain experts, it is important that the knowledge substrate be transparent. We cannot expect professors at medical schools to adopt technologies based on opaque knowledge when they are responsible for the competence of the physicians they train. In MVP, the need for transparency was addressed in three ways: (a) by encapsulating each disease model used to author instances of patients with that disease; (b) by organizing in human-readable, tabular form the types of knowledge that extend beyond what is captured by the patient-authoring interface; and (c) by graphically displaying traces of system functioning for purposes of system demonstration. We consider these visualization capabilities in turn. (Note that although all of the visualizations to be described were implemented in interactive interfaces, the reproduction quality of those screenshots was suboptimal, making it preferable to convey the material here using other expressive means. Examples of actual interfaces are available in McShane, Jarrell, et al. (2008) and Nirenburg et al. (2010a), as well as at https://homepages.hass.rpi.edu/mcsham2/Linguistics-for-the-Age-of-AI.html).

8.1.5.1 Authoring instances of virtual patients The virtual patients that users interact with are *instances* that are spawned from a single, highly parameterizable ontological model of patients experiencing the given disease. Authors of patient instances—who could be professors in medical schools, system developers, or even students preparing practice cases for their study partners—create patient instances by selecting particular values for variable features in the model. All patient models include basic information including name, age, gender, height, weight, and select personality traits. Beyond that, the nature of the model depends on the nature of the disease being modeled.

To illustrate the patient-authoring process, we use the esophageal disease called achalasia, introduced earlier. As a reminder, it involves the opposite physiological abnormality as GERD—namely, a *hyper*tensive LES. We switch from GERD to achalasia for two reasons. First, this provides a glimpse into a different class of disease: unlike GERD, achalasia has an unknown etiology, so the disease model derives from population-level clinical observations. Second, the achalasia model is more easily encapsulated using visualizations.

Each patient-authoring session opens with a short description of the disease to refresh the memory of patient authors. Methods of progressive disclosure—for example, displaying a portion of explanatory texts in a relatively small window with scroll bars offering the rest—permit users of different profiles to interact with the interface efficiently. The explanatory texts are not only remedial reminders. Instead, they describe key aspects of the modeling strategy to everyone interacting with the interface—including, importantly, specialists whose mental model might be different from the one implemented in the system.

Table 8.5 shows patient-authoring choices involving stage duration, physiological properties, and symptoms for achalasia. Property values in plain text are fixed across patient instances, whereas those in square brackets are variable. The actual value shown in each set of brackets is the editable default. In the dynamic interface, the legal range of values was shown by rolling over the cell. This amount of variability allows for a wide range of patient profiles while still ensuring that disease progression remains within clinically observed patterns. However, given other teaching goals or new clinical evidence, the choice of variable versus fixed features could be changed with no need to alter the simulation engine. There is more variability in symptom profiles than in the physiological model itself, reflecting the clinical observation that different patients can perceive a given, test-confirmed physiological state in very different ways.

The patient author also indicates—using pull-down menus and editable text fields presented in tables similar to table 8.5—how the virtual patient will respond to treatments, should they be administered at each stage of the disease. Details aside, two of the treatment options—pneumatic dilation and Heller myotomy—have three potential outcomes: unsuccessful, successful with regression, and successful with no regression. In the case of success with regression, the rate of regression is selected by the patient author. The third treatment option, BoTox, always involves regression, but the rate can vary across patients. These authoring options are provided with extensive explanations because only specialists are expected to remember all the details.

To summarize, the patient-authoring interface for each disease provides patient authors with an encapsulated version of the disease model along with the choice space for patient parameterization. It does not repeat all the information about each disease available in textbooks or attempt to elucidate every detail of implementing the simulation engine. The grain size of description—including which aspects are made parameterizable and which physiological causal chains are included in the model—is influenced by the judgment calls of the domain experts participating in system development.

8.1.5.2 The knowledge about tests and interventions Throughout this chapter, we have been presenting ontological knowledge about clinical medicine in tables, which are a method of knowledge representation understandable by domain experts, knowledge engineers, and programmers alike. Specifically, such tables are readable enough to be vetted by domain experts and formal enough to be converted into the ontological metalanguage by knowledge engineers for programmers. Among the kinds of table-based knowledge that are not displayed in the patient-creation interface are those shown in tables 8.6–8.8.

Table 8.6 includes the test name (in some cases, multiple tests can measure the same property), sample results, and a specialist's interpretation of those results. Results are

Table 8.5
Patient-authoring choices for the disease achalasia

	Start	t0	t1	t2	t3	t4
Stage duration (in months)		[12]	[12]	[12]	[12]	[12]
Physiological Properties						
Ratio of relaxing to contracting neurons in the distal esophagus	100/100	80/100	60/100	40/100	20/100	10/100
Basal LES pressure (torr)	[25]	[start-value] +8	[start-value] +16	[start-value] +24	[start-value] +32	[start-value] +40
Residual LES pressure (torr)	0	8	16	24	32	40
Residual LES diameter (cm)	2.0	1.5	1.0	.5	0.0	0.0
Amplitude of contraction during peristalsis	80	65	40	30	20	10
Efficacy of peristalsis	peristalsis	peristalsis	peristalsis	Intermittent	aperistalsis	aperistalsis
Diameter of distal esophagus (cm)	2	2.8	3.6	4.2	5	6
Retained esophageal content (on the abstract scale {0,1})	0	[.1]	[.3]	[.55]	[.85]	[1]
Emptying delay (min)	0	1	5	10	30	35,000[12]
Symptoms						
Difficulty swallowing distal	0	0.1	[1]	[2]	[3]	[4]
Do solids stick?	No	no	Yes	Yes	yes	yes
Do liquids stick?	No	no	No	Yes	yes	yes
Weight loss	0	0	[0]	[0]	[.1]	[.2]
Chest pain	0	0	[.1]	[.3]	[.5]	[.7]
Regurgitation (times/month)	0	0	[0]	[10]	[40]	[70]

Table 8.6
Examples of ontological knowledge about tests relevant for achalasia

Test	Sample results (presented informally)	Specialist's interpretation
EGD or BARIUM-SWALLOW	LES diameter <= .5	"Narrowing of the LES with a pop upon entering the stomach"
EGD or BARIUM-SWALLOW	LES diameter .5–1.5	"Subtle narrowing of the LES"
EGD or BARIUM-SWALLOW	Diameter of lower body of esophagus: 3–4 cm	"Slightly dilated esophagus"
EGD or BARIUM-SWALLOW	Diameter of lower body of esophagus: >= 4 cm	"Moderately dilated esophagus"
ESOPHAGEAL-MANOMETRY	LES pressure at rest: > 45 torr	"Hypertensive LES"
ESOPHAGEAL-MANOMETRY	LES pressure at rest: 35–45 torr	"High-normal LES pressure"
BARIUM-SWALLOW	Duration of swallowing: 1–5 mins	"Slight delay in emptying"
BARIUM-SWALLOW	Duration of swallowing: > 5 mins	"Moderate-severe delay in emptying"

Table 8.7
Examples of knowledge that supports clinical decision-making about achalasia, which is used by the virtual tutor in the MVP system

PROPERTY	Values (presented in plain English for readability)
SUFFICIENT-GROUNDS-TO-DIAGNOSE	All three of the following conditions: 1. Either a bird's beak (a visual test finding) or a hypertensive LES 2. Aperistalsis 3. Negative esophagogastroduodenoscopy (EGD) for cancer (i.e., a pertinent negative)
SUFFICIENT-GROUNDS-TO-TREAT	Definitive diagnosis

Table 8.8
MVP. Knowledge about the test results expected at different stages of the disease achalasia. Used by the tutoring agent in MVP. The test results in italics are required to definitively diagnose the disease.

Test Name	t1	t2	t3	t4
EGD	Dilated esophagus & no tumor at the GI junction		Dilated esophagus *& narrowing and pop upon entering LES (i.e., a hypertensive LES)* & retained debris *& no tumor at the GI junction*	
Esophageal Manometry	Incomplete relaxation of the LES & high-normal LES pressure	Incomplete relaxation of the LES & high-normal or hypertensive LES & intermittent peristalsis	Incomplete relaxation of the LES *& hypertensive LES* *& aperistalsis*	

Table 8.9
Inventory of under-the-hood panes that are dynamically populated during MVP simulation runs

Physiology	Interoception	Thoughts	Knowledge learned	TMRs
A list of disease-relevant property value pairs, with values being highlighted every time they change during the simulation. Reflects an omniscient view of the patient's physiology.	A list of the virtual patient's perceived symptoms as property value pairs. Every time a symptom appears or changes, a new entry is posted.	Dynamically populated traces of the patient's decision functions, rendered in plain English for readability. E.g., "I don't know the risks of EGD. I'd better ask about them."	Traces of words and concepts learned through dialog. The words are mapped to concepts, and the concepts are placed appropriately in the ontological hierarchy.	Text meaning representations of the virtual patient's interpretations of the user's inputs during the simulated doctor-patient interactions.[13]

expressed informally for readability. The actual results are written in the ontological metalanguage.

Our point in presenting these tables is to emphasize that it is important to make the collaboration between the domain experts, knowledge engineers, and programmers explicit, organized, and easily modifiable. In other words, the knowledge representation must serve many masters.

8.1.5.3 Traces of system functioning Dynamic traces of system functioning are shown in what we call the under-the-hood panes of MVP. The inventory of panes is shown in table 8.9, along with brief descriptions of what they contain. The panes are presented as columns since this is how they are rendered in the demonstration system—that is, all of them can be viewed at the same time. (As a reminder, sample screenshots are shown at https://homepages.hass.rpi.edu/mcsham2/Linguistics-for-the-Age-of-AI.html.)

The under-the-hood panes of the MVP environment are key to showing that the simulations are real—there is no hand-waving; nothing is hidden. They can also be used to pedagogical ends, for example, by allowing students to view the physiological changes during disease progression and the effects of medical interventions.

8.1.6 To What Extent Can MVP-Style Models Be Automatically Learned from Texts?

In the current climate of big data and machine learning, a natural question is, *To what extent can models like these be automatically learned from texts?*[14] The answer: Only very partially. Full models cannot be automatically learned, or even cobbled together by diligent humans, from the literature because they do not exist in the literature. However, we think that some *model components* could be automatically extracted. We define model components as ontologically grounded property value pairs that contribute to full models. Learnable properties have the following characteristics:

- They are straightforward and concrete, such as LES-PRESSURE (measurable by a test) and SENSITIVITY-TO-CAFFEINE (knowable based on patient reports). Learnable properties cannot be abstract, like our MODIFIED-TOTAL-TIME-IN-ACID-REFLUX or MUCOSAL-RESISTANCE, because abstract properties will certainly have no equivalents in published texts.
- They are known to be changeable over time, based on our ontological knowledge of the domain. For example, since we know that new medications and tests are constantly being invented, we know that the properties TREATED-BY-MEDICATION and ESTABLISHED-BY-TEST must have an open-ended inventory of values. By contrast, we do not expect to have to change the fact that heartburn can be a symptom of GERD or that HEARTBURN-SEVERITY is best modeled as having values on the abstract scale {0,1}.
- They describe newly discovered causal chains that can replace clinical bridges in a current model. By contrast, if the model already includes causal chains that fully or partially overlap, their modification is likely to be too complex to be learned automatically without inadvertently perturbing the model.[15]

Table 8.10 shows some examples of properties—associated with their respective concepts—whose values we believe could be learned from the literature.

In order for a model component to be fully learned, the property and the fillers for its domain and range must be ontological entities, not words of language. LEIAs can, in principle, produce these using NLU. For example, all of the following text strings, and many more, will result in text meaning representations that include the knowledge GASTROESOPHAGEAL-REFLUX-DISEASE (HAS-TREATMENT PROTON-PUMP-INHIBITOR):

- A proton pump inhibitor treats <can treat, can be used to treat, can be prescribed to treat, is often prescribed to treat> GERD.
- GERD is <can be> treated <cured> by (taking) a proton pump inhibitor.
- Doctors <Your doctor may> recommend <prescribe> (taking) a proton pump inhibitor to treat GERD symptoms.
- If you have GERD, you might <may> be advised to take a proton pump inhibitor.

Table 8.10
Examples of properties, associated with their respective concepts, whose values can potentially be automatically learned from the literature

Concept	Properties
DISEASE	HAS-EVENT-AS-PART, AFFECTS-BODY-PART, CAUSED-BY, HAS-SYMPTOMS, HAS-DIAGNOSTIC-TEST, HAS-TREATMENT
DIAGNOSTIC-TEST	MEASURES-PROPERTY, NORMAL-RESULT, ABNORMAL-RESULT, SIDE-EFFECTS, PAIN-INDUCED
MEDICAL-TREATMENT	HAS-EVENT-AS-PART, EFFICACY, HAS-RISKS, PAIN-INDUCED

Establishing the functional equivalence of these strings would not be done by listing. Instead, it would be done by combining our general approach to natural language understanding with methods for paraphrase detection and ontologically grounded reasoning.[16]

Let us consider just three examples of how natural language understanding could support the automatic learning of disease model components. Assume that the LEIA is seeking to automatically learn or verify the correctness of the previously discussed fact GASTROESOPHAGEAL-REFLUX-DISEASE (HAS-TREATMENT PROTON-PUMP-INHIBITOR). As we said, all the inputs above provide this information, albeit some more directly than others. The input *GERD is treated by a proton pump inhibitor* perfectly matches the lexical sense for the verb *treat* that is defined by the structure DISEASE *is treated by* MEDICATION, and the analyzer can generate exactly the text meaning representation we are seeking: GASTROESOPHAGEAL-REFLUX-DISEASE (HAS-TREATMENT PROTON-PUMP-INHIBITOR).

In other cases, the basic text meaning representation includes additional information that does not affect the truth value of the main proposition. For example, the potential modality scoping over the proposition *GERD can be treated by a proton pump inhibitor* does not affect the truth value of the main proposition, which is the same as before and matches the expectation we seek to fill.

In still other cases, the meaning we are looking for must be inferred from what is actually written. For example, the input *Your doctor may recommend a proton pump inhibitor* does not explicitly say that a proton pump inhibitor treats GERD, but it implies this based on the general ontological knowledge that a precondition for a physician advising a patient to take a medication is DISEASE (HAS-TREATMENT MEDICATION). Because a LEIA's language understanding system has access to this ontological knowledge, it can be taught to make the needed inference and fill in our slot as before. It should be noted that these types of reasoning rules are not spontaneously generated—they must be recorded, like any other knowledge. However, once recorded, they can be used for any applicable reasoning need of the agent.

When we were investigating what information could be extracted from medical texts in service of disease-model development, we focused on two genres that offer different opportunities for knowledge extraction: case studies and disease overviews.

Case studies do not present all disease mechanisms. Instead, they typically begin with a broad overview of the disease to serve as a reminder to readers who are expected to be familiar with it. Then they focus on a single new or unexpected aspect of the disease as manifest in one or a small number of patients. For example, Evsyutina et al.'s (2014) case study reports that a mother and daughter both suffer from the same rare disease, achalasia, and suggests that this case supports previous hypotheses of a genetic influence on disease occurrence. The new findings are typically repeated in the abstract, case report, and discussion sections, offering useful redundancy to improve system confidence.

A LEIA could set to the task of comparing the information in a case study with the ontologically grounded computational model as follows. First it could semantically analyze the case study, focusing on the TMR chunks representing the types of learnable property values listed above. (This focusing means that the system need not achieve a perfect analysis of every aspect of the text: it knows what it is looking for.) Then, it could compare the learned property values with the values in the model. Continuing with our example of mother-daughter achalasia, our current model of achalasia has no filler for the value of CAUSED-BY since, when we developed the model, the cause was not definitively known (it still is not; the genetic influence remains to be validated). Automatically filling an empty slot with a new filler can be carried out directly, with no extensive reasoning necessary. However, the nature of that slot filler must be understood: in the context of a case study, it represents an instance, not a generic ontological fact. The system has two sources of evidence that this information is an instance: (a) the individuals spoken about are instances, so the features applied to them are also instances (compare this with assertions about people in general), and (b) the genre of case study sets up the expectation that reported information will be at the level of an instance.

Such analysis could, for example, be folded into an application to alert clinicians to new findings in a snapshot formalism like the one shown below (invented for illustration):

Journal article: "Meditation as medication for GERD"

Contribution type: Case study
Author: Dr. Joseph Physician
Date: *Some future date*

GERD Therapies:
Non-medical: lifestyle modifications, **MEDITATION-new**
Mild: H2 blocker, PPI QD
Severe: PPI BID

This presentation style encapsulates the following expectations:

1. Clinicians know, without explanation, that one of the ontological properties of diseases is that they have therapies.
2. When providing new information, it is useful to provide old information as the backdrop, with a clear indication of whether the new information adds to or overwrites the old information.
3. Clinicians understand that information provided in case studies represents instances and not across-the-board generalizations.
4. Modern-day users understand that entities can be clicked on for more information (e.g., which lifestyle modifications are being referred to).
5. Terseness is appreciated by busy people operating within their realm of specialization.

Let us turn now to the other genre from which model information can be extracted: disease overviews. Disease overviews typically present a stable inventory of properties of interest, often even introduced by subheadings, such as causes of the disease, risk factors, physiological manifestations, symptoms, applicable tests and procedures, and so on. Not surprisingly, these categories align well with the knowledge elements we seek to extract from texts, shown in table 8.10. The natural language processing of disease overviews would proceed as described for case studies. However, we envision applications for this processing to be somewhat different. For example, an application could respond to a clinician's request for a thumbnail sketch of a disease by reading overviews, populating the inventory of key property values, and presenting them in a semiformal manner, such as a list of concept-property-value triples.

To wrap up this section on learning components of disease models, note how different the sketched approaches are from statistically oriented knowledge extraction. Our goal would be to speed up, and dynamically enhance, cognitively inspired disease models, not extract uninterpreted text strings into templates that have no connection to ontologies or related cognitive models.

8.1.7 To What Extent Can Cognitive Models Be Automatically Elicited from People?

Text processing is only one of the available methods of reducing the role of knowledge engineers in the process of domain modeling. Another is to guide domain experts through the process of recording components of disease models using a mixed-initiative computer system.[17] The results can then seed the collaborative process between the experts and knowledge engineers.[18]

The strategy for the methodology we describe below, OntoElicit, was informed by two things: lessons learned from developing the first several disease models for MVP through unstructured and semistructured interviews with domain experts, and our past work on a mixed-initiative knowledge elicitation system in a different domain—machine translation.

Let us present just a passing introduction to the latter. The Boas system (McShane et al., 2002; McShane & Nirenburg, 2003) was designed to quickly gather machine-tractable knowledge about lesser-studied languages from native speakers of those language without the assistance of linguists or system developers. The results of the knowledge-elicitation process had to directly feed into a machine translation system from that language into English. By "directly feed into" we mean that, once the user supplied the requested information, he or she pushed a button, waited a minute, and ended up with a translation system. Since developers were completely out of the loop once they delivered the environment to the user, the elicitation process and associated interface had to ensure that the necessary knowledge would get recorded in the right way and that the environment itself provided users with sufficient pedagogical support. The informants, for their part, were not expected to have any formal linguistic knowledge, just the ability to read and write the language in question, as well as a functional knowledge of English. The automatically elicited knowledge was not identical to what could be crafted if knowledge engineers were involved in the process, but it was sufficient to enable basic machine translation capabilities to be configured in this way. Change the domain from language to medicine, the experts from native speakers to physicians, and the goal from machine translation to seeding ontological models for clinical medicine, and Boas smoothly morphs into OntoElicit.

The knowledge elicitation methods of OntoElicit, shown below, will look familiar to readers as they share much in common with the patient-authoring interfaces and knowledge representation schemes described earlier.

In OntoElicit, domain experts are asked to divide the disease into any number of conceptual stages correlating with important events, findings, symptoms, or the divergence of disease paths among patients. They are also asked to indicate the typical duration of each stage as a range (x–y in table 8.11) with a default value (d). Next, they are led through the process of describing the relevant physiological and symptom-related properties during each stage. They can either record all information directly in a table like table 8.11 or be led through a more step-by-step process that results in a summary like table 8.11. Following a practice we invented for Boas, we call the former the *fast lane* and the latter the *scenic route*. Both paths offer links explaining the *why* and *how* of the associated decision-making, as well as examples.

In describing tests and their results, the expert indicates the test name, alternative names, which physiological properties are measured, clinically relevant ranges of results, the specialist's interpretation of those ranges (e.g., "Suggestive of disease X"), clinical guidelines regarding ordering the test, and diseases for which the test is appropriate.

For interventions, including medications, the expert indicates which properties and/or symptoms are affected by the intervention, the possible outcomes of the intervention, possible side effects, and, if known, the percentage of the population expected to have each outcome and side effect.

Table 8.11
Fast-lane elicitation strategy for recording information about physiology and symptoms

	Properties	Start value	Stage 1	Stage 2	…
Duration			x–y (d)	x–y (d)	
Physiology	P1	x–y (d)	x–y (d)	x–y (d)	
	P2	x–y (d)	x–y (d)	x–y (d)	
	…				
Symptoms	S1	x–y (d)	x–y (d)	x–y (d)	
	S2	x–y (d)	x–y (d)	x–y (d)	
	…				

Note: "x–y" indicates the acceptable range of values; (d) indicates the default. If "x" and "y" are nonnumerical, they are presented as a list.

As concerns recording knowledge about clinical practices—that is, the knowledge to support automatic tutoring—two different functionalities must be supported: checking the validity of a clinical move, which is relatively simple and relies on the knowledge of *preconditions of good practice*, and advising what to do next, which can range from simple to very complex.

The knowledge about preconditions of good practice is readily encoded using ontological properties. For example, for each disease, we record values for properties such as SUFFICIENT-GROUNDS-TO-SUSPECT, SUFFICIENT-GROUNDS-TO-DIAGNOSE, and SUFFICIENT-GROUNDS-TO-TREAT (e.g., clinical diagnosis or definitive diagnosis). Similar inventories of properties are used for tests, treatments, making definitive diagnoses, and so on. The content of this knowledge is both broader and deeper than that available in published "best practices" guides. OntoElicit uses tables for eliciting this information (see table 8.12), with the experts providing prose descriptions of property fillers. These descriptions are then converted—like all other aspects of acquired knowledge—into formal, ontologically grounded structures by knowledge engineers and programmers.

As concerns clinical knowledge about what to do next, things can get complicated quickly. Many clinical moves must be decided (a) in the face of competing conditions, (b) with different preferences of different stakeholders (e.g., the patient, the physician, the insurance company), and (c) using incomplete knowledge of relevant property values. For those cases, we have experimented with the use of Bayesian networks that are constructed with the help of influence diagrams.[19] The knowledge encoded in influence diagrams represents an expert's opinion about the utility scores (i.e., the preference level, or "goodness") of different combinations of property values associated with each possible decision. One of the main reasons why we chose to work with influence diagrams is that the kind of information required of experts is of a nature that they can readily conceptualize. In essence, they are asked: *Given this combination of property values, how good is solution Y? Given this other combination of property values, how good is solution Z?* and so on. The properties and values are familiar to our experts because they are the same

Table 8.12
Sample precondition of good practice. Domain experts supply the descriptive fillers and knowledge engineers convert it into a formal representation.

DISEASE	ACHALASIA
PROPERTY	SUFFICIENT-GROUNDS-TO-SUSPECT
Descriptive filler	solid and liquid dysphagia or regurgitation
Formal encoding	(or (and ((SOLIDS-STICK HUMAN YES) (LIQUIDS-STICK HUMAN YES)) (REGURGITATION-FREQUENCY HUMAN (> 0)))))

ones used to build the other models in the system. Knowledge engineers help experts to organize the problem space into subproblems, as applicable, and to develop a case-specific methodology of filling out the utility tables in the most efficient way.

Although the nature of information required of experts in an influence-diagram-driven methodology is straightforward, one problem is that the number of features involved in making a complex decision can be large, easily driving the number of feature-value permutations into the tens or hundreds of thousands. As in all aspects of modeling, we approach this problem using realistic strategies including the following:

1. We organize the knowledge optimally—for example, covering as many variables as possible using local decisions whose output contributes to a more general decision.
2. We simplify the problem space and judge whether the results are sufficient to yield realistic, accurate functioning—for example, not including every property we can think of but, instead, focusing on those considered to have the most impact by clinicians.
3. We work toward automating the process of knowledge acquisition—for example, using functions to provide values for many of the feature-value combinations once a pattern of utility scores has been recognized.[20]

As regards incorporating aspects of influence diagram creation into OntoElicit, our thinking is that experts could, in fact, be led through the process of decomposing the problem into the main variables in the decision versus the variables in local decisions.

We have not yet experimented with how far we can push a mixed-initiative elicitation strategy in the domain of clinical medicine. However, considering that we covered a lot of ground with the Boas predecessor, and considering that the realm of language description is arguably no easier than clinical medicine, we believe that this approach has great potential to be useful.

This wraps up our discussion of the MVP application which was, as we mentioned, implemented at a prototype level. We now move to a model that has not yet been implemented but relies largely on the same knowledge substrate as MVP.

8.2 A Clinician's Assistant for Flagging Cognitive Biases

Cognitive bias is a term used in the field of psychology to describe distortions in human reasoning that lead to empirically verified, replicable patterns of faulty judgment. Cognitive biases result from the inadvertent misapplication of necessary human abilities: the ability to simplify complex problems, make decisions despite incomplete information (called decision-making under uncertainty), and generally function under the real-world constraints of limited time, information, and cognitive capacity (cf. Simon's [1957] theory of bounded rationality). Factors that contribute to cognitive biases include, nonexhaustively: overreliance on one's personal experience as heuristic evidence; misinterpretations of statistics; overuse of intuition over analysis; acting from emotion; the effects of fatigue; considering too few options or alternatives; the illusion that the decision-maker has more control over how events will unfold than he or she actually does; overestimation of the importance of information that is easily obtainable over information that is not readily available; framing a problem too narrowly; and not recognizing the interconnectedness of multiple decisions. (For further discussion see, e.g., Kahneman, 2011; Korte, 2003.)

Even if one recognizes that cognitive biases could be affecting decision-making, their effects can be difficult to counteract. As Heuer (1999, Chapter 9) writes, "Cognitive biases are similar to optical illusions in that the error remains compelling even when one is fully aware of its nature. Awareness of the bias, by itself, does not produce a more accurate perception. Cognitive biases, therefore, are, exceedingly difficult to overcome."

However, the fact that a problem is difficult does not absolve us from responsibility for solving it. Biased thinking can have detrimental consequences, particularly in a high-stakes domain like clinical medicine. We hypothesize that at least some errors in judgment caused by some cognitive biases could be reduced if LEIAs serving as clinician advisors were able to detect potentially biased decisions and generate explanatory alerts to their human collaborators. Even such partial solutions to very difficult problems have the potential to offer rewards at the societal level.

The bias-related functionalities we will address and the psychological phenomena they target are summarized in table 8.13.[21]

Table 8.13
Functionalities of a bias-detection advisor in clinical medicine

Advisor functionalities	Targeted decision-making biases
Memory support: Supplying facts the clinician requests using text generation, structured presentation of knowledge (e.g., checklists), process simulation, and so on	• Depletion effects
Detecting and flagging potential clinician biases	• Illusion that more features are better • False intuitions • Jumping to conclusions • Small sample bias • Base-rate neglect • Illusion of validity • Exposure effect
Detecting and flagging potential patient biases	• Framing sway • Halo effect • Exposure effect • Effects of evaluative attitudes

In discussing each class of bias, we will present (a) a theory of how to model cognitive support to avoid the bias, which involves the selection of properties and values to be treated (e.g., bias types), detection heuristics, decision functions, and knowledge support, and (b) the descriptive realization of the theory as a set of models compatible with LEIA modeling overall.

8.2.1 Memory Support for Bias Avoidance

Memory lapses are unavoidable in clinical medicine due to not only the large amount of knowledge that physicians must manipulate but also *depletion effects*—that is, the effects of fatigue. We believe that depletion effects could be decreased with timely, ergonomically presented reminders, cribs, and checklists (Gawande, 2009) that reflect particular aspects of the knowledge already available in a LEIA's expert models. This type of cognitive assistance would be user initiated, meaning that the user must recognize his or her own potential to misremember or misanalyze something in the given situation, as might happen under conditions of sleep deprivation (Gunzelmann et al., 2009). Consider just a few situations in which a LEIA's knowledge could be leveraged to counter clinician memory lapses. Let us use as our example primary care physician Dr. Allegra Clark.

Example 1. It's the end of the day, Dr. Clark is tired, and she forgets some basic ontological properties of a disease or treatment. She queries the LEIA with an English string such as, *What are the symptoms of achalasia?* The LEIA semantically analyzes this input, converting it into the following text meaning representation.

```
REQUEST-INFO-1
   AGENT      PHYSICIAN-1
   THEME      ACHALASIA.CAUSES-SYMPTOM-1
```

This TMR says that this input is requesting the fillers of the CAUSES-SYMPTOM property of ACHALASIA. The LEIA can answer the question by looking up the needed information in its ontology, the relevant portion of which is shown below.

```
ACHALASIA
   CAUSES-SYMPTOM   sem   CHEST-PAIN, DYSPHAGIA, REGURGITATE, WEIGHT-LOSS
```

Example 2. Dr. Clark wants to order the test called EGD (esophagogastroduodenoscopy) but forgets what preconditions must hold to justify this. She queries the LEIA with *What's needed to diagnose achalasia?* As before, the LEIA translates the input into a TMR and understands that the answer will be the filler of the property SUFFICIENT-GROUNDS-TO-DIAGNOSE in the ontological concept ACHALASIA. Table 8.14 shows a subset of properties of the disease ACHALASIA that relate to diagnosis and treatment. For presentation purposes, the property values in the right-hand column are presented in plain English rather than the ontological metalanguage.

Table 8.14
Four clinical properties of the esophageal disease achalasia, with values written in plain English for readability

PROPERTY	Values
SUFFICIENT-GROUNDS-TO-DIAGNOSE	All three of the following: 1. Either a bird's beak (a visual test finding) or a hypertensive lower esophageal sphincter (LES) 2. Aperistalsis 3. Negative esophagogastroduodenoscopy (EGD) for cancer (i.e., a pertinent negative)
SUFFICIENT-GROUNDS-TO-SUSPECT	Either: 1. Dysphagia (difficulty swallowing) to solids and liquids 2. Regurgitation
SUFFICIENT-GROUNDS-TO-TREAT	Definitive diagnosis
PREFERRED-ACTION-WHEN-DIAGNOSED	Either: 1. HELLER-MYOTOMY (a surgical procedure) 2. PNEUMATIC-DILATION (an endoscopic procedure)

Example 3. Dr. Clark knows that the disease achalasia can have different manifestations in different patients but forgets the details and asks the LEIA to display the ontologically grounded disease model for achalasia. As explained earlier, all disease models are available in the human-inspectable formats shown in section 8.1.5.1, which the LEIA displays.
Example 4. A patient asks Dr. Clark for a prognosis, but she is too tired, too rushed, or not familiar enough with the disease to provide a well-motivated answer. The LEIA could help by permitting her to run one or more simulations of virtual patients that are constrained by the known features of the human in question. This will make the sample simulations as predictive as possible given the coverage and accuracy of the underlying models.

The above four examples should suffice to convey our main point: the knowledge structures and simulation capabilities already developed for the MVP application can be directly reused to help clinicians to counteract memory lapses or knowledge gaps. For this category of phenomena, developing models that take into account biases involves anticipating the requests of clinicians and optimizing the presentation of already available knowledge to make it easily interpretable by them. The initiative for seeking this class of bias-avoidance support lies in the hands of the clinician-users. By contrast, solutions for the remaining two groups of phenomena will proactively seek to detect decision-making biases on the part of both participants in clinician-patient interactions.

8.2.2 Detecting and Flagging Clinician Biases

Diagnosing a patient typically begins with a patient interview and a physical examination. Next, the clinician posits a hypothesis and then attempts to confirm it through medical testing or trial therapy (e.g., lifestyle changes or medication). Confirming a hypothesis by testing leads to a definitive diagnosis, whereas confirming a hypothesis by successful therapy leads to a clinical diagnosis. Unintentionally biased decision-making by the clinician can happen at any point in this process.

The "need more features" bias. When people, particularly domain experts, make a decision, they tend to think that it will be beneficial to include more variables to personalize or narrowly contextualize it. As Kahneman (2011, p. 224) writes, "Experts try to be clever, think outside the box, and consider complex combinations of features in making their predictions. Complexity may work in the odd case, but more often than not it reduces validity. Simple combinations of features are better."[22] One point at which clinicians might erroneously—and at great expense—believe that more feature values are necessary is during diagnosis: they might not recognize that they already have sufficient information to diagnose a disease. For many diseases, clear diagnostic criteria exist, like that shown in the first row of table 8.14. If the patient chart shows sufficient evidence to diagnose a disease, but the clinician has not posited the diagnosis and has ordered more tests, a LEIA could issue an alert about the possible oversight.

Jumping to conclusions. The opposite of seeking too many features is jumping to conclusions, as by diagnosing a disease without sufficient evidence. Typically, each disease has a constellation of findings that permit a clinician to definitively diagnose it. For example, the disease achalasia can be definitely diagnosed by the combination of italicized test results shown in the third and fourth disease stages shown in table 8.15. Positing a diagnosis prior to obtaining the full set of definitive values could be incorrect. Whenever a clinician posits a diagnosis, a LEIA could double-check the patient's chart for the known property values and issue an alert if not all expected property values are attested.

False intuitions. Without entering into the nuanced debate about the nature and formal validation of expert intuition—as pursued, for example, in Kahneman and Klein (2009)—we define skilled intuition as the recognition of constellations of highly predictive property values based on sufficient past experience. Nobody can have reliable intuitions (a) about unknowable situations, (b) in the absence of reliable feedback, or (c) without sufficient experience.

We can operationalize the notion of intuition in at least two ways. The simpler way is to leverage only and exactly the knowledge recorded in tables like the ones above, which would assume that they exhaust valid medical knowledge. A more sophisticated approach would be to incorporate a LEIA's knowledge of the past history of the physician into its decision-making about the likelihood that the clinician is acting on the basis of false intuition. If a clinician has little past experience, then the LEIA will be justified in flagging

Table 8.15
Knowledge about expected test results during progression of achalasia

	Stage 1	Stage 2	Stage 3	Stage 4
EGD	Dilated esophagus & no tumor at the GI junction		Dilated esophagus & *narrowing and pop upon entering LES (i.e., hypertensive LES)* & retained debris & *no tumor at the GI junction*	
Esophageal manometry	Incomplete relaxation of the LES & high-normal LES pressure	Incomplete relaxation of the LES & high-normal or hypertensive LES & intermittent peristalsis	Incomplete relaxation of the LES & *hypertensive LES* & *aperistalsis*	
Barium swallow	Delayed emptying		*Bird's beak* & dilated esophagus & retained debris & retained barium	

Note: The combination of italicized results is sufficient to posit a definitive diagnosis of this disease.

seemingly false moves. However, if a clinician who has vast past experience with patients of a similar profile starts to carry out what appears to be an unsubstantiated move, the LEIA might better query him about the reason for the move and potentially learn this new constellation of findings and their predictive power. This aspect of *system-initiated learning by being told* is a core functionality of LEIAs.

The illusion of validity. The illusion of validity describes a person's clinging to a belief despite evidence that it is unsubstantiated. Kahneman (2011, p. 211) reports that the discovery of this illusion occurred as a result of his practical experience with a particular method of evaluating candidates for army officer training. A study demonstrated that the selected method was nonpredictive—that is, the results of the evaluation had no correlation with the candidate's ultimate success in officer training—but the evaluators still clung to the idea that the method was predictive because they believed that it *should* be predictive.

The illusion of validity can be found in clinical medicine when a physician refuses to change an early hypothesis despite sufficient counterevidence. (He or she might, for example, rerun tests or continue a failed medication trial.) The definition of *sufficient* counterevidence depends on (a) the strength of the constellation of features suggesting the diagnosis; (b) the strength of the constellation of features suggesting a *different* diagnosis, recorded in corresponding tables for other diseases; and (c) the trustworthiness of tests, whose error rates must be recorded in the ontology. A LEIA could detect overzealous pursuit of a hypothesis using decision functions that combine these three factors.

Base-rate neglect. Base-rate neglect is a type of decision-making bias that, applied to clinical medicine, can refer to losing sight of the expected probability of a disease for a given type of patient in a given circumstance. For example, a patient presenting to an emergency room in New York is highly unlikely to have malaria, whereas that diagnosis would be very common in sub-Saharan Africa. Although physicians are trained to think about the relative likelihood of different diagnoses, remembering all of the relative probabilities of given different constellations of signs and symptoms can be quite challenging. A LEIA could help with this by flagging situations in which a clinician is pursuing a diagnostic hypothesis that is unlikely given the available data.

For example, esophageal carcinoma can result from gastroesophageal reflux disease (GERD) but typically only if GERD is not sufficiently treated for a long time and if the person smokes, drinks alcohol, lives or works in an industrial environment, or has had exposure to carcinogenic materials. These likelihood conditions are recorded in the ontology as complex fillers for the property SUFFICIENT-GROUNDS-TO-SUSPECT for the disease ESOPHAGEAL-CARCINOMA, as pretty-printed below.

ESOPHAGEAL-CARCINOMA
  SUFFICIENT-GROUNDS-TO-SUSPECT
    Both
      - (GERD (EXPERIENCER MEDICAL-PATIENT-1) (DURATION (> 5 (measured-in YEAR)))
      - At least one of
        • (MEDICAL-PATIENT-1 (AGENT-OF SMOKE))
        • (MEDICAL-PATIENT-1 (AGENT-OF (DRINK (THEME ALCOHOL) (FREQUENCY (> .3)))))
        • (MEDICAL-PATIENT-1 (AGENT-OF (RESIDE (LOCATION INDUSTRIAL-PLACE))))
        • (MEDICAL-PATIENT-1 (AGENT-OF (WORK (LOCATION INDUSTRIAL-PLACE))))
        • (MEDICAL-PATIENT-1 (EXPERIENCER-OF (EXPOSE (THEME CARCINOGEN) (FREQUENCY (> .3)))))

If a clinician hypothesizes esophageal carcinoma for a twenty-year-old person with a three-month history of GERD, the LEIA should issue a warning that there appears to be insufficient evidence for this hypothesis, and it will show the clinician the conditions under which the hypothesis is typically justified.

The small sample bias. A person's understanding of the frequency or likelihood of an event can be swayed from objective measures by the person's own experience and by the ease with which an example of a given type of situation—even if objectively rare—comes to mind (Kahneman 2011, p. 129). The small sample bias can lead to placing undue faith in personal experience. For example, if the widely preferred medication for a condition happens to fail one or more times in a physician's personal experience, the physician is prone to give undue weight to those results—effectively ignoring population-level statistics—and prefer a different medication instead. This is where the *art* of medicine becomes fraught with complexity. While personal experience should not be discounted, its importance should not be inflated since it could be idiosyncratic. As Kahneman (p. 118) writes, "The exaggerated faith in small samples is only one example of a more general illusion—we pay more attention to the content of messages than to information about their reliability, and as a result end up with a view of the world around us that is simpler and more coherent than the data justify."

A LEIA could automatically detect the small sample bias in clinicians' decisions by comparing three things: (a) the clinician's current clinical decision, (b) the LEIA's memory of the clinician's past decisions when dealing with the particular disease, and (c) the objective, population-level preference for the selected decision compared to other options. For example, suppose that three of a clinician's recent patients with a particular disease did not respond sufficiently to the preferred treatment or developed complications from it. If the clinician then stops recommending that treatment and, instead, opts for a less preferred one, the LEIA can issue a reminder of the population-level preference for the originally selected treatment and point out there is a danger of a small sample bias. Of course, the actual reason for the switch in treatment preferences might be legitimate. For example, if the treatment involves a procedure carried out by a specialist, then perhaps a highly skilled specialist was replaced by a less skilled one—which is an eventuality that must be modeled as well.

The exposure effect. The exposure effect describes people's tendency to believe frequently repeated statements even if they are false because, as Kahneman (2011, p. 62) says, "familiarity is not easily distinguished from truth." This is biologically grounded in the fact that if you have encountered something many times and are still alive, it is probably not dangerous (p. 67). The LEIA can detect potential cases of the exposure effect using a function whose arguments include the following.

- A new ontological property, HYPE-LEVEL, that applies to interventions—drugs and procedures. Its values reflect the amount of advertising, drug company samples, and so on to which a clinician is exposed. If this is unknown for a particular clinician, a population-level value will be used, based on the amount of overall advertising and sample distribution.
- The objective "goodness" of an intervention, as compared with alternatives, at the level of population, which is a function of its relative efficacy, side effects, cost, and so on.
- The objective "goodness" of an intervention, as compared with alternatives, for the specific patient, which adds patient-specific features, if known, to the above calculation.
- The actual selection of an intervention for this patient in this case.
- The clinician's past history of prescribing, or not prescribing, this intervention in relevant circumstances. For example, a clinician might (a) be continuing to prescribe an

old medication instead of a better new one due to engrained past experience, (b) insist on a name brand if a generic has been made available, or (c) prefer one company's offering over a similar offering from another company despite high additional costs to the patient; and so on.

8.2.3 Detecting and Flagging Patient Biases

The gold standard of modern medical care is patient-centered medicine. In the patient-centered paradigm, the physician does not impose a single solution on the patient but, rather, instructs, advises, and listens to the patient with the purpose of jointly arriving at an optimum solution. The patient's goals might be summarized as "Talk to me, answer my questions, and solve my problem in a way that suits my body, my personal situation, and my preferences." The doctor's goals might be summarized as "Make an accurate diagnosis. Have a compliant patient who is informed about the problem and makes responsible decisions. Launch an effective treatment."

To best serve the patient, the doctor should be aware of psychological effects on decision-making that might negatively impact the patient's decisions. If a patient makes a decision that the doctor considers suboptimal, the doctor can attempt to understand why by modeling what he or she believes the patient knows, believes, fears, prioritizes, and so on and by hypothesizing the decision function that might have led to the given decision. For example, imagine that a doctor suggests that a patient, Matthew, take a medication that the doctor knows to be highly effective and that has infrequent, mild side effects about which the doctor informs Matthew. In response to the doctor's suggestion, Matthew refuses, saying he doesn't want to take that kind of medication. When the doctor asks why, Matthew responds in a vague manner, saying that he just has a bad feeling about it. Rather than try to force Matthew or badger him for a better explanation, the doctor—in the role of psychologist gumshoe—can break down the decision process into inspectable parts and constructively pursue them in turn. Let us consider the process in more detail.

A person who is considering advice to take a medication will likely consider things like the following: the list of potential benefits, risks, and side effects; the cost, in terms of money, time, emotional drain; the patient's trust in the doctor's advice; and the patient's beliefs in a more general sense—about medication use overall, being under a doctor's care, and so on.

Returning to our example, suppose the drug that the doctor recommended was hypothetical drug X, used for headache relief. Suppose also that the doctor describes the drug to Matthew as follows: "It is very likely that this drug will give you significant relief from your headaches and it might also improve your mood a little. The most common side effect is dry mouth, and there is a small chance of impotence. Unfortunately, the drug has to be injected subcutaneously twice a day." From this, Matthew will have the following information to inform his decision-making. We include conditional flags (described below) in the structures as italicized comments.

HEADACHE-RELIEF	[intensity: high, likelihood: high]	
MOOD-LIFT	[intensity: low, likelihood: moderate]	
IMPOTENCE	[likelihood: low]	*FLAG for male patients*
DRY-MOUTH	[likelihood: high]	*FLAG for wind instrument players*
COST-EFFORT	[high]	*FLAG because injectable*
COST-EMOTIONAL	[potentially high]	*FLAG if needle-phobic*

In addition, both patients and doctors know that the following can affect health care decisions:

COST-FINANCIAL	*FLAG if no health insurance*
TRUST-IN-PHYSICIAN	*FLAG if the doctor feels the patient doesn't trust him*
MEDICATION-AVERSION	*FLAG for certain individuals and socio-ethnic groups*

Finally, doctors know that patients can be affected by various decision-making biases such as the following, each of which can be considered a standing (always available) flag for the doctor as he or she attempts to understand the patient's thought processes:

- *The exposure effect.* People are barraged by drug information on the internet and in TV and radio ads, with the latter rattling off potential side effects at a pace. From this, the patient's impression of a medication might involve a vague but lengthy inventory of side effects that the doctor did not mention, and these might serve as misinformation for the patient's decision-making.
- *The effect of small samples.* The patient might know somebody who took this medication and had a bad time with it, thus generalizing that it is a bad drug, despite the doctor's description of it.
- *The effect of evaluative attitudes.* The patient might not like the idea of taking any medication, or some class of medications, due to a perceived stigma (e.g., against antidepressants). Or the patient might be so opposed to a given type of side effect that its potential overshadows any other aspect of the medication.
- *Depletion effects.* The patient might be tired or distracted when making a decision and therefore decide that refusing a proposed intervention is the least-risk option. Or fatigue might have caused lapses in attention so that the patient misremembers the doctor's description of the medication.

A LEIA could assist the physician in trying to understand the patient's decision-making by making the relevant flags explicit. For example, if our patient, Matthew, has good health insurance and a medical history of having given himself allergy injections for years, it is possible that the impotence side effect is an issue, but it is unlikely that the financial cost or fear of injections is a detractor. Since the LEIA will have access to Matthew's online medical records, it can make such contextual judgment calls and give the doctor advice about which features might be best to pursue first. Even things like a patient's trust in the doctor can, we believe, be detected to some degree by the doctor-patient dialog. For example, if Matthew argues with the doctor, or asks a lot of questions, or frequently voices disagreement, it is possible that low trust is affecting his decision-making.

Another factor that might affect a patient is the *halo effect*, which is the tendency to make an overall positive or negative assessment of a person on the basis of a small sample of known positive or negative features. For example, you might believe that a person who is kind and successful will also be generous, even though you know nothing about this aspect of the person's character. As Kahneman (2011, p. 83) says, "The halo effect increases the weight of first impressions, sometimes to the point that subsequent information is mostly wasted." We will suggest that an extended notion of the halo effect—in which it can apply also to objects and events—can undermine good decision-making by patients. On the one hand, our patient Matthew might like his doctor so much he agrees to the latter's advice before learning a sufficient amount about what is recommended to make a responsible, informed decision. On the other hand, he might dislike his doctor so much that he refuses advice that would actually be beneficial. Extending the halo effect to events, Matthew might be so happy that a procedure has few risks that he assumes that it will not involve any pain and will have no side effects—both of which might not be true. By contrast, Matthew might be so influenced by the knowledge that the procedure will hurt that he loses sight of its potential benefits. Doctors should detect halo effects in order to ensure that patients are making the best, most responsible decisions for themselves. It would be no better for Matthew to blindly undergo surgery because he likes his doctor than for him to refuse lifesaving surgery because he is angry with him or her.

In order to operationalize the automatic detection of halo effects, we can construct *halo-property nests* like the ones shown in table 8.16. These are inventories of properties that form a constellation with respect to which a person might evaluate another person, thing, or event.

Each value or range of values for a property has a positive-halo, negative-halo, or neutral-halo score. If a patient knows about a given property value that has a positive-halo score

Table 8.16
Example of halo-property nests

OBJECT or EVENT	Nest of PROPERTIES
MEDICAL-PROCEDURE	RISK, PAIN, SIDE-EFFECTS, BENEFITS
PHYSICIAN	INTELLIGENCE, SKILL-LEVEL, AFFABILITY, KINDNESS, TRUSTWORTHINESS

(e.g., low risk) but doesn't know about any of the other property values in the nest, it is possible that he or she will assume that the values of the other properties have the same halo-polarity score (e.g., low pain, low side effects, high benefits). This can explain why a patient who knows little about a procedure might accept or decline it out of hand. Understanding this potential bias can help a doctor to tactfully continue a knowledge-providing conversation until the patient actually has all the information needed to make a good decision. The agent's role in the process is to trace the hypothetical decision-making process of the patient, determine whether or not he or she knows enough feature values to make a good decision, and, if not, notify the doctor.

The final class of decision-making biases to which a patient might be subject pertains to the nature of the doctor-patient dialog. The way a situation is presented or a question is asked can impact a person's perception of it and subsequently affect related decision-making. For example, if someone is asked, "I imagine you hurt right now?" they will have a tendency to seek corroborating evidence by noticing something that hurts, even if just a little (*the confirmation bias*). If someone is asked, "Your pain is very bad, isn't it?" they are likely to overestimate the perceived pain, having been primed with a high pain level (*the priming effect*). And if someone is told, "There is a 20% chance that this will fail," they are likely to interpret it more negatively than if they were told, "There's an 80% chance that this will succeed" (*the framing sway*).

The agent could help doctors be aware of, and learn to avoid, the negative consequences of such effects by automatically detecting and flagging relevant situations. The detection methods involve recording constructions in the lexicon that can predictably lead to biased thinking. Table 8.17 shows some examples.

The semantic descriptions of such constructions (recorded in the lexicon) must include the information that the DISCOURSE-FUNCTION of the construction is the listed bias (e.g., SEEK-CONFIRMATION). So, for example, the semantic representation of tag-question constructions like "You don't VP, do you?" will be

Table 8.17
Examples of constructions that can lead to biased thinking

Example	Associated bias
You don't smoke, do you? I assume you don't eat before sleeping.	SEEK-CONFIRMATION
Do you have sharp pain in your lower abdomen?	SUGGESTIVE-YES/NO
Do you drink between 2 and 4 cups of coffee a day?	PRIME-WITH-RANGE
There's a 10% chance the procedure will fail.	NEGATIVE-FRAMING-SWAY
There's a 90% chance the procedure will succeed.	POSITIVE-FRAMING-SWAY

REQUEST-INFO

AGENT	HUMAN	("speaker")
BENEFICIARY	HUMAN	("hearer")
THEME	^main-clause	(the meaning of the main clause)
DISCOURSE-FUNCTION	SEEK-CONFIRMATION	

The values of DISCOURSE-FUNCTION can be incorporated into rules for good clinical negotiation. For example, a doctor is more likely to convince a patient to agree to a lifesaving procedure by framing the side effects, risks, and so on using a positive framing sway rather than a negative one. Similarly, a doctor is more likely to get a patient to provide maximally objective ratings of symptom severity by asking neutral questions ("Do you have any chest pain?") rather than questions framed as SUGGESTIVE-YES/NO or PRIME-WITH-RANGE. The agent can match the most desired utterance types with its assessment of the doctor's goal in the given exchange using the tracking of hypothesized goals and plans (e.g., "convince patient to undergo procedure").

When considering the utility of LEIAs in advising doctors, it is important to remember that the psychological effects we have been discussing are typically not recognized by people in the course of normal interactions. So it is not that we expect LEIAs to discover anything that doctors do not already know or could not learn in principle. Instead, we think that LEIAs could point out aspects of decision-making and interpersonal interactions that, for whatever reason, the doctor is unaware of in the heat of the moment. We think that LEIAs could be particularly useful to doctors who have less experience overall, who have little experience with a particular constellation of findings, who are under the pressures of time and/or fatigue, or who are dealing with difficult nonmedical aspects of a case, such as a noncompliant patient.

8.3 LEIAs in Robotics

It is broadly recognized that progress in social robotics is predicated on improving robots' ability to communicate with people. But there are different levels of communication. While robots have, for example, been able to react to vocal commands for quite some time, this ability does not invoke the kind of fundamental, broad-coverage NLU described throughout this book. Instead, the language utterances understood by robots have been tightly constrained, with most research efforts focused on enabling robots to learn skills through demonstration (e.g., Argall et al., 2009; Zhu & Hu, 2018). The robotics community has not willfully disregarded the promise of language-endowed robots; rather, it has understandably postponed the challenge of NLU, which, in an embodied application, must also incorporate extralinguistic context (what the robot sees, hears, knows about the domain, thinks about its interlocutor's goals, and so on). Integrating the language capabilities of LEIAs into robotic systems is the obvious next step forward.

Typical robots have some inventory of physical actions they can perform, as well as objects they can recognize and manipulate. A LEIA-robot hybrid can acquire a mental model of these actions and objects through dialog with human collaborators. That is, people can help LEIA-robots to understand their world by naming objects and actions; describing actions in terms of their causal organization, prerequisites, and constraints; listing the affordances of objects; and explaining people's expectations of the robots. This kind of understanding will enhance LEIA-robots' ability to understand their own actions and the actions of others and to become more humanlike collaborators overall. Clearly, this kind of learning relies on semantically interpreting language inputs, and it mirrors a major mode of learning in humans—learning through language. In this section we describe our work on integrating a LEIA with a robot in an application system.

The system we describe is a social robot collaborating with a human user to learn complex actions. The experimental domain is the familiar task of furniture assembly that is widely accepted as useful for demonstrating human-robot collaboration on a joint activity. Roncone et al. (2017) report on a Baxter robot supplied with high-level specifications of procedures implementing chair-building tasks, represented in the hierarchical task network (HTN) formalism (Erol et al., 1994). In that system, the robot uses a rudimentary sublanguage to communicate with the human in order to convert these HTN representations into low-level task planners capable of being directly executed by the robot. Since the robot does not have the language understanding capabilities or the ontological knowledge substrate of LEIAs, it cannot learn by being told or reason explicitly about the HTN-represented tasks. As a result, those tasks have the status of uninterpreted skills stored in the robot's procedural memory.

We undertook to develop a LEIA-robot hybrid based on the robot just described. The resulting system was able to

- learn the semantics of the initially uninterpreted basic actions;
- learn the semantics of operations performed by the robot's human collaborator from natural language descriptions of them;
- learn, name, and reason about meaningful groupings and sequences of actions;
- organize those sequences of actions hierarchically; and
- integrate the results of learning with knowledge stored in the LEIA-robot's semantic and episodic memories.

To make clear how all this happens, we must start from the beginning. The LEIA-robot brings to the learning process the functionalities of both the LEIA and the robot. Its robotic side can (a) visually recognize parts of the future chair (e.g., the seat) and the tools to be used (e.g., screwdriver) and (b) perform basic programmed actions, which are issued as non-natural-language commands such as GET(LEFT-BRACKET), HOLD(SCREWDRIVER), RELEASE(LEFT-BRACKET). The hybrid system's LEIA side, for its part, can generate ontologically grounded meaning representations (MRs) from both user utterances and physical actions.[23] The interactive learning process that combines these capabilities is implemented in three modules.

Learning module 1: *Concept grounding.* The LEIA-robot learns the connection between its basic programmed actions and the meaning representations of utterances that describe them. This is done by the user verbally describing a basic programmed action at the same time as launching it. For example, he or she can say, "You are fetching a screwdriver" while launching the procedure GET(SCREWDRIVER). The LEIA-robot generates the following TMR while physically retrieving the screwdriver.

```
CHANGE-LOCATION-1
    THEME          SCREWDRIVER-1
    AGENT          ROBOT-1
    TIME           find-anchor-time
    EFFECT         BESIDE-1
    textstring     fetch
    lex-sense      get-v1

BESIDE-1
    DOMAIN         SCREWDRIVER-1
    RANGE          HUMAN-1
```

Learning module 2: *Learning legal sequences of known basic actions.* The robot learns legal sequences of known basic actions by hierarchically organizing the TMRs for sequential event descriptions. It recognizes these sequences as new complex actions (ontological events), which it names and records in its ontology. Since the full process of chair assembly is far too long to present here (see Nirenburg & Wood, 2017, for details), we illustrate this process (in table 8.18) by tracing the robot's learning how to assemble the third of the four chair legs.

Learning module 3: *Memory management for newly acquired knowledge.* Newly learned process sequences (e.g., ASSEMBLE-RIGHT-BACK-LEG) and objects (e.g., RIGHT-BACK-LEG) must be incorporated in the LEIA-robot's long-term semantic and episodic memories. For each newly learned concept, the memory management module first determines whether this concept should be (a) added to the LEIA-robot's semantic memory or (b) merged with an existing concept. To make this choice, the agent uses an extension of the concept-matching algorithm reported in English and Nirenburg (2007) and Nirenburg et al. (2007). This algorithm is based on unification, with the added facility for naming concepts and determining their best position in the ontological hierarchy. Details aside, the matching algorithm works down through the ontological graph—starting at the PHYSICAL-OBJECT or PHYSICAL-EVENT node, as applicable—and identifies the closest match that does not violate any recorded constraints. Nirenburg et al. describe the eventualities that this process can encounter.

To recapitulate, the system described here concentrates on robotic learning through language understanding. This learning results in extensions to and modifications of the three

Table 8.18
Learning while assembling the right back leg

The user says,	"We are building the right back leg."
The LEIA-robot carries out a mental action:	It generates a TMR for that utterance.
The user says,	"Get another foot bracket."
The user launches the associated robotic action by inputting	GET(BRACKET-FOOT)
The LEIA-robot carries out a sequence of physical actions:	First, it undertakes the asserted GET(BRACKET-FOOT) action. Then it carries out the action it typically performs next: RELEASE(BRACKET-FOOT).
The LEIA-robot carries out a mental action:	It learns to associate this complex event with the TMR for "Get another foot bracket."
The user says,	"Get the right back bracket."
The user launches the associated robotic action by inputting	GET(BRACKET-BACK-RIGHT)
The LEIA-robot performs the associated physical and learning actions, as before.	
The user says,	"Get and hold another dowel."
The user launches the associated robotic actions by inputting	GET(DOWEL), HOLD(DOWEL)
The LEIA-robot performs the associated physical and learning actions.	
The user says,	"I am mounting the third set of brackets on a dowel."
The LEIA-robot carries out a mental action:	It generates a meaning representation of this utterance.
The user carries out a physical action:	He affixes the foot and the right back brackets to the dowel.
The LEIA-robot learns through demonstration:	It observes this physical action and generates a meaning representation of it.
The user says,	"Finished."
The LEIA-robot carries out a mental action:	It generates a meaning representation of this utterance.
The user says,	"Release the dowel."
The user launches the associated robotic action by inputting	RELEASE(DOWEL)
The LEIA robot performs the associated physical and learning actions.	
The user says,	"Done assembling the right back leg."
The LEIA robot carries out a sequence of mental actions:	(a) It generates a meaning representation for that utterance. (b) It learns the action subsequence for ASSEMBLE-RIGHT-BACK-LEG. (c) It learns the following ontological concepts in their meronymic relationship: RIGHT-BACK-LEG (HAS-OBJECT-AS-PART BRACKET-FOOT, BRACKET-BACK-RIGHT, DOWEL) (d) It learns that RIGHT-BACK-LEG fills the HAS-OBJECT-AS-PART slot of CHAIR.

kinds of memory in a LEIA-robot: explicit semantic memory (i.e., ontology); explicit episodic memory (i.e., a recollection of what happened during the learning session); and the implicit (skill-oriented) procedural memory. We expect these capabilities to allow the robot to (a) perform complex actions without the user having to spell out a complete sequence of basic and complex actions; (b) reason about task allocation between itself and the human user; and (c) test and verify its knowledge through dialog with the user, avoiding the need for the large number of training examples often required when learning is carried out by demonstration only.

The work on integrating linguistically sophisticated cognitive agents with physical robots offers several advantages over machine learning approaches. First, LEIA-robots can explain their decisions and actions in human terms, using natural language. Second, their operation does not depend on the availability of big-data training materials; instead, we model the way people learn, which is largely through natural language interactions. Third, our work overtly models the LEIA-robot's memory components, which include the implicit memory of skills (the robotic component), the explicit memory of concepts (objects, events, and their properties), and the explicit memory of concept instances, including episodes, which are represented in our system as hierarchical transition networks. The link established between the implicit and explicit layers of memory allows the robot to reason about its own actions.

Scheutz et al. (2013) discuss methodological options for integrating robotic and cognitive architectures and propose three "generic high-level interfaces" between them—the perceptual interface, the goal interface, and the action interface. In our work, the basic interaction between the implicit robotic operation and explicit cognitive operation is supported by interactions among the three components of the memory system of the LEIA-robot.

There are several natural extensions to this work. After the robot's physical actions are grounded in ontological concepts, the robot should be able to carry out commands or learn new action sequences by acting directly on the user's utterances, without the need for direct triggering of those physical actions through software function calls. In addition, the incorporation of text generation and dialog management capabilities would allow the robot to take a more active role in the learning process (as by asking questions) as well as enrich the verisimilitude of interactions with humans during joint task performance. Yet another direction of work, quite novel for the robotics field, would be to enable the robot to adapt to particular users, leveraging the sort of mindreading discussed earlier in this chapter.

8.4 The Take-Home Message about Agent Applications

We expect that some readers will have skipped over the details of the applications presented in this chapter. That is fine as long as the point behind all those details is not lost.

The main argument against developing deep NLU systems, particularly using knowledge-based methods, has been that it requires deep and broad, high-quality knowledge, which is expensive to acquire. Yes, it *is* expensive to acquire, but it is needed not only for language processing but also to enable virtual and robotic agents to function intelligently in many kinds of applications. That is, the knowledge problem is not restricted to matters of language; it is at the core of many of AI's most imposing challenges.

The program of knowledge-based AI presented in this book has not infrequently been dubbed ambitious—a term that tends to carry at least some degree of skepticism. That skepticism is not surprising: if very few people are pursuing one of the most compelling problems science has to offer, there must be some reason why. We hypothesize that a substantial contributing factor is that people simply don't enjoy, and/or don't receive sufficient personal and professional benefits from, doing the kinds of knowledge engineering we illustrate. However, personal preferences and personal cost-benefit analyses should not be confused with more objective assessments of the potential for knowledge engineering to foster progress in the field of AI at large.

We do not expect that this book will turn every reader into an optimistic champion of knowledge-based NLU or, more broadly, knowledge-based AI. However, as we have shown in this chapter and those that precede it, it would be unsound to dismiss our commitment to this paradigm as deriving from unrealistic ideas about how much work it requires. Only time will tell how AI will unfold over the decades to come, but we are making our bets with eyes wide open.

9

Measuring Progress

Measuring progress is an important aspect of developing models and systems. But for knowledge-based approaches, evaluation is not a simple add-on to system development. Instead, inventing useful, practical evaluation methodologies is best viewed as an ongoing research issue. And, as a research issue, we should expect the process to be marked by trial and error. However, the associated successes, failures, and lessons learned are as central to the program of work as are the models and implementations themselves.

The reason why new evaluation approaches and metrics are needed is because the ones that have been adopted as the standard within the world of empirical NLP are a poor fit. In the chapter entitled "Evaluation of NLP Systems," published in *The Handbook of Computational Linguistics and Natural Language Processing* (Clark, Fox, & Lappin, 2010), Resnik and Lin (2010) do not even touch on the evaluation of knowledge-based or theoretically oriented NLP systems, writing (italics ours):

> It must be noted that the design or application of an NLP system is sometimes connected with a broader scientific agenda; for example, cognitive modeling of human language acquisition or processing. In those cases, the value of a system resides partly in the attributes of the theory it instantiates, such as conciseness, coverage of observed data, and the ability to make falsifiable predictions. Although several chapters in this volume touch on scientific as well as practical goals (e.g., the chapters on computational morphology, unsupervised grammar acquisition, and computational semantics), *such scientific criteria have fallen out of mainstream computational linguistics almost entirely in recent years in favor of a focus on practical applications, and we will not consider them further here.* (p. 271)

However, work outside the mainstream is ongoing, and those pursuing it have to take on the evaluation challenge. Jerry Hobbs, in "Some Notes on Performance Evaluation for Natural Language Systems" (2004), makes the following observations: Evaluation through demonstration systems is not conclusive because it can "dazzle observers for the wrong reasons"; evaluation through deployed systems that use emergent technologies is also not conclusive because such systems can fail to be embraced for reasons unrelated to the promise of the technology; component-level evaluations are needed, but the competence and performance of systems must be considered separately, because "a system that represents significant progress in competence may be a disaster in performance for some trivial reason"; and, since there tends to be little uniformity of goals, foci, coverage, and applications across knowledge-based systems, head-to-head comparisons are well-nigh impossible.

Apart from the difficulty of formulating useful, representative evaluation suites, another important issue is cost. In their historical overview of evaluation practices in NLP, Paroubek et al. (2007, p. 26) point out that in the 1980s "the issue of evaluation was controversial in the field. At that time, a majority of actors were not convinced that the benefits outweighed the cost." They proceed to describe the subsequent emphasis on formal evaluation as a "trend reversal" but, curiously, do not overtly link this effect to its cause.

It seems clear that the focus on evaluation only became possible because the field at large adopted a system-building methodology based on statistical machine learning approaches that allowed individual developers to use standardized, straightforward, and inexpensive evaluation regimens of a particular kind. So this trend reversal toward an emphasis on evaluation says much more about the history of mainstream NLP than about the cost-benefit analysis of formal evaluations in principle.

Our point is not that knowledge-based programs of R&D should be absolved of providing evidence of progress—certainly not![1] However, the measures of progress adopted must be appropriate to the approaches and systems they evaluate, genuinely useful, and not so demanding of resources that they overwhelm the overall program of R&D.[2] This is the spirit of our ongoing efforts to measure progress on NLU within the broader program of work on developing humanlike LEIAs.

9.1 Evaluation Options—and Why the Standard Ones Don't Fit

Whatever methods one uses to build NLP capabilities, the top-level choice in evaluation is whether to evaluate an end application or a component functionality. Many believe that end-system evaluation is the gold standard. And, in fact, the evaluation practices that have become the standard in mainstream, empirical NLP originally grew out of such applications as information retrieval and information extraction. However, end-system evaluation is not always ideal. For example, if NLU capabilities are incorporated into a more comprehensive system, such as a robotic assistant, then the entire system—not only the NLU portion of it—needs to be at an evaluable stage of development, which can take a long time. Moreover, all NLP-specific capabilities required by the end system must also be developed and integrated prior to evaluation. Yet another drawback of end-system evaluations is that they are unlikely to address all language phenomena treated by the system, meaning that the evaluations say *something* about the system's capabilities but far from *everything interesting and useful*. Finally, error attribution can be difficult in an end system that is comprised of many diverse parts.

An alternative evaluation option focuses on individual components. The empirical NLP community has developed the practice of creating tasks that foster field-wide competitions targeting one or another linguistic phenomenon. The tasks—which are described in extensive guidelines—are formulated by individuals representing the community at large. Often, those individuals also oversee the compilation and annotation of corpora to support the training and evaluation of the supervised machine learning systems that will compete on the task. Among the many phenomena that have been addressed by task descriptions are coreference resolution, named-entity recognition, case role identification, and word-sense disambiguation. The fruits of this considerable, and expensive, task-formulation effort are made available to the community for free—something that stimulates work on the topic, allows for head-to-head comparisons between systems, and avoids the replication of effort across research teams. In short, these resources are of great utility to those whose goals and methods align with them.

However, to properly understand the role of tasks and task-oriented resources in the field, one must acknowledge not only their benefits but also their limitations. The task descriptions typically contain extensive listings of rule-in/rule-out criteria (e.g., Chinchor, 1997; Hirschman & Chinchor, 1997). The ruled-in instances are called *markables* because they are what annotators will mark in a corpus. Instances that are ruled out (not marked) are considered outside of purview.

The task descriptions reflect rigorous analysis by linguists, who must consider not only linguistic complexity but also the expected capabilities of annotators (often, college students), the speed/cost of annotation, the need for high interannotator agreement, and the anticipated strengths and limitations of the machine learning methods that are expected to be brought to bear. In many cases, the task description specifies that systems participating in an evaluation competition will be provided with annotated corpora not only for the training stage but also for the evaluation stage, which significantly distances the task from the full, real-world problem. And the more difficult instances of linguistic phenomena are usually excluded from purview because they pose problems for annotators and system developers alike. In sum, using the word *task* to describe such enterprises is quite appropriate; so is using such tasks to compare results obtained by different machine learning methods. It is important, however, to thoughtfully interpret—having read the task specifications—what the scores on associated evaluations mean. After all, 90% precision on a task does not mean 90% precision on automatically processing all examples representing the given linguistic phenomenon.

The reason for detailing the nature of mainstream NLP tasks was to make the following point. When we evaluate a LEIA's domain-neutral NLU capabilities using stages 1–5 of processing (*not* Situational Reasoning) over unrestricted corpora, our evaluation suites are no less idiosyncratic. They, too, cover only a subset of eventualities, and for the same reason—the state of the art is too young for any of us to do well on the very hardest of language inputs, and we all need credit for midstream accomplishments. There are, however, significant differences between our pre-situational (stages 1–5) evaluations and mainstream NLP tasks.

1. Whereas mainstream tasks are formulated by community-wide representatives, we need to formulate our own. Community-level task formulation has three advantages that we do not share:
    a. The community gives its stamp of approval regarding task content and design, absolving individual developers of having to justify it.
    b. The task description exists independently and can be pointed to, without further discussion, by individual developers reporting their work, which facilitates the all-important publication of results.
    c. The community takes on the cost of preparing the task and all associated resources, so there is little to no cost to individual developers.
2. Mainstream tasks involve manually preselecting markables before systems are run, effectively making the difficult examples go away. We, by contrast, expose LEIAs to all examples but design them to operate with self-awareness. Just as people can judge how well they have understood a language input, so, too, must LEIAs. Relying on a model of metacognitive introspection that uses simpler-first principles (see section 2.6), LEIAs can automatically select the inputs that they believe they can treat competently. It is these inputs that are included in our evaluation runs. Requiring LEIAs to treat every example would be tantamount to requiring mainstream NLP tasks—and the associated annotation efforts—to treat every example as a markable. The field overall, no matter the paradigm (be it statistical NLP or knowledge-based NLU), is just too young for a *treat everything* requirement to be anything but futile. Recalling the discussion in section 1.6.3 (which juxtaposes NLU and NLP), we rightfully cheer for every individual behavior demonstrated by robots, not expecting them to be fully humanlike today. We need to shift the collective mindset accordingly when it comes to processing natural language.

    Note: We must reiterate that the evaluation setups we are talking about treat NLU outside the full cognitive architecture, applying only those knowledge bases and processors that cover the open domain (i.e., those belonging to stages 1–5 of LEIA operation). The above juxtaposition with mainstream NLP tasks is meant to stress that evaluating pre-situational, open-domain NLU by LEIAs is very different from evaluating full NLU within an end application. Within end applications, LEIAs have to treat every input but can take advantage of (a) specialized domain knowledge, (b) Situational Reasoning (stage 6), and (c) the ability to decide how precise and confident an analysis must be to render it actionable.
3. Whereas NLP task suites include a manually annotated gold standard against which to evaluate system performance, most of our evaluation experiments—namely, those

requiring TMR generation—have involved checking the system's output after it was produced. The reason why is best understood by considering the alternatives.

a. If people were asked to manually create gold standard TMRs on the basis of the ontology alone (i.e., without the lexicon), this gold standard would be suboptimal for evaluation because of the possibility of ontological paraphrase. That is, the system might generate a perfectly acceptable TMR that did not happen to match the particular paraphrase listed in the gold standard. This is similar to the problem of accounting for linguistic paraphrase when evaluating machine translation systems.[3]
b. If people were told to use the lexicon and ontology together to create gold standard TMRs, then they would be carrying out a very inefficient replication of the automatic process. It is for good reason that the time and cost of annotation has always been at the center of attention in statistical NLP. We cannot collectively afford to spend unbounded resources on evaluation—particularly if they would be as ill-used as under this scenario.

4. Whereas mainstream task formulation involves teams of people carrying out each aspect of manual data preparation (with interannotator agreement being an important objective), we do not have comparable resources and so must find alternative solutions.
5. Whereas mainstream task-oriented evaluations are black box and geared at generating numerical results to facilitate comparisons across systems, ours are glass box and only partially numerical. Our emphasis is on understanding the reasons for particular outcomes, which is necessary to assess the quality of our models, to determine the success of the model-to-system transition, and to chart directions for future development.

The following are among the approaches to evaluation that knowledge-based efforts can adopt.

1. *Carrying out evaluations that target specific phenomena within small domains.* This has been done, for example, in the work of James Allen and collaborators (e.g., Allen et al., 2006, 2007; Ferguson & Allen, 1998).
2. *Wearing two hats: scientific and technological.* In their role as scientists, developers carry out cognitively inspired, rigorous descriptive work, but in their role as technologists, they select simplified subsets of phenomena for use in application systems that are evaluated using the traditional NLP approach. This appears to be the choice of the dialog specialist David Traum (compare Traum, 1994, for scientific work with Nouri, Artstein, Leuski, & Traum, 2011, for application-oriented work).
3. *Building theories but not applying them in computational systems.* This approach—which is typical, for example, of computational formal semanticists—has been criticized on the grounds that NLP must involve actual computation (see, e.g., Wilks, 2011). However, its motivation lies in the promise of contributing to future system building.

4. *Pursuing hybrid evaluations.* Hybrid evaluations combine aspects of the above approaches, which we find most appropriate for evaluating NLU by LEIAs.

The sections to follow describe our team's experience with evaluation. It includes five component-level (i.e., microtheory-oriented) evaluation experiments (section 9.2) and two holistic ones (section 9.3). We describe the experiments in some detail because we believe that our experience will be of use to others undertaking evaluation as part of R&D in NLU. All the evaluations we describe were carried out on unrestricted corpora, with varying rules of the game that we will specify for each evaluation. In all cases, the experiments validated that our system worked essentially as expected. But the real utility of the experiments lay in the lessons learned—lessons that would have been unavailable had we not actually implemented our models, tested them on real inputs, and observed where they succeeded and failed. Introspection, no matter how informed by experience, just does not predict all the ways people actually use language.

The most important lesson learned was that, with higher-than-expected frequency, the interpretation of an input can seem to work out well (i.e., receive a high-confidence score) yet be incorrect. For example, an agent cannot be expected to guess that *kick the bucket* or *hit the deck* have idiomatic meanings if those meanings are not recorded in the lexicon, since it is entirely possible to strike a bucket with one's foot and slap a deck with one's hand. However, once such meanings are recorded, agents can include the idiomatic readings along with the direct ones in the analysis space. Although adding a lexical sense or two would be a simple fix for many attested errors, what is needed is a much more comprehensive computational-semantic lexicon than is currently available. Building such a lexicon is an entirely doable task, but, in the current climate, it is unlikely to be undertaken at a large scale because the vast majority of resources for human knowledge acquisition field-wide are being devoted to corpus annotation. So the "seems right but is wrong" challenge to NLU systems operating in the open domain will remain for the foreseeable future.

9.2 Five Component-Level Evaluation Experiments

Over the past several years we formally evaluated our microtheories for five linguistic phenomena: nominal compounding; multiword expressions;[4] lexical disambiguation and the establishment of the semantic dependency structure; difficult referring expressions; and verb phrase ellipsis. Each of these evaluation experiments played a minor part in a published report whose main contribution was the microtheory itself—that is, the description of a model, grounded in a theory, that advances the fields of linguistics and computational cognitive modeling. However, it was important to include the description of an evaluation experiment to show that the microtheories were actually computational. The challenge in each case was to carve out a part of the microtheory that could be teased apart relatively cleanly from all the other interdependent microtheories required for comprehensive NLU.

This section presents a sketch of each of those evaluations. We do not repeat the numerical results for three reasons: they are available in the original papers; their precise interpretation requires a level of detail that we are not presenting here; and we don't believe that a theoretically oriented book, which should have a reasonably long shelf life, should include necessarily fleeting progress reports.

It is worth noting that all of these evaluation setups were deemed reasonable by at least those members of the community who served as reviewers for the respective published papers. Our hope is that these summaries highlight the unifying threads across experiments without losing the aspects of those original reports that made them convincing.

9.2.1 Nominal Compounding

Our evaluation of the microtheory of nominal compounding (McShane et al., 2014) focused on lexical and ontological constructions that both disambiguate the component nouns and establish the semantic relationship between their interpretations. That is, if two nouns can be interpreted using the expectations encoded in a recorded construction, then it is likely that they should be interpreted using that construction. These constructions were described in section 6.3.1.

For example, the nouns in the compound *bass fishing* are ambiguous: bass [BASS-FISH, STRING-BASED-INSTRUMENT], fishing [FISHING-EVENT, SEEK]. Combining these meanings leads to four interpretations:

- Carrying out the sport/job of fishing in an attempt to catch a type of fish called a *bass*;
- Carrying out the sport/job of fishing in an attempt to catch a stringed musical instrument called a *bass*;
- Seeking (looking for) a type of fish called a *bass*; or
- Seeking (looking for) a stringed musical instrument called a *bass*.

However, only one of these interpretations, the first, matches a recorded NN construction, namely, FISH + *fishing* → FISHING-EVENT (THEME FISH). By analyzing *bass fishing* according to this construction, the system simultaneously selects a meaning of *bass*, a meaning of *fishing*, and the relationship between them. The existence of this construction asserts a preference for this interpretation as the default. We must emphasize that this is still only a tentative, default interpretation that may be discarded when the analysis of the nominal compound is incorporated into the clause-level semantic dependency structure. In the reported evaluation, we assessed how often this default interpretation was correct.

This evaluation did not address all aspects of the microtheory of nominal compounding, such as processing compounds containing three or more nouns or compounds in which one or both of the words are unknown and need to be learned on the fly. This would have invoked not only new-word learning capabilities but also all the aspects of analysis contributing to it, such as clause-level lexical disambiguation and coreference resolution.

The corpus used for evaluation was the *Wall Street Journal* (1987; hereafter, WSJ). String-search methods identified sentences of potential interest, and those candidates remained in the evaluation corpus if they met all of the following criteria:

1. The sentence could be analyzed, with no technical failures, by the CoreNLP preprocessor, the CoreNLP syntactic dependency parser, and the LEIA's semantic analyzer. If there was a failure, then the given sentence was automatically excluded from purview. It is not feasible to turn the evaluation of a particular microtheory into the evaluation of every system component—particularly those, like CoreNLP, that we import.
2. The NN string was recognized by the parser as a compound, it contained exactly two nouns, and neither of those was a proper noun or an unknown word.
3. The semantic analyses of both the NN and the verb that selected it as an argument were headed by an ontological concept rather than a modality frame, a call to a procedural semantic routine, or a pointer to a reified structure. This made the manual inspection of the system's results reasonably fast and straightforward.
4. The NN served as an argument of the main verb of the clause, which permits clause-level disambiguation using selectional constraints. If the NN was, for example, located in a parenthetical expression or used as an adjunct, then disambiguation would rely much more heavily on reference resolution and extrasentential context.

This pruning of candidate contexts was carried out automatically, with supplementary manual inspection to weed out processing errors (e.g., not recognizing that a compound contained three, not two, nouns). After this pruning, 72% of the examples initially extracted were deemed within purview of the evaluation, resulting in 935 examples.

The manual checking of the system's results was carried out by a graduate student under the supervision of a senior developer. The manual vetting involved reading the portion of the TMR(s) that represented the meaning of the compound and determining whether it was correct in the context.

The evaluation results overall were positive: the system returned the appropriate decision when it could be expected to do so. What is most interesting is what the system got wrong and why. There were three main sources of errors—lexical idiosyncrasy, polysemy/ambiguity, and metaphorical usage—which we describe in turn.

Lexical idiosyncrasy. Most mistakes involved lexically idiosyncratic compounds—that is, ones whose meanings need to be explicitly recorded in the lexicon rather than dynamically computed using standard expectations. For example (in plain English rather than the ontological metalanguage):

1. *Talk program* was incorrectly analyzed as a social event whose purpose was either conversation or lecturing—as might be plausible, for example, as an activity for nursing home residents to keep them socially active. The intended meaning, however, was a radio or TV program that involves talking rather than, say, music or drama.

2. *College education* and *public education* were incorrectly analyzed as teaching about college and society, respectively, using the construction that would have been correct for *science education* or *history education.*
3. *Pilot program* was analyzed as a social event that benefits airplane pilots, which is actually plausible but is not the meaning (*feasibility study*) that was intended in the examples.
4. *Home life* was analyzed as the length of time that a dwelling could be used, employing a construction intended to cover compounds like *battery life* and *chainsaw life.*

In some cases, these errors pointed to the need to further constrain the semantics of the variables in the construction that was selected. However, more often, the compounds simply needed to be recorded as constructions in the lexicon along with their not-entirely-predictable meanings.

Polysemy/ambiguity. In some cases, a compound allowed for multiple interpretations, even though people might zero in on a single one due to its frequency, their personal experience, or the discourse context. When the system recognized ambiguities, it generated multiple candidate interpretations. To cite just a few examples:

1. *Basketball program* was analyzed as a program of activities dedicated either to basketballs as objects (maybe they were being donated) or to the game of basketball.
2. *Oil spill* was analyzed as the spilling of either industrial oil or cooking oil.
3. *Ship fleet* was analyzed as a set of sailing ships or spaceships.

A curious ambiguity-related error was the analysis of *body part* as a part of a human, since that compound was recorded explicitly in the lexicon during our work on the MVP application. In a particular corpus example, however, it referred to a car part.

As mentioned earlier, the meaning of compounds must be incorporated into the meaning of the discourse overall, and this is part of our full microtheory of compounding. However, in order to keep this experiment as simple and focused as possible, we put the system at an unfair disadvantage. We forced it to accept the default NN interpretation that was generated using a construction without allowing it to reason further about the context; yet we penalized it if that default interpretation was incorrect! This is a good example of the trade-offs we must accept when, for purposes of evaluation, we extract particular linguistic phenomena from the highly complex, multistage process of NLU.

Metaphorical usage. Metaphorical uses of NNs are quite common. For example, in (9.1) both *rabbit holes* and *storm clouds* are used metaphorically.

(9.1) He also alerts investors to key financial *rabbit holes* such as accounts receivable and inventories, noting that sharply rising amounts here could signal *storm clouds* ahead. (WSJ)

In some cases, automatically detecting metaphorical usage is straightforward, as when the NN is preceded by the modifier *proverbial*.

(9.2) "They have taken the proverbial *atom bomb* to swat the fly," says Vivian Eveloff, a government issues manager at Monsanto Co. (WSJ)

In other cases, it can be difficult to detect that something other than the direct meaning is intended.

In addition to non-compositionality and residual ambiguity, our work on NN compounding has revealed other challenges. For example, certain classes of compounds are very difficult to semantically analyze, even wearing our finest linguistic hats. A star example involves the headword *scene*, used in compounds such as *labor scene*, *drug scene*, and *jazz scene*. The meanings of these compounds can only be adequately described using full ontological scripts—a different script for each kind of scene. Anything less, such as describing the word *scene* using an underspecified concept like SCRIPT-INDICATOR, would just be passing the buck.

Even if NNs are not as semantically loaded as *scene* compounds, many more than one might imagine are not fully compositional and, therefore, must be recorded as fixed expressions. In fact, most of the NNs that we recorded as headwords in the lexicon prior to the evaluation study were analyzed correctly, which suggests that our lexicalization criteria are appropriate. Of course, occasionally we encountered an unforeseen point of ambiguity, as in the case of *body part*, referred to earlier.

To summarize the NN compounding experiment:

- It validated the content and utility of the portion of the microtheory tested.
- The system worked as expected; that is, it faithfully implemented the model.
- The lexicon needs to be bigger: there is no way around the fact that language is in large part not semantically compositional.
- The experimental setup did not address the need for contextual disambiguation of nominal compounds.
- It can be difficult to automatically detect certain kinds of mistakes when the wrong interpretation seems to work out fine, as in the case of NNs being used metaphorically.

Spoiler alert: This list of experimental outcomes will largely be the same for the rest of the experiments we describe here.

9.2.2 Multiword Expressions

As explained in section 4.3, there is no single definition of *multiword expression* (MWE). For the evaluation reported in McShane, Nirenburg, and Beale (2015), we defined *MWEs of interest* as those lexical senses whose syn-struc zones included one or more specific words that were not prepositions or particles. For example, *cast-v3* (*X {cast} a spell on/over Y*) requires the direct object to be the word *spell*; similarly, *in-prep15* (*X be in surgery*) requires the object of the preposition to be the word *surgery*. Neither the inventory of MWEs covered in the lexicon nor the lexicon entries themselves were modified before evaluation: all evaluated MWEs were recorded during regular lexical acquisition over prior decades.

The evaluation worked as follows. The system automatically identified 382 MWEs of interest in our lexicon and then used a string-based (nonsemantic) method to search the *Wall Street Journal* corpus of 1987 for sentences that might contain them—"might" because the search method was underconstrained and cast a wide net. The requirements of that search were that all lexically specified roots in the MWE occur within six tokens of the headword. For example, to detect candidate examples of the MWE *something {go} wrong with X*, the word *something* had to be attested within six tokens preceding *go/went/goes*, and the words *wrong* and *with* had to be attested within six tokens following *go/went/goes*. This filtering yielded a corpus of 182,530 sentences, which included potential matches for 286 of our 382 target MWEs. We then selected the first 25 candidate hits per MWE, yielding a more manageable set of 2,001 sentences, which were syntactically parsed. If the syntactic parse of a sentence did not correspond to the syntactic requirements of its target MWE—that is, if the actual dependencies returned by CoreNLP did not match the expected dependencies recorded in our lexicon—the sentence was excluded. (Recall that the initial candidate extraction method was quite imprecise—we did not expect it to return exclusively sentences containing MWEs.) This pruning resulted in 804 sentences that syntactically matched 81 of our target MWEs. We then randomly selected a maximum of 2 sentences per target MWE, resulting in an evaluation corpus of an appropriate size: 136 sentences.

These 136 sentences were semantically analyzed in the usual way. The analyzer was free to select any lexical sense for each word of input, either using or not using MWE senses. To put the lexical disambiguation challenge in perspective, consider the following:

- The average sentence length in the evaluation corpus was 22.3 words.
- The average number of word senses for the headword of an MWE was 23.7. This number is so high because verbs such as *take* and *make* have over 50 senses apiece due to the combination of productive meanings and light-verb usages (e.g., *take a bath*, *take a nap*, *take sides*).
- The average number of word senses for each unique root in the corpus was 4.

To summarize, the system was tasked with resolving the syntactic and semantic ambiguities in these inputs using an approximately 30,000-sense lexicon that was not tuned to any particular domain. One developer manually inspected the system's results and another carried out targeted double-checking and selective error attribution.

Since the TMRs for long sentences can run to several pages, we used a TMR-simplification program to automatically extract the minimal TMR constituents covered by the candidate

MWE. For example, in (9.3) and (9.4), the listed TMR excerpts were sufficient to determine that the MWEs (whose key elements are in italics) were treated correctly.[5]

(9.3) The company previously didn't *place* much *emphasis on* the development of prescription drugs and relied heavily on its workhorse, Maalox. (WSJ)

```
EMPHASIZE-1⁶
   AGENT    FOR-PROFIT-CORPORATION-1
   THEME    DEVELOP-1
```

(9.4) "I'm sure nuclear power is good and safe, but it's impossible in the Soviet bloc," says Andrzej Wierusz, a nuclear-reactor designer who *lost* his *job* and was briefly jailed after the martial-law crackdown of 1981. (WSJ)

```
ASPECT-1
   SCOPE    WORK-ACTIVITY-1
   PHASE    end

WORK-ACTIVITY-1
   AGENT    HUMAN-1
```

The questions posited in the evaluation were:

- Did the system correctly identify sentences in which an MWE was used?
- Did it correctly compute the meaning of the MWE portion of those sentences?

Note that the latter does not require correctly disambiguating all words filling variable slots in MWEs, since that would complicate the evaluation tenfold, forcing it to cover not only lexical disambiguation overall but also coreference resolution.

In most cases, these evaluation criteria meant that the EVENT head of the TMR frame representing the MWE's meaning needed to be selected correctly—such as EMPHASIZE in (9.3). But in some cases, multiple elements contribute to the core meaning of an MWE, so all of them needed to be correct. For example, in (9.4) the combination of the ASPECT frame—with its "PHASE end" property—and the WORK-ACTIVITY frame represents the core meaning of the MWE.

The decision about correctness was binary. If the needed TMR head was (or heads were) correct, then the MWE interpretation was judged correct; if not, the MWE interpretation was judged incorrect.

In many cases, the system correctly analyzed more than what was minimally needed for this evaluation. For example, in (9.3) it correctly disambiguated the fillers of the AGENT and THEME case roles. It selected FOR-PROFIT-CORPORATION as the analysis of the ambiguous word *company* (which can also refer to a set of people), and it selected DEVELOP-1 as the analysis of *development* (which can also refer to a novel event or a residential area). However, to reiterate, we did not require that case role fillers be correctly disambiguated in order to mark an MWE interpretation as correct because this can require much more than clause-level heuristics. For example, in (9.5), the MWE analysis was correct: *to look forward to X* means (roughly) to *want* the event or state of affairs X to occur, which is represented in the TMR by the highest value of volitive modality scoping over X.

(9.5) We *look forward to* the result.

MODALITY-1	
SCOPE	ANY-NUMBER-1
VALUE	1
ATTRIBUTED-TO	SET-1
TYPE	volitive

However, the filler of one of the slots in this TMR—the SCOPE of the modality—is probably not correct. This TMR says that what was looked forward to was some number (ANY-NUMBER), which is one sense of *result*, whereas what is probably being looked forward to is some state of affairs—another sense of *result*. However, in all fairness, the system *could* be correct since the sentence could be uttered in a math class by students waiting for their resident genius to solve a problem. Lacking extra-clausal heuristic evidence, the system arrived at comparable scores for both analyses and randomly selected between them.[7]

Examples (9.6) and (9.7) offer further insights into why we did not fold into the evaluation the disambiguation of case role fillers. All four salient case roles in these examples of the MWE *X {pose} problem for Y* were analyzed incorrectly, even though analysis of the MWE was correct.

(9.6) The changing image did however *pose* a *problem* for the West. $_{(WSJ)}$

(9.7) But John McGinty, an analyst with First Boston Corp., said he believed dissolution of the venture won't *pose* any *problem* for Deere. $_{(WSJ)}$

Two of the errors—the analyses of *the West* and *Deere*—were due to the mishandling of proper names (something handled by the CoreNLP tool set, whose preprocessing results we import). One error—the analysis of *the changing image*—could not be correctly disambiguated using the sentence-level context provided by our examples: that is, *image* can be a pictorial representation or an abstract conceptualization. And the final error—the analysis of *dissolution of the venture*—results from a failure to simultaneously recognize the metaphorical usage of *dissolve* and select the correct sense of the polysemous noun *venture*. These examples underscore just how many different factors contribute to making NLU as difficult as it is.

In some cases, the system did not select the MWE sense of a lexical item that it should have preferred. Instead, it analyzed the input compositionally. The reasons were not always apparent, apart from the fact that the scoring bonus for MWE analyses over compositional ones is relatively minor. Clearly, other preferences in the analysis of the sentence overall had a deciding role.

To reiterate a point made earlier, the system had to select from an average of 23.7 word senses for each MWE head, each having their own inventories of expected syntactic and semantic constraints, which competed to be used in the analysis of each input. So, although our approach to NLU and the system implementation are as transparent as they can be, the effects of combinatorial complexity cannot always be untangled.

Many of the errors in processing MWEs can be obviated through additional knowledge acquisition: namely, by acquiring more MWEs, by adding more senses to existing MWEs, and by more precisely specifying the rule-in/rule-out constraints on MWEs. This was discussed in section 4.3.5.

As promised, we can summarize the results of this experiment using the same points as for the nominal compounding experiment:

- The experiment validated the content and utility of the portion of the microtheory tested.
- The system worked as expected.
- The lexicon needs to be bigger and some of its entries need to be more precisely specified.
- Some problems, such as residual ambiguity, need to be resolved by methods that were not invoked for the experiment.
- It is difficult to automatically detect certain kinds of mistakes when the wrong interpretation seems to work fine, as in the case of metaphorical usage.

9.2.3 Lexical Disambiguation and the Establishment of the Semantic Dependency Structure

The experiment reported in McShane, Nirenburg, and Beale (2016) focused on the system's ability to carry out lexical disambiguation and establish the semantic dependency structure.[8] As always, we attempted to give the system a fair opportunity to demonstrate its capabilities while neither overwhelming it with complexity nor reducing the endeavor to a toy exercise. The system was required to

1. disambiguate head verbs: that is, specify the EVENT needed to express their meaning as used in the context; and
2. establish which case roles were needed to link that EVENT to its semantic dependents.

We did not evaluate the disambiguation of the fillers of case role slots for the same reason as was described earlier: this often requires coreference resolution and/or other aspects of discourse analysis that would have made the evaluation criteria impossibly complicated.

The evaluation corpus included four Sherlock Holmes stories: "A Scandal in Bohemia," "The Red-Headed League," "A Case of Identity," and "The Boscombe Valley Mystery" (hereafter referred to collectively as *S-Holmes*). We selected these because they are freely available from Project Gutenberg (EBook #1661) and, to our knowledge, nobody has recorded linguistic annotations of these works, so there can be no question that the system operated on unenhanced input.

We first selected an inventory of verbs of interest from our lexicon, all of which had the following two properties: (a) they had at least two senses, so that there would be a disambiguation challenge, and (b) those senses included syntactic and/or semantic constraints that allowed for their disambiguation. Ideally, all lexical senses would include such disambiguating constraints, but this is not always possible. A frequent confounding case involves pairs of physical and metaphorical senses that take the same kinds of arguments. For example, if person A attacks person B, A might be physically assaulting or criticizing B, something that can only be determined using additional knowledge about the context.

The system automatically selected, and then semantically analyzed, 200 sentences containing verbs that corresponded to the selection criteria. We then manually checked the correctness of the resulting TMRs. One developer carried out this work with selective contributions from another. The evaluation involved not only identifying errors but also attempting to trace them back to their source so that they could be fixed to improve future system functioning. We did not amend the lexicon or ontology in any way to prepare for this evaluation.

The experimental setup included challenges of a type that are often filtered out of mainstream NLP evaluation suites. For instance, some examples did not contain sufficient information to be properly disambiguated, as by having semantically underspecified pronouns fill key case roles; other examples reflected what might be considered nonnormative grammar. However, considering the importance of automatically processing nonstandard language genres (texting, email, blogs), we felt it appropriate to make the system responsible for all encountered phenomena.

The two main sources of errors, beyond singletons that are of interest only to developers, were lexical lacunae and insufficiencies of the experimental design, which we describe in turn.

Lexical lacunae. Most disambiguation errors resulted from the absence of the needed lexical sense in the lexicon. Often, the missing sense was part of an idiomatic construction that had not yet been acquired, such as *draw the blinds* in (9.8).

(9.8) The *drawn* blinds and the smokeless chimneys, however, gave it a stricken look. $_{\text{(S-Holmes)}}$

In other cases, the needed semantic representation (sem-struc) was available in the lexicon but it was not associated with the needed syntactic realization (syn-struc). For example, for the system to correctly process (9.9), the lexicon must permit *announce* to take a direct object. However, the sense available in the lexicon—which did semantically describe the needed meaning of *announce*—required a clausal complement.

(9.9) She became restive, insisted upon her rights, and finally *announced* her positive intention of going to a certain ball. $_{\text{(S-Holmes)}}$

A trickier type of lexical lacuna involves grammatical constructions that are not sufficiently canonical (at least in modern-day English) to be recorded in the lexicon. For example, the verb *pronounce* in (9.10) is used in the nonstandard construction *X pronounces Y as Z*.

(9.10) "I found the ash of a cigar, which my special knowledge of tobacco ashes enables me to *pronounce* as an Indian cigar." $_{\text{(S-Holmes)}}$

Such sentences are best treated as unexpected input, to be handled by the recovery procedures described in section 3.2.4.

Insufficiencies of experimental design. Since we did not invoke the coreference resolution engine for this experiment, we should have excluded examples containing the most underspecified pronominal case role fillers: *it*, *they*, *that*, and *this*. (By contrast, personal pronouns that most often refer to people—such as *he*, *she*, *you*, and *we*—are not as problematic.) An argument like *it* in (9.11) is of little help for clause-level disambiguation.

(9.11) I walked round it and *examined it* closely from every point of view, but without noting anything else of interest. $_{\text{(S-Holmes)}}$

For this example, the system selected the abstract event ANALYZE, which expects an ABSTRACT-OBJECT as its THEME. It should have selected the physical event VOLUNTARY-VISUAL-EVENT, which expects a PHYSICAL-OBJECT as the THEME. Of course, if this experiment had included coreference and multiclause processing, then the direct object of *examined* would corefer with the previous instance of *it*, which must refer to a physical object since it can be walked around.

Were the results of, and lessons learned from, this experiment the same as for the previous ones? Indeed, they were.

- The experiment validated the content and utility of the portion of the microtheory tested.
- The system worked as expected.
- The lexicon needs to be bigger.
- Some problems, such as residual ambiguity, need to be resolved by methods that were not invoked for the experiment.
- It is difficult to automatically detect certain kinds of mistakes when the wrong interpretation seems to work fine (as in the case of metaphorical usage).

One additional note deserves mention. Since most errors were attributable to missing or insufficiently precise verbal senses, and since we used the verbs in our lexicon to guide example selection, we could have avoided most mistakes by using a different experimental setup. That is, before the evaluation we could have optimized the inventory of lexical senses for each selected verb, particularly by boosting the inventory of recorded multiword expressions. This would have required some, but not a prohibitive amount of, acquisition time. It would likely have substantially decreased the error rate, and it would likely have better highlighted the system's ability to manipulate competing syntactic and semantic constraints during disambiguation. However, an experiment of this profile would have less realistically conveyed the current state of our lexicon since we would not have done that enhancement for all of its verbs, not to mention all of the verbs in English. It would be hard to argue that either of these task formulations is superior to the other given that both would confirm the core capability of lexical disambiguation that was being addressed.

9.2.4 Difficult Referring Expressions

McShane and Babkin (2016a) describe the treatment and evaluation of two classes of referring expressions that have proven particularly resistant to statistical methods: broad referring expressions (e.g., pronominal *this*, *that*, and *it*; see section 5.3) and third-person personal pronouns (see section 5.2).

Broad referring expressions are difficult not only because they can refer to spans of text of any length (i.e., one or more propositions) but also because they can refer to simple noun phrases, and the system does not know a priori which kind of sponsor it is looking for. Third-person personal pronouns, for their part, are difficult because semantic and/or pragmatic knowledge is often required to identify their coreferents. We prepared the system to treat difficult referring expressions by defining lexico-syntactic constructions that predicted the coreference decisions. These constructions do not cover a large proportion of instances in a corpus (i.e., they have low recall), but they have proven useful for what they do cover.

For the evaluation, the system had to (a) automatically detect, in an unrestricted corpus, which instances of difficult referring expressions matched a recorded construction and then (b) establish the coreference link predicted by that construction. This evaluation is more difficult to summarize than others because each construction was evaluated individually. That is why select evaluation results were reported in the sections that introduced the microtheories themselves (sections 5.2.2 and 5.3).

For the development and evaluation portions of this experiment, we used different portions of the English Gigaword corpus (Graff & Cieri, 2003; hereafter, Gigaword). For the first time, we compiled a gold standard against which the system would be evaluated. This involved two steps. First, the system identified the examples it believed it could treat confidently (since they matched recorded constructions). Then two graduate students and one undergraduate student annotated those examples according to the following instructions:

[NE]	If the selected entity is not actually a referring expression (e.g., pleonastic *it*), type [NE] before the example.
[ ]	If there is a single perfect or near-perfect antecedent, surround it with brackets.
[Mult]	If there is more than one possible antecedent, type [Mult] before the example and use multiple sets of brackets to indicate the options.
[Close]	If an available text string is close to the needed antecedent but not a perfect match, type [Close] before the example and use brackets to show the best available antecedent.
[Impossible]	If no text string captures the meaning of the antecedent, type [Impossible] before the example.
[Prob]	If there is some other problem with the context (e.g., it is unintelligible) type [Prob] before the example.

Annotators were shown a few worked examples but given no further instructions. This contrasts with the mainstream NLP annotation efforts that involve extensive guidelines that are painstakingly compiled by developers and then memorized by annotators.

When the annotation results were in, senior developers manually reviewed them (with the help of the program KDiff3[9]) and selected which ones to include in the gold standard. Often we considered more than one result correct. Occasionally, we added an additional correct answer that was not provided by the annotators. As expected, there was a considerable level of interannotator disagreement, but most of those differences were inconsequential. For example, different annotators could include or exclude a punctuation mark, include or exclude a relative clause attached to an NP, include or exclude the label [Close], or select different members of a coreference chain as the antecedent. We did not measure interannotator agreement because any useful measure would have required a well-developed approach to classifying important versus inconsequential annotation decisions—something that we did not consider worth the effort. To evaluate the system, we semiautomatically (again, with the help of KDiff3) compared the system's answers to the gold standard, calculated precision, and carried out error analysis toward the goal of system improvement.

As with previous experiments, this one validated the content and utility of the portion of the microtheory tested, and the system worked as expected. It pointed to the need for additional knowledge engineering on the constructions themselves, particularly on specifying rule-out conditions. This evaluation differed from previous ones in that we first created a gold standard and then tested system results against it. That process was more expensive and time-consuming than our previous approaches to vetting system outputs, but it was not prohibitively heavy because the decision-making about the coreference relations was relatively straightforward. However, as we will see in the next section, applying the same gold standard–first methodology to the task of VP ellipsis was a different story entirely.

9.2.5 Verb Phrase Ellipsis

To date, we have carried out two evaluations of different iterations of our model for VP ellipsis. The first, reported in McShane and Babkin (2016b), treated only elided VPs, whereas the second, reported in McShane and Beale (2020), treated both elided and overt-anaphoric VPs—the latter realized as *do it*, *do this*, *do that*, and *do so*. We use the former experiment for illustration because it involved a more formal evaluation setup and, therefore, offers more discussion points for this chapter on evaluation.

The 2016 system—called ViPER (Verb Phrase Ellipsis Resolver)—had to

1. identify instances of VP ellipsis in an unconstrained corpus (Gigaword);
2. determine which instances it could treat using its repertoire of resolution strategies; and
3. identify the text string that served as the sponsor.

As our original report explains, this definition of *resolution* is partial in that it does not account for the important semantic decisions that we describe in section 5.5. However, the ability of this module to detect which contexts can be treated by available resolution strategies and to point out the sponsor counts as a significant contribution to the very demanding challenge of full VP ellipsis resolution.

The aspect of this experiment that is most salient to this chapter involves the repercussions of our decision to create a gold standard first, by annotating examples in the way that is traditional for mainstream (machine learning–oriented) NLP tasks. We anticipated a lot of eventualities and incorporated them into the annotation instructions. For example:

- Some of the examples that the system selected to treat might not actually be elliptical.
- The sponsor might be outside the provided context.
- There might be no precisely correct sponsor in the linguistic context at all.
- There might be multiple reasonable sponsor selections.

A few examples will serve to illustrate tricky cases, with their complexities indicated in square brackets.

(9.12) [Either of the previous mentions in the chain of coreference is a valid sponsor.]
However, Beijing still [rules the country with harsh authoritarian methods] in the provinces and will [continue to do so] for as long as it can __. (Gigaword)

(9.13) [The direct object could be included (which is more complete) or excluded (which sounds better).]
Nuclear power may [[give] NASA's long-range missions] the speed and range that combustion engines can not __, but research is sputtering for lack of funds. (Gigaword)

(9.14) [The first conjunct, 'go out and', may or may not be considered part of the sponsor.]
"We had to [go out and [play the game]] just like they did __." (Gigaword)

(9.15) [The actual sponsor is the noncontiguous 'pull off'; we did not allow for noncontiguous sponsors in order to avoid complexity, but this decision had some negative consequences.]
"I feel I can [[pull] that shot off]; that's just one of those I didn't __." (Gigaword)

(9.16) [The sponsor can, itself, be elided. Here, the actual resolution should be 'let them disrupt us'.]
"They can disrupt you if you [let them], and we didn't ___." (Gigaword)

It would have taken a very detailed, difficult-to-master set of annotation rules to ensure that annotators were highly likely to make the same sponsor selection.

Given our lenient annotation conventions, for 81% of the examples in the evaluation suite (320 out of 393), all student annotators agreed on the sponsor, and that answer was considered correct (i.e., it was not vetted by senior developers). For the other 73 examples, senior developers had to decide which answer(s) qualified as correct. Then, in order to make the evaluation results as useful as possible, we created guidelines to judge ViPER's answers as correct, incorrect, or partially correct.

Correct required that the answer be exactly correct. *Incorrect* included three eventualities:

- Sentences that ViPER thought were elliptical but actually were not;
- Sentences whose sponsor was not in the provided context but ViPER pointed to a sponsor anyway; or
- Cases in which ViPER either did not identify the head of the sponsor correctly or got too many other things wrong (e.g., the inclusion or exclusion of verbs scoping over the sponsor head) to qualify for partial credit.

The second eventuality is actually the most interesting since it represents a case that would never make it into traditional evaluation tasks—that is, the case in which the answer is not available and the system is required to understand that. In traditional evaluation setups, examples that are deemed by task developers to be too difficult or impossible are excluded from the start.

Partial credit covered several eventualities, all of which involved correctly identifying the verbal head of the sponsor but making a mistake by including too many other elements (e.g., modal scopers) or excluding some necessary ones. Getting the sponsor head right is actually a big deal because it shows not only that the system can identify the sponsor clause but also that it understands that the example is, in principle, within its ability to treat.

It is important to note that many of the problems of string-level sponsor selection simply go away when the full NLU system is invoked, since actual VP resolution is done at the level of TMRs (semantic analyses), not words.

ViPER's methods for identifying treatable cases of VP ellipsis and identifying the sponsor worked well, as the evaluation numbers reported in the paper show. For reasons described earlier, the system itself chose what to treat and what not to treat, and we made no attempt to calculate its recall over the entire corpus. This would actually not have been trivial because our VP-detection process did not attempt full recall—ellipsis detection being a difficult problem in its own right. Instead, our goal in developing detection methods was to compile a useful corpus with relatively few false positives.

In terms of system operation, this experiment yielded no surprises. However, we did learn to think thrice before undertaking any more annotation-first approaches to evaluating system operation for a problem as complex as VP ellipsis. At least in this case, the game was not worth the candle. We would have obtained the same information about the problem space and system operation if developers had reviewed system results without a precompiled gold standard. In fact, when it came time to do our next evaluation of VP ellipsis resolution (along with overt-anaphoric VP resolution), we did not create a gold standard first. Instead, developers vetted the results, and we called the process a *system-vetting experiment* rather than an evaluation (McShane & Beale, 2020). In fact, the latter approach was not only faster and cheaper but also more useful than the evaluation just described because we were not bound to one round of experimentation for reasons of cost/practicality. Instead, we iteratively developed and vetted the model and system in a way that best served our scientific and engineering goals.

9.3 Holistic Evaluations

The evaluation reported in McShane et al. (2019)—as well as a follow-up, unpublished evaluation that we present for the first time here—attempted to assess the system's ability to semantically interpret sentences from an open corpus using processing stages 1–5. As a reminder, this covers all modules before Situational Reasoning. Constraining the scope to non-situational semantics was necessary because neither our agents, nor any others within the current state of the art, have sufficiently broad and deep knowledge to engage in open-domain situational reasoning. The evaluations were carried out using a portion of the COCA corpus (Davies, 2008–). These experiments, like previous ones, required the agent to select those examples that it thought it could treat correctly.

The biggest challenges in applying our NLU engine to the open domain are incomplete coverage of the lexicon and incomplete coverage of our microtheories. These limitations are key to understanding the evaluation processes and outcomes, so let us consider them in more detail.

Incomplete lexicon. As a reminder, the lexicon that LEIAs currently use contains approximately 30,000 senses, which include individual words, multiword expressions, and constructions. This size is substantial for a deep-semantic, knowledge-based system, but it is still only a fraction of what is needed to cover English as a whole. In formulating our first holistic experiment, we attempted to account for this limitation by having the system select sentences that seemed to be fully covered by the lexicon. That is, we made the clearly over-simplifying assumption that if the lexicon contained the needed word in the needed part of speech, then there was a good chance that the needed sense was among the available options.

This assumption turned out to be false more often than anticipated, but it wasn't completely unfounded. The knowledge engineers who acquired the bulk of the lexicon some two decades ago were instructed to embrace, rather than back away from, ambiguity. And, in fact, the lexicon amply represents ambiguity. On average, prepositions have three senses each, conjunctions have three, and verbs have two. Ninety-eight verbs have more than five senses each, and the light verbs *make* and *take* have over forty and thirty senses, respectively. Nouns, adjectives, and adverbs average slightly over one sense each. The fact that not *all* senses of all words were acquired from the outset reflects competing demands on acquisition time, not an intentional avoidance of ambiguity. After all, for an open-domain lexicon (unlike a lexicon crafted for a narrowly defined application), there is no advantage to omitting word senses that have a reasonable chance of appearing in input texts.

The fact that a lexicon can contain a lot of senses but still lack the one(s) needed was amply demonstrated in these experiments. Consider just one example. The lexicon contains nineteen senses of *turn*, covering not only the core, physical senses (rotate around an axis and cause to rotate around an axis) but also a large number of multiword expressions in one or more of their senses: for example, *turn in*, *turn off*, *turn around*, *turn away*. Each of these is provided with syntactic and semantic constraints to enable automatic disambiguation. However, although nineteen well-specified senses sounds pretty good, our experiment used practically none of these and, instead, required three senses that the lexicon happened to lack:

- *Turn to*, meaning 'to face (physical)': *She turns to Tripp.* $_{\text{(COCA)}}$
- *Turn to*, meaning 'to seek emotional support from': *People can turn to a woman.* $_{\text{(COCA)}}$
- *Turn to food*, meaning 'overeat in an attempt to soothe one's emotions': *I'd always turn to food.* $_{\text{(COCA)}}$

This means that lexicon lookup is not a reliable guide for determining whether a given input is or is not treatable. Relying on lexicon lookup is tantamount to a child's overhearing a conversation about parse trees and assuming that the trees in question are the big leafy things. The upshot is that the system often thinks it is getting the answer right when, in fact, it is mistaken. Later we will return to the important consequences of this both for system evaluation and for lifelong learning by LEIAs.

Incomplete coverage of microtheories. The second coverage-related complication of holistic evaluations involves microtheories. As readers well understand by now, although our microtheories attempt to cover all *kinds* of linguistic phenomena, they do not yet cover all *realizations* of each one—that will require more work.[10] In our first holistic experiment, we did not directly address the issue of incomplete coverage of microtheories. This resulted in the system's attempting to analyze—and then analyzing incorrectly—inputs containing realizations of linguistic phenomena that we knew were not yet covered.

So, for our second holistic evaluation, we improved the example selection process by formalizing what each microtheory did and did not cover, and we used this knowledge to create a set of sentence extraction filters. This added a second stage to the task of selecting sentences to process as part of the evaluation. First the system extracted sentences that seemed to be covered by the lexicon. Then it filtered out those that contained phenomena that our microtheories do not yet cover.

These filters are not just an engineering hack; they are the beginning of a *microtheory of language complexity*. We do not call it *the* microtheory of language complexity because it reflects a combination of objective linguistic reality and idiosyncratic aspects of our environment.[11] The full inventory of extraction filters combines unenlightening minutiae that we will not report with points of more general interest, which we describe now.

The *intrasentential punctuation mark filter* rules in sentences with intrasentential punctuation marks that are either included in a multiword expression (e.g., *nothing ventured, nothing gained*) or occur in a rule-in position recorded in a list (e.g., commas between full clauses, commas before or after adverbs). It rules out sentences with other intrasentential punctuation marks, which can have a wide variety of functions and meanings, as illustrated by (9.17)—(9.19).

(9.17) She squeezed her eyelids shut, damming the tears. (COCA)

(9.18) Working light tackle, he had to give and take carefully not to lose it. (COCA)

(9.19) Now, think, she thought. (COCA)

The *relative spatial expression filter* excludes sentences containing relative spatial expressions because their meanings (e.g., *to the far left of the table*) can only be fully grounded in a situated agent environment. We are currently developing the associated microtheory within a situated agent environment, not as an exclusively linguistic enterprise.

The *set-based reasoning* and *comparative* filters exclude complex expressions that require constructions that are not yet covered in the lexicon.

(9.20) The second to last thing she said to him was, …. (COCA)

(9.21) In these stories he's always ten times smarter than the person in charge. (COCA)

The *conditional filter* rules in conditionals whose *if*-clause uses a present-tense verb and no modality marker (the *then*-clause can contain anything). It rules out counterfactuals, since counterfactual reasoning has not yet made it to the top of our agenda.

The *multiple negation filter* excludes sentences with multiple negation markers since they can involve long-distance dependencies and complex semantics.

(9.22) Except around a dinner table I had never before, at an occasion, seen Father not sit beside Mother. (COCA)

The *no-main-proposition filter* detects nonpropositional sentences that must necessarily be incorporated into the larger context.

(9.23) As fast as those little legs could carry him. $_{\text{(COCA)}}$

(9.24) Better even than Nat and Jake expected. $_{\text{(COCA)}}$

Note that the system *can* process the latter when it has access to multiple sentences of context, but in the reported experiments it did not.

The *light verb filter* excludes some inputs whose main verb is a light verb: *have*, *do*, *make*, *take*, and *get*. Specifically, it rules in inputs that are covered by a multiword expression that uses these verbs, and it rules out all others. The reason for this filter is that we know that the lexicon lacks many multiword expressions that contain light verbs. And, although the lexicon contains a fallback sense of each light verb that can formally treat most inputs, the analyses generated using those senses are often so much vaguer than the meaning intended by the input that we would evaluate them as incorrect.

For example, in our testing runs, use of the fallback sense led to overly vague interpretations of *take a cab*, *make the case*, and *get back to you*, all of which are not fully compositional and require their own multiword lexical senses. Note that this exclusion is not actually as strict as it may seem because constructions that the lexicon does contain actually cover large semantic nests. For example *have* $+NP_{EVENT}$ means that the subject is the AGENT of the EVENT, which handles inputs like *have an argument*, *have an affair*, and *have a long nap*.

Let us pause to recap where we are in our story of holistic evaluations. Both holistic evaluation experiments encountered the same problem related to lexical lacunae: the lexicon could contain the needed word in the needed part of speech, but not the needed sense (which was often part of a multiword expression). As concerns the coverage of microtheories, the first experiment made clear that we needed to operationalize the agent's understanding of what each microtheory did and did not cover. We did that for the second experiment using the kinds of sentence extraction filters just illustrated.

In the first holistic experiment, which did not use the microtheory-oriented sentence extraction filters, there was a high proportion of difficult sentences that were beyond the system's capabilities. Some of the problems reflected how hard NLU can be, whereas others pointed to suboptimal decisions of experimental design. Starting with the problem that NLU is very difficult, consider the following sets of examples, whose challenges are described in brackets. As applicable, constituents of interest are italicized.

(9.25) [Compositional analysis failed due to a multiword expression not being in the lexicon.]

a. She is *long gone from* the club. $_{\text{(COCA)}}$
b. I *got a good look at* that shot. $_{\text{(COCA)}}$
c. The Knicks *can live with* that. $_{\text{(COCA)}}$
d. But once Miller *gets on a roll*, he can make shots from almost 30 feet. $_{\text{(COCA)}}$
e. I *can't say enough about* him. $_{\text{(COCA)}}$
f. *This better be good.* $_{\text{(COCA)}}$
g. You *miss the point*. $_{\text{(COCA)}}$

(9.26) [A nonliteral meaning was intended but not detected.]
He not only hit the ball, he *hammered* it. (COCA)

(9.27) [It would be difficult, even for humans, to describe the intended meaning given just the single sentence of context.]
a. Training was a way of killing myself without dying. (COCA)
b. The supporting actor has become the leading man. (COCA)
c. This is about substance. (COCA)
d. The roots that are set here grow deep. (COCA)

(9.28) [The intended meaning relies more on the discourse interpretation than on the basic semantic analysis.]
a. It takes two to tango. (COCA)
b. And he came back from the dead. (COCA)

(9.29) [It is unclear what credit, if any, to give to a basic semantic interpretation when a large portion of the meaning involves implicit comparisons, implicatures, and the like.]
a. She's also a woman. (COCA)
b. How quickly the city claimed the young. (COCA)
c. They sat by bloodline. (COCA)
d. I think he is coming into good years. (COCA)
e. Fathers were for that. (COCA)

(9.30) [Without knowing or inferring the domain—the examples below refer to sports—it is impossible to fully interpret some utterances.]
a. The Rangers and the Athletics have yet to make it. (COCA)
b. He hit his shot to four feet at the 16th. (COCA)
c. We stole this one. (COCA)
d. I wanted the shot. (COCA)

As concerns the experimental setup for the first holistic evaluation, two of our decisions were suboptimal. First, we required the TMR for the entire sentence to be correct, which was too demanding. Often, some portion nicely demonstrated a particular functionality, while some relatively less important aspect (e.g., the analysis of a modifier) was wrong. Second, we focused exclusively on examples that returned exactly one highest-scoring TMR candidate (for practical reasons described below). We did not consider cases in which multiple equally plausible candidates were generated—even though this is often the correct solution when sentences are taken out of context. For example, the system correctly detected the ambiguity in, and generated multiple correct candidates, for (9.31) and (9.32).

(9.31) [The fish could be an animal (FISH) or a foodstuff (FISH-MEAT).]
He stared at the fish. (COCA)

(9.32) [*Walls* could refer to parts of a room (WALL) or parts of a person undergoing surgery (WALL-OF-ORGAN)]
He glanced at the walls. (COCA)

There are two reasons—both of them practical—for excluding sentences with multiple high-scoring candidate interpretations. First, since sentences can contain multiple ambiguous strings, the number of TMR candidates can quickly become large and thus require too much effort to manually review. Second, we would have needed a sophisticated methodology for assigning partial credit because, not infrequently, some but not all of the candidates are plausible.

Despite all the linguistic complications and tactical insufficiencies, our first holistic experiment yielded quite a number of satisfactory results, as shown by the following classes of examples.

(9.33) [Many difficult disambiguation decisions were handled properly. For example, this required disambiguating between sixteen senses of *look*.]
He looked for the creek. (COCA)

(9.34) [Many highly polysemous particles and prepositions were disambiguated correctly.]
a. She rebelled against him. (COCA)
b. He stared at the ceiling. (COCA)

(9.35) [Modification (*old*, *white*) and sets (*couple*) were treated properly.]
An old white couple lived in a trailer. (COCA)

(9.36) [Multiword expressions were treated properly.]
He took me by surprise. (COCA)

(9.37) [Dynamic sense bunching allowed the system to underspecify an interpretation rather than end up with competing analyses. For *ask* the system generalized over the candidates REQUEST-INFO, REQUEST-ACTION, and PROPOSE, positing their closest common ontological ancestor, ROGATIVE-ACT, as the analysis.]
I didn't ask him. (COCA)

(9.38) [New-word learning functioned as designed: the unknown word *uncle* was learned to mean some kind of HUMAN since it filled the AGENT slot of ASSERTIVE-ACT.]
The uncle said something to him. (COCA)

Turning to the second holistic experiment, it was different in three ways, the second of which (the use of filters) was already discussed.

1. *Example extraction*. The system sought examples of each verbal sense in the lexicon, still requiring that all other words used in the sentence be covered by the lexicon as well. This bunched results for easier comparative review.

2. *Filters*. We implemented the sentence extraction filters described earlier to automatically weed out linguistic phenomena that were known to not yet be covered by our microtheories.
3. *Defining "correct."* We developed a more explicit definition of a correct TMR, such that a correct TMR represented *a* (possibly not the only available) correct interpretation. In assessing correctness, we tried hard not to allow ourselves to question every decision knowledge engineers made when building the lexicon. For example, all words expressing breeds of dogs are mapped to the concept DOG since we never concentrated on application domains for which distinctions between dog breeds were important. So, a TMR that analyzed *poodle* as DOG would be considered correct. In short, our definition of *correct* allowed for underspecifications deriving from knowledge-acquisition decisions, but it did not allow for actual mistakes. If the lexicon was lacking a needed word sense, the fact that the agent used the only sense available does not make it right. After all, this is an evaluation of semantic analysis; it is not code debugging.

As in the first holistic evaluation, we excluded sentences that offered multiple high-scoring analyses, and we evaluated sentence-level analyses on the whole. The rationale was as before: to avoid introducing excessive complexity into the evaluation setup.

Before turning to the successes of the second holistic evaluation, let us consider some failures. No surprise: most of them were due to missing lexical senses, including multiword expressions and constructions. Here are just a few examples for illustration:

(9.39) [The lexicon contains the direct/physical meaning of the underlined word but not the conventional metaphor, which also must be recorded.]

a. Christians in Egypt worry about the ascent of Islamists. (COCA)

b. Shaw had won the first battle. (COCA)

(9.40) [Although the lexicon contains several nominal senses of *line*, *clothesline* was not among them.]

They wash their clothes, and they hang them on a line. (COCA)

(9.41) [The system found a sense of *wait out* in the lexicon, but that sense expected the complement to be an EVENT (e.g., *wait out the storm*), not an OBJECT. When the complement is an OBJECT, the expression is actually elliptical: the named OBJECT is the AGENT of an unspecified EVENT whose meaning must be inferred from the context. Our lexicon already contains treatments of lexemes requiring a similar ellipsis detection and resolution strategy (e.g., covering *A raccoon caused the accident*); it just happens to lack the one needed for this input.]

They waited out the bear. (COCA)

(9.42) [*Would always* is a multiword expression indicating that the event occurred repeatedly in the past. The lexicon lacked this multiword sense at the time of the evaluation run.]

My mother would always worry. (COCA)

(9.43) [This entire sentence is idiomatic, conveying a person's inability to think of the precise word(s) needed to express some thought. Analyzing it compositionally is just wrong.]
I wish I had the word. (COCA)

In many cases, the meaning of a sentence centrally required understanding implicatures. That is, the sentence did not merely give rise to implicatures; instead, arriving at the basic meaning required going beyond what was stated. We did not give the system credit for implicature-free interpretations in such cases, even if they contained correct aspects of the full meaning.

(9.44) [This does not simply mean that individuals representing the IRS will serve as collaborating agents with *them* on something. It means working out a way for the people to pay off their tax burden to the IRS.]
The IRS will work with them. (COCA)

(9.45) [This sentence does not involve a single instance of writing a single sentence, as the basic analysis would imply. Instead, it means that he is an excellent writer, an interpretation that relies on a construction (cf. *She plays a mean horn*; *He makes a delicious pizza pie*).]
He writes a great sentence. (COCA)

(9.46) [Neither of the auxiliary senses of *can* in the lexicon (indicating ability and permission) is correct for this sentence. Here, *can* means that they have written, and have the potential to write in the future, such emails.]
They can also write some pretty tough e-mails. (COCA)

(9.47) [This elliptical utterance requires the knowledge that *Parks and Recreation* is a department that is part of the city government.]
I work for the city, Parks and Recreation. (COCA)

(9.48) [This implies that golfers do not ride in golf carts, not that they simply walk around in principle.]
Golfers have always walked in competitive tournaments. (COCA)

(9.49) [A full interpretation requires identifying which features of a felon are salient.]
I was no better than a felon. (COCA)

(9.50) [This refers to particular political actions (perhaps protesting or contacting voters), not strolling around.]
I will walk for candidates. (COCA)

(9.51) [Compositionally, this means that members of the cabinet are coagents of voting. However, there is a political sense—relevant only if the indicated people represent appropriate political roles—that means *to vote the same as a higher-positioned politician*.]
The cabinet voted with Powell. (COCA)

It should be clear by now why counting things is a poor yardstick for evaluation. There's something not quite fair about marking an open-domain system wrong for not inserting golf carts into the interpretation of *Golfers have always walked in competitive tournaments*—especially when many human native speakers of English probably don't know enough about golf to understand what was meant either.

So, as in the first holistic evaluation, in the second one we oriented around qualitative rather than quantitative analysis, focusing on (a) what the system *did* get right, as proof of concept that our approach and microtheories are on the right track, and (b) lessons learned.

We halted the experiment after collecting fifty sentences that the system processed correctly, since by that point we had learned the big lessons and found ourselves just accumulating more examples of the same. Those fifty sentences are listed below, with selective, highly abbreviated comments on what makes them interesting.

(9.52) [*Told* was disambiguated from six senses of *tell*.]
Shehan told him about the layoffs. (COCA)

(9.53) [There are multiple propositions.]
Dawami says neighbors told her they heard Hassan beat the girl. (COCA)

(9.54) [There are multiple propositions and volitive modality from *hope*.]
I told him I hope he wins. (COCA)

(9.55) [The TMR is explanatory: TEACH (THEME INFORMATION (ABOUT BUDDHISM))]
Monks teach you about Buddhism. (COCA)

(9.56) [*Poetry* is described as LITERARY-COMPOSITION (HAS-STYLE POETRY). *Write* was disambiguated from eight senses.]
I write poetry. (COCA)

(9.57) [*Worsen* is described as a CHANGE-EVENT targeting the relative values of evaluative modality in its PRECONDITION and EFFECT slots.]
You'd worsen the recession. (COCA)

(9.58) [*About* was disambiguated due to the inclusion of the multiword expression *worry about*.]
I worried about him. (COCA)

(9.59) [Turin was correctly analyzed as CITY (HAS-NAME 'Turin').]
He worried about Turin's future. (COCA)

(9.60) [*On* was disambiguated due to the inclusion of the multiword expression *work on*.]
I'll work on the equipment. (COCA)

(9.61) [*And* creates a DISCOURSE-RELATION between the meanings of the propositions. The instances of *we* are coreferred.]
We work and we eat. (COCA)

(9.62) [*Fast* is analyzed as (RAPIDITY .8).]
I work fast. (COCA)

(9.63) [*Championship* is an unknown word that was learned as meaning some sort of EVENT.]
The team won the championship. (COCA)

(9.64) [The multiword sense for *watch out* is used. There is obligative modality from *should*.]
Everybody should watch out. (COCA)

(9.65) [The causative sense of *wake up* is correctly integrated with the obligative modality from *should*.]
They should wake you up. (COCA)

(9.66) [The instances of *I* are correctly coreferred. *Wake up* is described using the *end* value of ASPECT scoping over a SLEEP event whose EXPERIENCER is the HUMAN indicated by *I*.]
I slept till I woke up. (COCA)

(9.67) [*Wake up* is analyzed as above. *With* is correctly interpreted as BESIDE.]
Jack wakes up with Jennifer. (COCA)

(9.68) [The conjunction structure is analyzed as a set. The proper names are correctly analyzed as two cities and a state with their respective names.]
He has visited Cincinnati, Tennessee, and Miami. (COCA)

(9.69) [The analysis explicitly points to all the concepts relevant for reasoning: COME (DESTINATION (PLACE (LOCATION-OF CRIMINAL-ACTIVITY))).]
They visited the crime scenes. (COCA)

(9.70) [The modification and nominal compound are treated correctly.]
Ghosts visit a grumpy TV executive. (COCA)

(9.71) [The multiword expression *turn down* allows the system to disambiguate among twenty-six verbal senses of *turn*.]
Thoreen turned down the offer. (COCA)

(9.72) [This sense of *use* is underspecified, instantiating an EVENT whose INSTRUMENT is the set ONION, GARLIC, CUMIN. Although people would probably infer that seasoning food is in question, this is not a necessary implicature: this sentence could also refer to gardening or even painting.]
You use onion, garlic, and cumin. (COCA)

(9.73) [This uses the multiword expression *turn off*.]
He turns off the engine. (COCA)

(9.74) [This uses the multiword expression *turn on*.]
Somebody turned on a television. (COCA)

(9.75) [The question is interpreted as a request for information: namely, the agent of the proposition.]
Who trained them? (COCA)

(9.76) [Two different senses of *agent* work here: an intelligence agent and the agent of an event more generically. This instance of residual ambiguity launched an automatic sense-bunching function that generalized to their most common ancestor, HUMAN. It left a trace of that generalization in the TMR, in case the agent later chooses to seek a more specific interpretation using discourse-related reasoning.]
We train our agents. (COCA)

(9.77) [This uses the multiword expression *track down*. Also note that the lexicon acquirer chose to attribute null semantics to *always* because it rarely, actually, means *always*! For example, *He is always teasing me* does not literally mean all the time. One can disagree with this acquisition decision, but it was a conscious, documented decision that we did not overturn for this experiment.]
Chigurh always tracks him down. (COCA)

(9.78) [This uses *belief* modality from *think* and the multiword expression *find out*.]
I think we found out. (COCA)

(9.79) [This uses the multiword expression *think about*.]
They think about the road. (COCA)

(9.80) [*For* was correctly disambiguated (from among eighteen senses) as PURPOSE.]
We stayed for lunch. (COCA)

(9.81) [*Forever* is analyzed as 'TIME-END never'.]
Joseph would stay there forever. (COCA)

(9.82) [This uses the MWE *stand up for* and the property ASPECT with the value *end* scoping over the main event, PROTECT (from *stand up for*).]
Maeda had stood up for Mosley. (COCA)

(9.83) [*Speed up* is described as a CHANGE-EVENT whose PRECONDITION and EFFECT have different relative values of SPEED.]
They speed up. (COCA)

(9.84) [This uses the multiword expression *sign off on*.]
The courts must sign off on any final accounting. (COCA)

(9.85) [*Showered* is described in the lexicon as BATHE-HUMAN (INSTRUMENT SHOWER).]
He showered. (COCA)

(9.86) [This uses the multiword expression *shoot back at*.]
Nobody had shot back at them. (COCA)

(9.87) [The nominal compound *lab space* is analyzed using a generic RELATION since the candidate meanings of the nouns do not match any of the more narrowly defined ontological patterns supporting compound analysis.]
They share lab space. (COCA)

(9.88) [*Policy* is described in the lexicon as a necessary procedure—that is, PROCEDURE scoped over by obligative modality with a value of .7]
They would shape policy. (COCA)

(9.89) [*Settle in* is correctly disambiguated as INHABIT.]
They settled in Minsk. (COCA)

(9.90) [*Never* is described using epistemic modality with a value of 0 scoping over the proposition.]
He'll never send the money. (COCA)

(9.91) [*See* is disambiguated from thirteen available senses.]
I'll see you on the freeway. (COCA)

(9.92) [Highly polysemous *see* and *at* are correctly disambiguated; there is modification of a proper noun and interpretation of a nominal compound (*cafeteria door*).]
She saw an injured Graves at the cafeteria door. (COCA)

(9.93) [*Legislator* is described in the lexicon as POLITICIAN (MEMBER-OF LEGISLATIVE-ENTITY).]
Legislators scheduled hearings. (COCA)

(9.94) [*Ran* is disambiguated from twelve available senses.]
I ran to the door. (COCA)

(9.95) [This uses the MWE *run a campaign*; *but* instantiates the discourse relation CONTRAST between the meanings of the clauses; the lexical description of *prefer* uses evaluative modality; and the proper names are correctly handled.]
Biss has run a good campaign, but we prefer Coulson. (COCA)

(9.96) [This uses the multiword expression *rise up*.]
We will rise up. (COCA)

(9.97) [This uses the lexical sense for the middle voice of *ring*.]
A dinner bell rang. (COCA)

(9.98) [This uses the multiword expression *station wagon*.]
We rented a station wagon. (COCA)

(9.99) [*Release* is correctly analyzed as INFORM.]
The NCAA releases the information. (COCA)

(9.100) [*Refuse* is analyzed as an ACCEPT event scoped over by epistemic modality with a value of 0—that is, *refusing* is described as *not accepting*.]
USAID refused interviews with staff in Badakhshan. (COCA)

(9.101) [*Race* is correctly understood as a MOTION-EVENT with a VELOCITY of .8 (not an actual running race).]
She raced to the church. (COCA)

The automatically generated TMRs for these examples are available at https://homepages.hass.rpi.edu/mcsham2/Linguistics-for-the-Age-of-AI.html.

This pair of holistic experiments served its purposes. First, they validated that the system was working as designed and could generate impressive analyses of real, automatically selected inputs from the open domain. Second, they highlighted the need for the *microtheory of language complexity* and led to our developing the first version of that microtheory. Third, they gave us empirical evidence that has allowed us to make a major improvement in our system: a redesign of confidence assessment for TMRs.

Before these experiments, our confidence measures relied exclusively on how well the input aligned with the syntactic and semantic expectations recorded in our knowledge bases. However, as we have seen, when an analysis seems to work fine, the agent can fail to recognize that it is missing a word sense, multiword expression, construction, or piece of world knowledge needed for making implicatures. So we now understand that it is important to enable the system to do all of the following:

1. When operating in a particular application area (i.e., a narrow domain), the LEIA will need to distinguish between in-domain and out-of-domain utterances. As a first approximation, this will rely on frequency counts of words describing concepts participating in known ontological scripts.
2. The LEIA will need to apply additional reasoning to in-domain utterances, effectively asking the question, "Could the input have a deeper or different meaning?"
3. The LEIA will need to decrease overall confidence in out-of-domain analyses due to the fact that they are not being fully semantically and pragmatically vetted using the kinds of ontological knowledge a person would bring to bear.

All of these have important implications for lifelong learning, which is a core functionality if agents are to both scale up and operate at near-human levels in the future. That is, although an agent can learn outside its area of expertise, that learning will be of a different quality than learning within its area of expertise. This suggests that the most efficient approach to learning will involve starting from better-understood domains and expanding from there. It is worth noting that, although we have been in this business for a long time, we would not have realized how frequently this overestimation of confidence occurs—that is, how frequently the agent thinks it has understood perfectly when, in fact, it has not—had we not gone ahead and developed a system and evaluated it over unrestricted text. There are just no introspective shortcuts.

9.4 Final Thoughts

One of the strategic decisions used in all the reported experiments (the five devoted to individual microtheories and the two holistic ones) was to challenge the system with examples from the open domain but allow it to select the examples it believed it could treat effectively. The rationale for this independent-selection policy is that we are developing intelligent agents that will need to be able to collaborate with people whose speech is not constrained. This means that utterances will be variously interpretable. For example, a furniture-building robot will lose the thread if its human collaborators launch into a discussion of yesterday's sports results. So, given each input, each agent must determine what it understands and with what confidence. This is the same capability that each evaluated subsystem displayed when it selected treatable examples from an open corpus. It reflects the agent's introspection about its own language understanding capabilities.

As we saw, the biggest hurdle in correctly making such assessments—and our biggest lesson learned—involves the lexicon. It can be impossible for an agent to realize that it is missing a needed lexical sense (which might be a multiword expression or construction) when the analysis that uses the available senses seems to work fine. Three directions of R&D will contribute to solving this problem.

1. Redoubling our emphasis on learning by reading and by interaction with humans, which is the most practical long-term solution for resource acquisition. The best methodology will be to start with domains for which the agent has the most knowledge—and, therefore, can generate the highest-quality analyses—and spiral outward from there.
2. Consulting lists of potential multiword expressions (which can be generated in-house or borrowed from statistical NLP) during language processing. These lists will contain (potential) MWEs that are not yet recorded in our lexicon and, therefore, are not yet provided with semantic interpretations. However, they will serve as a red flag during processing, suggesting that the compositional analysis of the given input might not be correct. This should improve our confidence scoring, helping the agent to not be overconfident in analyses that might not be fully compositional.
3. Carrying out manual lexical acquisition. Although manual acquisition is too expensive to be the sole solution to lexical lacunae, it would be rash to exclude it from the development toolbox, particularly since it is no more time-consuming than many other tasks that are garnering resources in the larger NLP community, such as corpus annotation.

As should be clear, our agents can work in various modes. In application mode, they must do the best they can to process whatever inputs they encounter. In component-evaluation mode, they attempt to identify inputs that they believe they can interpret correctly. And in learning mode, they use their ability to assess what they do and do not understand to identify learnable information.

To return to the starting point of this chapter, there is no simple, all-purpose strategy for evaluating knowledge-based systems. Crafting useful, feasible evaluation suites is an ongoing research issue. Evaluations are useful to the extent that they teach us something that we would not have understood through introspective research practices and normal test-and-debug cycles. It is, therefore, not a stretch to say that bad evaluation results can be a blessing in disguise—as long as they lead to new insights and suggest priorities for future R&D.

Epilogue

This book presented an approach to operationalizing natural language understanding capabilities. The approach is rooted in the hypothesis that it is both scientifically fruitful and practically expedient to model intelligent agents after the humans they are intended to emulate. This involves endowing agents with a host of interconnected capabilities that represent our best functional approximation of *modeling what people do, how they do it*. This *how* is not at the level of brain or biology; it is at the level of folk-psychological explanation, introspection, and commonsense human reasoning. The multifaceted interconnectedness of cognitive processes does not lend itself to a reassuringly small and targeted program of R&D—and that is the rationale behind the large-scale program of R&D that we have dubbed Linguistics for the Age of AI.

Generations of linguists have acknowledged that the comprehensive analysis of language use must invoke world knowledge, general reasoning, linguistic reasoning, mindreading of the interlocutor, and the interpretation of nonlinguistic features of the real world. But, having acknowledged this, the vast majority have chosen to work on quite narrowly defined linguistic subproblems. This is understandable on two counts: first, modularization fosters the development of certain kinds of theories; and, second, linguists are not necessarily drawn to modeling all of the nonlinguistic capabilities that interact with linguistic ones.

In this book we have attempted to explain why holistically addressing a broad range of cognitive capabilities is the only realistic path toward cracking the problem of natural language understanding and artificial intelligence overall. It is not that we are *choosing* to solve dozens of problems rather than select a more manageable handful; instead, we are acknowledging the inevitability of this course of action. For example, dynamically tracking the plans and goals of one's interlocutors is a well-known basic challenge of AI, and one that linguists might not consider a first priority, but it is the key to fully interpreting elliptical and fragmented dialog turns. That is, any attempt to interpret incomplete utterances without understanding their function in the discourse would have nothing in common with what people do and, at best, it would be no more than a temporary stopgap to give a system the veneer of intelligence.

Let us return to the notion of *the knowledge bottleneck*, which has been a key reason why knowledge-based methods have been at the periphery of AI for decades. In order for agents to become truly humanlike, they must acquire knowledge about language and the world through a process of lifelong learning. From preschool to their college years and beyond, people learn largely by reading and interacting—in natural language!—with other people. Our goal should be to impart this capability to artificial agents. To bootstrap this process, agents must be endowed with an initial knowledge base of sufficient size and quality. The ontology/lexicon knowledge base that we developed in our lab is a good candidate. People will have to help the system learn by providing explanations, answering questions, and checking the quality of the system's output. Importantly, the labor requirements for this kind of project will not be unusual for NLP/NLU. In fact, the machine learning community has devoted extensive resources to the manual preparation of datasets over the past several decades, and it continues to do so with no signs of letting up. If those resources had, instead, gone toward building the kinds of knowledge bases we describe in this book, it is entirely possible that we would already be benefiting from highly functional LEIAs who would already be overcoming the knowledge bottleneck with an effective level of lifelong learning.

Knowledge-based and statistical methods are not in competition; they offer different methods of achieving different types of analysis results that can serve as input to intelligent systems. The most plausible path to human-level AI is to integrate the results of both of these approaches into hybrid environments. In fact, combining the best that each approach has to offer has been at the center of attention of AI researchers for quite some time. However, because of the high profile of machine learning, most of the thinking has focused on how to improve the results of current applications by adding a sprinkling of stored, human-acquired knowledge. We believe that this is the exact opposite of the most promising long-term direction. We hypothesize that the most fruitful path of integration will be for statistical methods both to support an agent's lifelong learning and to supply knowledge-based systems with high-quality modules for subtasks that lend themselves well to knowledge-lean approaches—such as named-entity recognition and syntactic parsing.

While writing this epilogue, we fortuitously attended a lecture by Jeffrey Siskind, who, like us, is working toward integrating multiple cognitive capabilities in AI systems. When asked to project into the future, he explained that he thinks about the future only to the level of his "great-grandstudents": anything beyond that is over the horizon. This resonated with us. The field of computational linguistics—with *linguistics* as a central topic of study—has been woefully underexplored over the past generation. This means that a tremendous amount remains to be done, and there is no telling what the field might look like in forty years' time. We hope that the program of research that we call Linguistics for the Age of AI will serve as a shot in the arm to the linguistics community, renewing excitement in the challenge of endowing agents with human-level language capabilities.

Notes

Chapter 1

1. For historical overviews of machine translation, see, e.g., Hutchins (1986) and Nirenburg et al. (2003). Portions of this discussion were originally published as Nirenburg & McShane (2016a).

2. A similar approach is a cornerstone of the current corpus-based paradigm in NLP, which usually involves the analysis not of text meaning in toto but either the treatment of the meaning of selected textual strings or no treatment of meaning at all—instead orienting exclusively around the syntactic and morphological properties of words.

3. This survey was published in Bar Hillel (1970).

4. For overviews of head-driven phrase structure grammar and lexical functional grammar—along with generative grammar and construction grammar—see Pustejovsky & Batiukova (2019, Chapter 3).

5. The automatic learning of knowledge by the agent has also started to be addressed in cognitively inspired approaches to multifunctional agent modeling. See, e.g., Forbus et al. (2007), Navigli et al. (2011), Nirenburg et al. (2007), and Wong et al. (2012). See the deep dive in section 1.6.2 for additional details about learning by LEIAs.

6. The approach to recording knowledge and computing meaning that we will describe was originally developed for natural language processing outside of a full agent architecture (Nirenburg & Raskin, 2004) but has easily accommodated agent-oriented extensions.

7. There have been ongoing debates between eliminativists and functionalists on the status of unobservables but, for our needs, a human-level explanation of behavior cannot be formulated in terms of neuronal activity. We prefer possibly "naive" explanations that may ultimately not be true in scientific terms but are expected to be accepted as explanations by regular people in regular circumstances, outside a philosopher's study or a psychologist's lab.

8. Although nonlinguistic channels of perception are largely outside the scope of this book, we have been working on them in earnest and include select mentions. For example, chapter 8 describes interoception by virtual patients (section 8.1.3.2) and vision by physical robots (sections 7.7.2 and 8.3), both modeled within the OntoAgent architecture.

9. For an analysis of how annotated corpora can support the development of linguistic theories, see de Marneffe & Potts (2017).

10. Readers of our draft manuscript offered interesting suggestions for additions to this literature review. However, we kept this review, like others throughout the book, highly selective in order not to stray from the main narrative.

11. FrameNet, described later in the chapter, is a computationally oriented resource deriving from early work in construction grammar led by Charles Fillmore.

12. This theory was actually used by Purver et al. (2011) as the theoretical substrate for an incremental parser. However, although that parser included semantics, it appears to be a version of upper-case semantics—i.e., the intended interpretations were provided manually.

13. See Chambers et al. (2002) for experiments exploring how the affordances of objects in a workspace affect subjects' interpretations of language inputs. For reviews of the literature, see Kruijff et al. (2007) and Chambers et al. (2004).

14. For a discussion of various approaches to semantic analysis in cognitive systems, see McShane (2017a).

15. Thesauri are quite diverse. To take two extremes, Apresjan's (2004) *New Explanatory Dictionary of Russian Synonyms* was manually created, is extremely rich in detail, and was intended for use by people. By contrast, Inkpen & Hirst's (2006) lexical knowledge base of near-synonym differences was compiled automatically, using a multistage process of machine learning, and was intended for use by systems.

16. Manning (2006) promotes the idea of bridging work between the NLP and Knowledge Representation and Reasoning (KR&R) communities, writing: "NLP people can do robust language processing to get things into a form that KR&R people can use, while KR&R people can show the value of using knowledge bases and reasoning that go beyond the shallow bottom-up semantics of most current NLP systems."

17. A good example is Leafgren's (2002) description of how different kinds of referring expressions are used in Bulgarian.

18. A recent contribution on the resource front is the Stanford Natural Language Inference corpus, which includes 570,152 labeled pairs of captions (Bowman et al., 2015). Users of the Amazon Mechanical Turk environment were presented with image captions—without the images themselves—and asked to write three alternative captions: one that was definitely true (an entailment), one that may or may not be true (a neutral statement), and one that was definitely false (a contradiction). The task instructions included explanations of the quartet:

> Two dogs are running through a field. (prompt)
> There are animals outdoors. (entailment)
> Some puppies are running to catch a stick. (neutral)
> The pets are sitting on a couch. (contradiction)

This annotated corpus has been used for experiments in machine learning related to natural language inference. We do not think that this corpus will facilitate knowledge acquisition for agent systems due to the unconstrained selection of *any* entailment, *any* neutral statement, and *any* contradiction.

19. Past theoretical work linking cognitive linguistics and computation includes, e.g., Feldman & Narayanan (2004) and Feldman (2006).

20. This short survey did not even touch on the field of neurolinguistics because agent development does not attempt the biological replication of a human brain, and it remains to be seen whether and how the results of neurolinguistics will ever inform computational cognitive modeling.

21. We do not include here the resources that will be described in upcoming chapters.

22. For a discussion of commonsense reasoning and knowledge, see Davis & Marcus (2015).

23. Note that research and development in the area of component integration has become increasingly prominent (e.g., Bontcheva et al., 2004; Shi et al., 2014). An additional benefit of the integrative approach is the possibility of viewing components of application systems as black boxes communicating with other components only at the input-output level, thus allowing integration of components built using potentially very different approaches.

24. Much thought has been given to understanding and repairing this state of affairs; see, e.g., Clegg & Shepherd (2007) for an analysis within the biomedical domain.

25. Paraphrase detection has become a well-investigated topic in itself; see Magnolini (2014) for a survey.

26. See S. Clark (2015) for an excellent overview, including historical references.

27. Work is underway to extend distributional semantics to exploit compositionality (Goyal et al., 2013).

28. A detailed discussion of phenomenology and its influence on our view of modeling agents is outside the scope of this book. See Zlatev (2010), Löwe & Müller (2011), Carruthers (2009), and Andler (2006) for relevant discussions.

29. This learning capability is realistic only when the agent is already endowed with a critical mass of knowledge of the world and language. In our work, we do not address early development stages. We concentrate on modeling the behavior of adult humans.

30. Comprehensive LEIA-based applications address other types of learning as well, such as the learning of ontological scripts (Nirenburg et al., 2018).

31. These ideas were first published in McShane (2017b).

32. "Detroit," Wikipedia, accessed December 10, 2016, https://en.wikipedia.org/wiki/Detroit

33. Some annotations have been removed for concise presentation.

34. To be done properly, such comparisons require (a) a near-developer's understanding of each environment, which is hardly ever achievable using published materials because the environments are constantly in flux; (b) a thoughtful, selective process of analysis, which could consume a level of effort equivalent to several doctoral theses; and (c) a presentation strategy that provides readers with all the necessary background about each environment referred to, which would constitute a separate book.

35. Further reading includes Stich & Nichols (2003) for folk psychology; Malle (2010) for attribution theories; Bello (2011) for mindreading; and Carruthers (2009) for metacognition.

36. An extended version of this work is reported in DeVault et al. (2011).

37. Throughout the book, examples drawn from the COCA corpus (Davies, 2008–) will be indicated by the subscript (COCA) following the example.

38. For a list of coreference annotation schemes for NPs, see Hasler et al. (2006).

39. Here we will focus on English, but the same principles apply to other language/culture pairs.

40. Dialog acts have also been called speech acts; locutionary, illocutionary, and perlocutionary acts; communicative acts; conversation acts; conversational moves; and dialog moves. See Traum (1999b, 2000) for references to the literature.

41. Another sample classification is found in Traum (1994, p. 57), which identifies four conversation act types that are relevant for different-sized chunks of conversation. Traum presents each with a sample of the associated dialog acts, as shown below (we normalize capitalization and spell out abbreviations):

turn-taking:	Take-turn, Keep-turn, Release-turn, Assign-turn
grounding:	Initiate, Continue, Ack[nowledgment], Repair, Req[uest]Repair, Req[uest] Ack[nowledgment], Cancel
core speech acts:	Inform, YNQ [i.e., ask question], Check, Eval[uate], Suggest, Request, Accept, Reject
argumentation:	Elaborate, Summarize, Clarify, Q&A, Convince, Find-Plan

42. The formatting of dialog act names (small caps, capitalization conventions, and singular vs. plural) is directly from Stolcke et al.'s (2000) table 2 (p. 341), which also includes corpus examples of each dialog act and its corpus-attested frequency.

43. For related literature see, e.g., the special issues of *Robotics and Autonomous Systems* (Coradeschi & Saffiotti, 2003) and *Artificial Intelligence* (Roy & Reiter, 2005), as well as Roy (2005), Scheutz et al. (2004), Gorniak & Roy (2005), and Steels (2008).

44. However, in field-wide competitions that pit knowledge-lean systems head-to-head, annotated corpora are often provided not only for the training portion but also for the evaluation portion. This means that such systems are actually not operating exclusively over observable data.

45. The Message Understanding Conferences (MUCs; Grishman & Sundheim, 1996) and Text Retrieval Conferences (TRECs; http://trec.nist.gov/) are noteworthy examples.

46. For surveys of the literature, see Lapata (2002), Girju et al. (2005), Lieber & Štekauer (2009), and Tratz & Hovy (2010).

47. Similarly, Lapata (2002) developed a probabilistic model covering only those two-noun compounds in which N1 is the underlying subject or direct object of the event represented by N2: e.g., *car lover.*

48. Accessed June 6, 2020, https://ufal.mff.cuni.cz/pcedt2.0/publications/t-man-en.pdf.

Chapter 2

1. Simplifications include removing most of the metadata that TMRs typically carry. We will not comment further on the specific simplifications of the different knowledge structures to be presented throughout.

2. Upper-case semantics refers to the practice, undertaken by some researchers in formal semantics and reasoning, of avoiding natural language challenges like ambiguity and semantic non-compositionality by

asserting that strings written using a particular typeface (often, uppercase) have a particular meaning: e.g., TABLE might be said to refer to a piece of furniture rather than a chart.

3. The TIME slot includes a call to a procedural semantic routine, *find-anchor-time*, that can attempt to concretize the time of speech if the agent considers that necessary. In many applications, it is not necessary: the general indication of past time—i.e., "before the time of speech"—is sufficient.

4. At the time of writing we are using the Stanford CoreNLP toolset (Manning et al., 2014). This toolset was updated several times during the writing of this book. The most recent version (version 4.0.0) was released late in the book's production process. We made an effort to update examples and associated discussions accordingly. We present screen shots of CoreNLP output for some examples in the online Appendix at https://homepages.hass.rpi.edu/mcsham2/Linguistics-for-the-Age-of-AI.html.

5. A more complete lexicon would include many more constructions, such as *eat one's hat*, *eat one's heart out*, *eat someone alive*, and so on.

6. One might ask, Why is there no concept for EAT in the ontology? This decision reflected the priorities and goals of knowledge acquisition at the time this corner of the ontology/lexicon pair was being acquired.

7. LEIAs also use several auxiliary knowledge bases, rule sets, and algorithms, which will be described in conjunction with their associated microtheories.

8. There has been much debate about the optimal inventory of case roles for NLP systems. Some resources use a very large inventory of case roles. For example, O'Hara & Wiebe (2009) report that FrameNet (Fillmore & Baker, 2009) uses over 780 case roles and provides a list of the most commonly used 25. Other resources underspecify the semantics of case roles. For example, PropBank (Palmer et al., 2005) uses numbers to label the case roles of a verb: Arg0 and Arg1 are generally understood to be the agent and theme, but the rest of the numbered arguments are not semantically specified. This approach facilitates the relatively fast annotation of large corpora, and the resulting annotations support investigation into the nature and frequency of syntactic variations of the realization of a predicate; however, it does not permit automatic reasoning about meaning to the degree that an explicit case role system does.

9. For early work on scripts, see Minsky (1975), Schank & Abelson (1977), Charniak (1972), and Fillmore (1985).

10. The scientific question related to this engineering solution for writing scripts is, Can we write all of this knowledge using a nonprogramming-oriented, static formalism (something that looks more typically ontological) and then write a program that automatically generates code to drive the actions of the agent? Phrased differently, can we write programs to write programs? We will leave this problem to the field of automatic programming or source-code generation. It is a fascinating problem whose full exploration would take our core research program too far afield. Instead of pursuing this, our knowledge engineers write scripts using semiformal knowledge representation strategies that include tables, slot-filler structures, and even diagrams (see chapter 8 for examples), and then they directly collaborate with the programmers who engineer all the required system behavior.

11. McShane et al. (2016) detail that NLU system.

12. For approaches to incremental syntactic parsing see, e.g., Ball (2011) and Demberg et al. (2013).

13. The Pens are the Pittsburgh Penguins, a professional ice hockey team.

14. This dichotomy might be considered mutatis mutandis, parallel to Newell's (1982) distinction between the knowledge level and the symbol level(s) in AI systems.

15. Discussing this very complex issue in any detail is beyond the scope of this book. For an introduction to the relevant topics, see Frigg & Hartman (2020) and Winther (2016).

16. Our conception of a model is strategically congruent with the views of Forbus (2018, Chapter 11), though we concentrate on modeling less observable phenomena than those in the focus of Forbus's presentation. Our views on the nature of theories, models, and systems have been strongly influenced by Bailer-Jones (2009), but for the purposes of our enterprise we do not see a need to retain the same fine grain of analysis of these concepts and their interrelationship as the predominantly philosophy-of-science angle of Bailer-Jones.

17. The psycholinguistic evidence that Wilks and Ide cite to support this position is irrelevant because systems do not operate the way people do. In fact, Wilks's own famous "theorem" to the effect that there is no linguistic theory, however bizarre, that cannot be made the basis of a successful NLP system (Wilks actually said "MT system") seems to argue for discounting psycholinguistic evidence for NLP.

18. Border collies are the real geniuses, with a default INTELLIGENCE value of 1.

19. For a comparison of upper ontologies see Mascardi et al. (2007) and references therein.

20. In reviewing the utility of a number of large medical ontologies for NLP, Hahn et al. (1999) report that MeSH shows "semantic opacity of relations and concepts … [and a] lack of formal concept definitions"; in SNOMED, "only a small part of all possible [automatically generated] combinations of axes correspond to consistent and reasonable medical concepts"; and in UMLS, "by merging concepts collected from different sources, a problematic mixture of the semantics of the original terms and concepts is enforced." It should be mentioned that these resources, like the very widely used WordNet, were actually developed for people, not NLP.

21. We believe that history will look back on this period of building wordnets with similar disillusionment, at least with respect to sophisticated NLP applications, for which disambiguation is crucial.

22. As Wilks et al. (1996, pp. 121–122) describe the situation, many practitioners of NLP consider lexicons "primarily subsidiary" resources expected to "fall out of" their theory of choice.

23. The languages are Catalan, Danish, Dutch, English, Finnish, French, German, Greek, Italian, Portuguese, Spanish, and Swedish. The work continues the earlier PAROLE project, which developed 20,000-sense morphological and syntactic lexicons for these languages.

24. The notion of a largely language independent lexicon was introduced in McShane et al. (2005a).

25. If, for example, the first LEIA lexicon was created for German, it is possible that the acquirer would create a special concept for WHITE-HORSE, since German has a word for that, *Schimmel*. If an English lexicon were bootstrapped from the German one, the acquirer would either ignore that lexical sense and concept or encode the multiword expression *white horse* to map to it.

26. We could, alternatively, have created a concept WEAPON-OF-MASS-DESTRUCTION, whose parent would be WEAPON and whose children would be CHEMICAL-WEAPON and BIOLOGICAL-WEAPON. We tend not to create special concepts for groups of things, however, since there are expressive means to group things, as needed, in the lexicon.

27. There has been extensive work on linguistic paraphrase in the knowledge-lean paradigm which involves, e.g., determining the distributional clustering of similar words in corpora (e.g., Pereira et al., 1993; Lin, 1998), using paraphrases for query expansion in question-answering applications (e.g., Ibrahim et al., 2003), and automatically extracting paraphrases from multiple translations of the same source text (Barzilay & McKeown, 2001).

28. One could, of course, create a multiword lexical sense for *go (somewhere) by plane*, which would map to AERIAL-MOTION-EVENT and avoid this particular case of ontological paraphrase. However, the point of this example is to show how ontological paraphrase can be reasoned about when it does occur.

Chapter 3

1. See de Marneffe et al. (2006) for a description of generating dependency parses from phrase structure parses.

2. Captured November 8, 2019. Figures 3.1 and 3.2 originally appeared in color.

3. Recall that ^, used in the sem-struc zones of lexicon entries, indicates *the meaning of.*

4. For early work on noncanonical input, see Carbonell & Hayes (1983).

5. Some annotations have been removed to keep the presentation concise.

6. You might wonder if we are making the task artificially more difficult by providing LEIAs with only written transcripts instead of the speech stream itself—after all, speech includes prosodic features that assist people in extracting meaning. No doubt, prosodic features could be very useful to LEIAs; however, the precondition for using them remains outstanding. Specifically, methods must be developed to automatically extract and interpret such features within the agent's ontological model.

7. There are formalism-related reasons why we cannot have a single template that would encompass all three.

8. For further description of this approach, albeit within a different implementation of the NLU system, see McShane et al. (2016).

9. For purposes of clarity, we focus on verbs as argument-taking heads, though other parts of speech can take arguments as well.

10. The LEIA *can* detect certain kinds of plays on idioms, but not at this stage; that occurs later on, during Extended Semantic Analysis.

11. This output was confirmed on January 25, 2020, using the online interface at the website corenlp.run.

12. Note that the example intentionally includes no coreferential expressions (e.g., no definite articles), so we cannot assume that the preceding linguistic context will aid in disambiguation.

13. For past work on semantic sense bunching and underspecification, see, e.g., Buitelaar (2000) and Palmer et al. (2004).

14. We use various sense-numbering conventions in the lexicon for internal bookkeeping.

Chapter 4

1. At the time of writing, we are working on a project whose objective is to record the algorithms and code base of not only our NLU system but also the OntoAgent cognitive architecture overall, using the graphic representations of the Unified Modeling Language™ (UML). Future publications reporting the results of that work will provide interested developers with the algorithms actually implemented in our current NLU engine.

2. Consider the following distinction, which might be missed by all but the most informed foodies: "Note the difference between smoked Scottish salmon and Scottish smoked salmon. It's possible that the wording of the latter has been deliberately used to cover the fact that the salmon has been smoked in Scotland, but not necessarily sourced from Scotland." From "Scottish or Scotch? A Guide to Interpreting Food Labels," The Larder, *The List*, May 1, 2009, https://food.list.co.uk/article/17265-scottish-or-scotch-a-guide-to-interpreting-food-labels/.

3. Garden-path sentences are grammatical sentences for which the hearer's initial interpretation of the first part ends up being incorrect when the sentence is completed. A classic example is *The horse raced past the barn fell.*

4. For earlier work on adjectives in this paradigm, see Raskin & Nirenburg (1998).

5. The "sad" meaning of *blue house* and the "colored blue" meaning of *blue person* would not be generated because they are instances of nonliteral language. Nonliteral interpretations are generated only if there is a trigger to do so, such as an incongruity during processing (see section 6.2), which is not the case here.

6. We do not adopt procedural semantics as our overarching lexical theory. That is, our approach would not associate common objects like *chair* with procedural routines to seek out their extensions (see Wilks et al., 1996, pp. 20–22 for a discussion of the various interpretations of "procedural semantics"). Instead, we apply the term to only that subclass of lexical phenomena that predictably requires a context-bound specification of their meaning, for which the lexicon explicitly points to the necessary routine.

7. *Evaluative* is actually a type of modality. The treatment of modality and aspect is described in sections 4.2.1 and 4.2.2. For clarity of presentation in this set of examples, we use a shorthand representation for the full modality and aspect frames.

8. If we were to use proper names instead, their TMR frames would be instances of HUMAN described using the feature HAS-PERSONAL-NAME. The TMRs omit indications of time.

9. SET-1 and SET-2 are generated by a lexical sense for the conjunction *and* in which *and* requires two noun phrases as its arguments.

10. We chose an example that describes the set as participating in an event because this most clearly illustrates why set expansion is needed: the agent must understand that each of the set elements is engaged in its own instance of the event. However, even if no event is mentioned, this type of set expansion best serves downstream reasoning. *Two gray wolves* generates two different instances of WOLF, each of which is described by "COLOR gray."

11. We do not present the formal meaning representations, since they are representationally complex and are not needed to convey our main point, which is conceptual.

12. It is stripped of metadata and the call to resolve the reference of *the*.

13. See Nirenburg & Raskin (2004) for motivation and a more formal description of this inventory.

14. One can also demand an action that is not agentive, such as when the interlocutor will be the experiencer: *Get better soon!* or *Catch the flu so you can get out of your final!*

15. According to Monti et al. (2018, p. 3), "Biber et al. (1999) argue that they [multiword expressions] constitute up to 45% of spoken English and up to 21% of academic prose in English. Sag et al. (2002) note that they are overwhelmingly present in terminology and 41% of the entries in WordNet 1.7 are reported to be multiword units."

16. We also do not adopt most of the theoretical principles they advocate, such as the lack of empty categories, the lack of transformations, and an inheritance network of constructions. But those aspects are not directly related to the definition of the word *construction*, which is our interest here.

17. Our approach to integrating all kinds of constructions into the lexicon resonates with Stock et al.'s (1993, p. 238) opinion that idioms (a subtype of constructions) should be integrated into the lexicon as "more information about particular words" rather than treated using special lists and idiosyncratic procedures.

18. Other names for multiword expressions, and subtypes of them, are *multiword units*, *fixed expressions*, *set expressions*, *phraseological units*, *formulaic language*, *phrasemes*, *idiomatic expressions*, *idioms*, *collocations*, and *polylexical expressions* (Monti et al., 2018, p. 3). Monti et al. provide a reference-rich overview of work on multiword expressions in language processing technologies, with an emphasis on machine translation.

19. For more extensive literature reviews, see Cacciari & Tabossi (1993) or McShane, Nirenburg, & Beale (2015). The analysis presented here draws from the latter.

20. This section draws from McShane et al. (2014).

21. Tratz & Hovy (2010) prioritized achieving interannotator agreement in a Mechanical Turk experiment, and this methodology influenced their final inventory of relations. Thanks to them for sharing their data.

22. This is just one of many possible control strategies, another being to launch all analysis functions, score them, and select the one with the highest score.

23. This section draws from Nirenburg & McShane (2016b).

24. The mid-2010s gave rise to a new wave of research on metaphor in the NLP community that follows the accepted knowledge-lean paradigm. But, as Shutova (2015) writes, "So far, the lack of a common task definition and a shared data set have hampered our progress as a community in this area. This calls for a unification of the task definition and a large-scale annotation effort that would provide a data set for metaphor system evaluation …" (p. 617). Whether that will occur is one question, and whether it will actually address the full complexity of metaphors is another.

25. Quoted from "The D Team" by Lee Edwards, *American Spectator*, April 30, 2013, https://spectator.org/33713_d-team/

26. A more comprehensive, but also more complex, approach would be to posit a disjunctive set allowing for either an event or a state.

Chapter 5

1. This process is called *grounding* in the robotics community. *Grounding* has a different meaning in the area of discourse and dialog processing: it refers to indicating—using utterances (e.g., *uh-huh*) or body language (e.g., nodding)—that one has understood what the other person said and meant.

2. There is no bucket involved in the idiomatic meaning of *kick the bucket*. There is also no kicking in the usual sense; however, when used in this idiom, *kick* means DIE, and DIE is referential.

3. This section draws from McShane (2009).

4. Work on associated topics has been carried out in the knowledge-lean paradigm. Boyd et al. (2005), Denber (1998), Evans (2001), and Li et al. (2009) describe methods for automatically detecting pleonastic *it*; Vieira & Poesio (2000) present a system for automatically processing definite descriptions; Bean & Riloff (1999) describe a method for identifying nonanaphoric noun phrases.

5. In the TMR, this two-part idiom will instantiate the property CONTRAST, whose domain is the meaning of the first proposition and whose range is the meaning of the second.

6. In the TMR, PLUMBER will fill the property HAS-SOCIAL ROLE, as follows: HUMAN-1 (HAS-PERSONAL-NAME 'Danny') (HAS-SOCIAL-ROLE PLUMBER).

7. *Has* is used here as an auxiliary verb, not a main verb.

8. The study reported in Elsner & Charniak (2010) shows that, given a coreference window of ten sentences using the MUC-6 coreference dataset, only about half of same-headed NPs were coreferential. However, this is a very large window of coreference, making this result not entirely surprising.

9. See McShane (2005, Chapter 7) for a discussion of such phenomena crosslinguistically.

10. For discussions of bridging see, e.g., Asher & Lascarides (1996) and Poesio, Mehta, et al. (2004).

11. Recasens et al. (2012) refer to some of these as near-identity relations and discuss their treatment in corpus annotation.

12. For LEIAs, the ambiguity of universally known definite descriptions is accounted for by multiple lexical senses. For example, one sense of *sun* requires *the* and means 'the Earth's sun,' and another sense does not require *the* and can indicate the sun of any planet. This is just another matter of lexical disambiguation.

13. See, e.g., Marcu (2000).

14. Some approaches orient around utterances rather than sentences.

15. For a discussion of treating ambiguity in corpus annotation, see Poesio & Artstein (2005).

16. For example, intransitive verbs like *sleep* require only a subject; transitive verbs like *refinish* require a subject and a direct object; and ditransitive verbs like *give* require a subject, a direct object, and an indirect object.

17. Bos & Spenader (2011) summarize NLP's avoidance of ellipsis as follows: "First, from a purely practical perspective, automatically locating ellipsis and their antecedents is a hard task, not subsumed by ordinary natural language processing components. Recent empirical work (Hardt, 1997; Nielsen, 2005) indeed confirms that VPE [verb phrase ellipsis] identification is difficult. Second, most theoretical work begins at the point at which the ellipsis example and the rough location of its antecedent are already identified, focussing on the resolution task" (pp. 464–465). The field of generative syntax, for its part, has studied ellipsis in earnest for decades but, in accordance with its purview and goals, it focuses exclusively on syntax and licensing conditions, not on semantics or resolution.

18. The CoreNLP developers call this tool dcoref, for deterministic coreference annotator. Other external coreference resolution systems could be used if they were to offer better quality of results or coverage of phenomena. In general, we welcome imported solutions as long as the overhead of incorporating them is not prohibitive.

19. Poesio et al. (2016, pp. 88–90) also provide a nice overview, within a broader description of the field. For an example of a contribution that integrates more knowledge into a still primarily machine-learning approach, see Ratinov & Roth (2012).

20. The CoreNLP output for several examples in this chapter, including (5.35) and (5.36), is presented in the online appendix at https://homepages.hass.rpi.edu/mcsham2/Linguistics-for-the-Age-of-AI.html.

21. The fact that we would more naturally verbalize the sponsor of *it* as "her being happy" or "the fact that she is happy" is inconsequential. Coreferences are actually established at the level of semantic interpretations (TMRs), not text strings.

22. The constructions described here were actually inspired by McShane's work on argument ellipsis in Russian and Polish (McShane 2000, 2005). In these languages, the ellipsis of referential direct objects is permitted only with substantial linguistic or contextual support, and that often involves parallelism. Not surprisingly, some of the same lexico-syntactic constructions that permit argument ellipsis in Russian and Polish quite confidently predict the coreference of overt arguments in English. This overlap is not only of theoretical interest; it also suggests that, viewed crosslinguistically, a knowledge-based approach to reference resolution is both feasible and economical in terms of the descriptive work needed.

23. This evaluation excluded *it* because we did not leverage semantic analysis for the evaluation and, therefore, the system could not detect pleonastic and idiomatic usages.

24. We tasked the system only with identifying the nominal head of the antecedent, not the entire noun phrase, which might include a determiner, adjectives, or relative clauses. The full NPs are indicated in the examples for clarity's sake.

25. Developing a comprehensive analysis system means that not all components are necessarily ready to go at a given time. Some of our evaluations have involved only a subset of overall system capabilities, as discussed in chapter 9.

26. *Span of text* is a syntactic description. Semantically, we are talking about multiple propositions.

27. For example, *Horatio proposed that everyone in the office should go on a group jog every morning, but his suggestion was met with collective horror.*

28. Byron (2004) is an exception but has narrow domain coverage. (It provides a nice review of the linguistic literature on broad referring expressions.) Machine translation systems must treat broad RefExes, but they have the advantage of not having to actually resolve them: they can replace a vague expression in source language by an equally vague one in the target language.

29. Parentheses indicate optionality, a forward slash indicates a choice, caps indicate category types, and underlining indicates coreferential categories.

30. The evaluation (reported in McShane & Babkin, 2016a) included twenty-seven contexts, for which twenty-five answers were correct, one was incorrect, and one was partially correct.

31. Past research has shown that discourse segments that serve as sponsors for broad RefExes are almost always contiguous with the broad RefEx's clause—i.e., discourse segments are not skipped over. See Byron (2004) for a review of that literature.

32. More formally, simple syntactic structures have none of the following dependencies in the CoreNLP dependency parse: *advcl*, *parataxis*, *ccomp*, *rcmod*, *complm*, *dep*, *conj* (with verbal arguments), *xcomp* (with a lexically recorded matrix verb as the governor), or *aux* (not involving a tense marker). We do not assume that all CoreNLP dependency parses will be error-free; however, its accuracy for detecting syntactically simple clauses using this algorithm is quite good.

33. For more on sentence trimming in the literature, see the references cited in McShane, Nirenburg, & Babkin (2015).

34. The full list is available at https://homepages.hass.rpi.edu/mcsham2/Linguistics-for-the-Age-of-AI.html.

35. For work on near-synonyms, see DiMarco et al. (1993) and Inkpen & Hirst (2006).

36. The first stage of work involved compiling a test list of verbs for which either the subject or the object was narrowly constrained, then compiling a list of typical fillers for that role. We used a total of 202 verbs with an average of nearly 60 keywords each. However, the average was pulled up by verbs like *eat/cook* and *die*, for which hundreds of food items and animals, respectively, were listed as keywords. For details, see https://homepages.hass.rpi.edu/mcsham2/Linguistics-for-the-Age-of-AI.html. We used the Gigaword corpus (Graff & Cieri, 2003) for the experiment.

37. Formally speaking, these descriptions unify.

38. We know of no definitive, comprehensive list of such entities. Automatically generated lists of this kind include numerous false positives.

39. Of course, metaphorical and other senses can be recorded as well.

40. By contrast, nonrestrictive modifiers—which are usually set off by commas—provide additional information but are not as essential to the meaning of the sentence: e.g., *Our neighbors' dog, Captain, comes over often.*

41. Accessed January 1, 2019, https://en.wikipedia.org/wiki/Lion.

42. HAS-OBJECT-AS-PART and PART-OF-OBJECT are inverse relations in the ontology.

43. The naming conventions for concepts are irrelevant; they are for human orientation only. The agent understands their meaning as the set of property fillers defined in the ontology.

44. As regards past work on VP ellipsis outside our group, Johnson (2001) offers a descriptive account aptly entitled "What VP ellipsis can do, what it can't, but not why." In the computational realm, if knowledge-lean systems treat VP ellipsis at all, they do not address the semantic issues and they do not offer confidence estimates for resolutions. Hardt (1997) reports a system for resolving VP ellipsis that required a manually corrected parse and did not pursue semantics. Most work on instance versus type coreference of internal arguments has been carried out in the paradigm of theoretical linguistics, which does not offer heuristics that can guide system building.

45. AECs can also be expressed using pronouns and other descriptions, but those cases are discussed in their respective sections.

46. The terms *strict* and *sloppy* coreference are used in the generative syntax literature (see, e.g., Fiengo & May, 1994) to refer to objects showing what we call *instance* and *type* coreference.

47. See McShane (2005) for further discussion of the reference-oriented effects of repetition structures.

Chapter 6

1. Recall that many procedural semantic routines involve coreference and were resolved during Basic Coreference Resolution.

2. This dovetails with the views of Lepore & Stone (2010) about metaphor—that is, metaphorical meanings do not need to be fully semantically interpreted or recorded.

3. This description may bring to mind the statistical approach called distributional semantics. Distributional semantics operates over uninterpreted text strings, not meanings, and therefore (a) it is very noisy due to lexical ambiguity, and (b) it does not yield formal representations to support the agent's downstream reasoning.

4. COLOR is a literal attribute, which means that its values are not ontological concepts. That is why they are not in small caps.

5. Recall that ontological scripts use a wide variety of knowledge representation methods, far beyond the simple slot-facet-filler formalism used for the nonscript portion of the ontology.

6. Section 7.9 describes automatic learning by reading, which can be useful for this kind of knowledge acquisition.

7. This is different from the property RELATION, which is the head of a large subtree of the ontology.

8. See Onyshkevych (1997) for a discussion of shortest-path algorithms.

9. For reasoning by analogy see, e.g., Forbus (2018), Gentner & Smith (2013), and Gentner & Maravilla (2018).

10. The idea of a metonym repository ascends at least to Fass (1997).

11. The linguistic literature has identified several other types of performance errors as well, such as spoonerisms (saying *The Lord is a shoving leopard* instead of *The Lord is a loving shepherd*) and cases of anticipation (saying *bed and butter* instead of *bread and butter*). These are not particularly common, so their treatment is not a priority.

12. For a book-length cognitive model of idiomatic creativity, along with extensive manual analyses of data, see Langlotz (2006). This model does not directly inform our work because it is not a *computational* cognitive model—it lacks heuristics that would make the human-oriented observations automatable. For an entertaining take on linguistic creativity overall, see Veale (2012).

13. In the lexicon, these will be recorded using the more involved formalism of that knowledge base. We use shorthand here for readability's sake.

14. Although theoretically oriented accounts have classified sources of idiomatic creativity—e.g., Langlotz's (2006, pp. 176–177) model distinguishes *institutionalized variants*, *occasional variants*, *pun variants*, and *erroneous variants*, among others—that grain size of analysis is unrealistic for computational models within the current state of the art.

15. This example uses generic *you*, which we refer to here, for simplicity's sake, as an instance of HUMAN.

16. Recall that many procedural semantic routines were already run by this time, during both Basic Semantic Analysis and Basic Coreference Resolution.

17. K. B. Cohen et al. (2008) present a rigorous linguistic analysis in a related vein. They studied the alternations in the argument structure of verbs commonly used in the biomedical domain, as well as their associated nominalizations. The work was aimed at improving information extraction systems.

18. Bakhshandeh et al. (2016) divide up linguistic phenomena related to comparisons and ellipsis differently than we do, and they treat them using supervised machine learning. Although their target knowledge structures are deeper than those used by most machine learning approaches, their direct reliance on annotated corpora, unmediated by a descriptive microtheory, makes their results not directly applicable to our work.

19. A simple example of leaving a procedural semantic routine unresolved involves the representation of the past tense. Every past-tense verb spawns a TMR whose TIME slot is filled with the call to the procedural semantic routine '<*find-anchor-time*', meaning 'before the time of speech'. If this procedure is run, the agent attempts to determine the time of speech/writing, which may or may not be available and may or may not matter.

20. Some points of comparison with the literature are as follows. Fernández et al. (2007) identify fifteen classes of what they call "non-sentential utterances" (NSUs), which they use in their work on automatically classifying NSUs using machine learning. Cinková (2009) describes an annotation scheme for detecting and reconstructing NSUs. And Schlangen & Lascarides (2003) identify twenty-four speech act types that can be realized with NSUs, grounding their taxonomy in the rhetorical relations defined by a theory of discourse structure called SDRT (Asher, 1993; Asher & Lascarides, 2003). Although some of the descriptive work and examples from these contributions can inform the development of our microtheory of NSUs, the goals pursued are so different that comparisons are quite distant. Cinková's corpus annotation work is targeted at developers pursuing supervised machine learning. Fernández et al.'s approach assumes a downstream consumer as well: "Our experiments show that, for the taxonomy adopted, the task of identifying the right NSU class can be successfully

learned, and hence provide a very encouraging basis for the more general enterprise of fully processing NSUs." As for Schlangen & Lascarides's approach, it relies on Minimal Recursion Semantics (Copestake et al., 2005), a minimalistic approach to lexical and compositional semantic analysis, which they describe as "a language in which partial descriptions of formulae of a logical language (the *base language*) can be expressed. This allows one to leave certain semantic distinctions unresolved, reflecting the idea that syntax supplies only partial information about meaning. Technically this is achieved via a strategy that has become standard in computational semantics (e.g., (Reyle, 1993)): one assigns labels to bits of base language formulae so that statements about their combination can remain 'underspecified'" (p. 65).

21. This work was originally described in McShane et al. (2005b).

22. See Ginzburg & Sag (2001) for more on interrogatives.

Chapter 7

1. Bello & Guarini (2010) discuss mindreading as a type of mental simulation.

2. Sample problems are available at https://cs.nyu.edu/faculty/davise/papers/WinogradSchemas/WSCollection.xml

3. Some others require textual coreferences that the agent has not yet been able to identify. This can happen, e.g., for difficult uses of demonstrative pronouns.

Chapter 8

1. In the realm of medical pedagogy, *virtual patients* have also been defined as physical manikins, as live actors who role-play with trainees, and as computer programs that allow trainees to work through static decision trees.

2. See Bailer-Jones (2009) for a discussion of modeling within the philosophy of science.

3. The abstract features used for cognitive modeling are similar to the intermediate categories in ontologies. Although regular people might not think of WHEELED-AIR-VEHICLES as a category, this can still be an appropriate node in an ontology.

4. See section 2.3.1 for an overview of how scripts are represented in the LEIA's ontology.

5. This description draws from McShane, Nirenburg, et al. (2007).

6. See section 4.1.1 for a discussion of property classes and their allowable values.

7. For a description of emotional effects on GERD, see Mizyed et al. (2009). For our incorporation of these factors into the clinical model, see McShane, Nirenburg, Beale, et al. (2013).

8. See section 2.8.2 for a discussion of why a direct mapping between a word and an ontological concept does not constitute upper-case semantics.

9. See section 2.3.1 for a discussion of inheritance in the ontology.

10. See section 4.1.1 for examples of scalar attributes.

11. Remember, diseases can also be dynamically generated if their preconditions are met.

12. This is an arbitrary large number that signifies "never."

13. For a review of text meaning representations, see the warm-up example in section 2.2.

14. A more detailed version of this analysis was published as McShane, Nirenburg, Jarrell, & Fantry (2015).

15. Analogous difficulties are well-known in the domain of ontology merging.

16. See section 2.8.4 for a discussion of paraphrase.

17. As Cooke (n.d.) reports, reviews and categorization schemes for knowledge elicitation and modeling "abound." But, as Ford & Sterman (1998, p. 309) write, "While many methods to elicit information from experts have been developed, most assist in the early phases of modeling: problem articulation, boundary selection, identification of variables, and qualitative causal mapping. ... The literature is comparatively silent, however, regarding methods to elicit the information required to estimate the parameters, initial conditions, and behavior relationships that must be specified precisely in formal modeling."

18. This approach conforms to all seven of Breuker's (1987, summarized in Shadbolt & Burton, 1995) KADS (Knowledge Acquisition and Domain Structuring) principles for the elicitation of knowledge and construction of a system, as detailed in Nirenburg, McShane, & Beale (2010b).

19. For more on influence diagrams, see Howard & Matheson (2005). For an example of their use in another medical domain, see Lucas (1996).

20. For other issues related to reducing the complexity of knowledge acquisition of influence diagrams see Bielza et al. (2010).

21. This material draws from McShane, Nirenburg, & Jarrell (2013).

22. This observation was first made in Paul Meehl's (1996 [1954]) highly influential work that compared statistical predictions to clinical judgments and found the former to consistently outperform the latter. A recent review of Meehl's work (Grove & Lloyd, 2006) concludes that his findings have stood the test of time.

23. We call these MRs rather than TMRs because they need not derive from textual (T) inputs.

Chapter 9

1. A useful comparison can be made with the practices accepted in the *Advances in Cognitive Systems* community (cogsys.org). In its publication guidelines, the community recognizes that formal evaluations are not realistic at every stage of every program of work. Accordingly, the guidelines do not require that all conference and journal papers include formal evaluations. Instead, they require that papers adhere to other quite specific standards, including defining an important problem related to human-level intelligence or cognition, specifying theoretical tenets, making explicit claims, and supporting those claims—be it by argumentation, demonstration, or evaluation.

2. Jones et al. (2012, pp. 83–84) argue that cognitive systems need to integrate *adaptivity*, *directability*, *explainability*, and *trustworthiness* and that "evaluation requirements should, as nearly as possible, not be achievable by a system unless it addresses all four dimensions." Although we appreciate the spirit of this position, we find it too rigid. For both practical and scientific reasons, developers need to be able to evaluate and report intermediate successes as well.

3. There are typically multiple correct and appropriate ways of rendering a given text in another language. No prefabricated gold standard against which translations are measured can cover this space of paraphrases. If a system's translation does not textually match the gold standard, it does not necessarily mean that it is unacceptable. This observation motivates the justified criticisms of BLEU (Papineni et al., 2002; Callison-Burch et al., 2006), a widely used metric for evaluating machine translation systems.

4. Although in this book we subsume *multiword expressions* under the broader rubric of *constructions* (see section 4.3), the paper we cite treated a subset of phenomena that were appropriately referred to as *multiword expressions*.

5. The full TMRs were available to evaluators and were consulted as needed.

6. The negation is taken care of by a modality frame available in the full TMR.

7. Randomly selecting among same-scoring semantic analyses is only one of many possible system settings. The analyzer could also be configured to return all candidate analyses that score within a threshold of the highest score.

8. The software system reported there was subsequently replaced by a different one that incorporates the two kinds of incrementality described in this book. However, since the evaluation targets the underlying algorithms and knowledge bases, which stayed the same across implementations, we would expect similar results from the current system.

9. Joachim Eibl, http://kdiff3.sourceforge.net

10. If the field of NLP had not turned away from the problem of computing meaning some twenty-five years ago, we can imagine that the computational linguistics community might have, by now, made good progress toward this goal. However, as it stands, problems that were already identified and partially addressed became sidelined over the years, with full computational solutions remaining elusive.

11. Language complexity has been addressed from various perspectives. For example, the book *Measuring Grammatical Complexity* (Newmeyer & Preston, 2014) focuses on complexity as it pertains to theoretical syntax.

References

Allen, J., Ferguson, G., Blaylock, N., Byron, D., Chambers, N., Dzikovska, M., Galescu, L., & Swift, M. (2006). Chester: Towards a personal medication advisor. *Journal of Biomedical Informatics*, *39*(5), 500–513.

Allen, J., Ferguson, G., Swift, M., Stent, A., Stoness, S., Galescu, L., Chambers, N., Campana, E., & Aist, G. (2005). Two diverse systems built using generic components for spoken dialogue (Recent progress on TRIPS). *Proceedings of the ACL Interactive Poster and Demonstration Sessions at the 43rd Annual Meeting of the Association for Computational Linguistics* (pp. 85–88). The Association for Computational Linguistics.

Allen, J., & Heeman, P. (1995). *TRAINS Spoken Dialog Corpus* (LDC95S25, CD). Linguistic Data Consortium.

Allen, J., Chambers, N., Ferguson, G., Galescu, L., Jung, H., Swift, M., & Taysom, W. (2007). PLOW: A collaborative task learning agent. *Proceedings of the 22nd National Conference on Artificial Intelligence* (Vol. 2, pp. 1514–1519). The AAAI Press.

Altmann, G. T. M., & Kamide, Y. (1999). Incremental interpretation at verbs: Restricting the domain of subsequent reference. *Cognition*, *73*(3), 247–264.

Andler, D. (2006). Phenomenology in artificial intelligence and cognitive science. In H. Dreyfus & M. Wrathall (Eds.), *The Blackwell companion to phenomenology and existentialism* (pp. 377–393). Blackwell.

Apresjan, Ju. D. (Ed.). (2004). *Novyj ob"jasnitel'nyj slovar' sinonimov russkogo jazyka* [*New explanatory dictionary of Russian synonyms*] (2nd ed.). Vienna Slavic Almanac.

Argall, B., Chernova, S., Veloso, M. M., & Browning, B. (2009). A survey of robot learning from demonstration. *Robotics & Autonomous Systems*, *57*, 469–483.

Asher, N. (1993). *Reference to abstract objects in discourse*. Kluwer.

Asher, N., & Lascarides, A. (1996). Bridging. In R. van der Sandt, R. Blutner, & M. Bierwisch (Eds.), *From underspecification to interpretation*. Working Papers of the Institute for Logic and Linguistics. IBM Deutschland, Heidelberg.

Asher, N., & Lascarides, A. (2003). *Logics of conversation*. Cambridge University Press.

Bailer-Jones, D. M. (2009). *Scientific models in philosophy of science*. University of Pittsburgh Press.

Baker, M., Hansen, T., Joiner, R., & Traum, D. (1999). The role of grounding in collaborative learning tasks. In P. Dillenbourg (Ed.), *Collaborative Learning: Cognitive and Computational Approaches* (pp. 31–63). Elsevier.

Bakhshandeh, O., Wellwood, A., & Allen, J. (2016). Learning to jointly predict ellipsis and comparison structures. *Proceedings of the 20th SIGNLL Conference on Computational Natural Language Learning (CoNLL)* (pp. 62–74). The Association for Computational Linguistics.

Ball, J. (2011). A pseudo-deterministic model of human language processing. In L. Carlson, C. Hoelscher, & T. F. Shipley (Eds.), *Proceedings of the Thirty-third Annual Conference of the Cognitive Science Society* (pp. 495–500). Cognitive Science Society.

Bar Hillel, Y. (1970). *Aspects of language*. Magnes.

Baral, C., Lumpkin, B., & Scheutz, M. (2017). A high level language for human robot interaction. *Proceedings of Advances in Cognitive Systems*, *5* (pp. 1–16). Cognitive Systems Foundation.

Barker, K., Agashe, B., Chaw, S., Fan, J., Friedland, N., Glass, M., Hobbs, J., Hovy, E., Israel, D., Kim, D. S., Mulkar-Mehta, R., Patwardhan, S., Porter, B., Tecuci, D., & Yeh P. (2007). Learning by reading: A prototype system, performance baseline, and lessons learned. *Proceedings of the 22nd AAAI Conference on Artificial Intelligence* (pp. 280–286). The AAAI Press.

Barzilay, R., & McKeown, K. (2001). Extracting paraphrases from a parallel corpus. *Proceedings of the 39th Annual Meeting of the Association for Computational Linguistics* (pp. 50–57). The Association for Computational Linguistics.

Beale, S., Nirenburg, S., & McShane, M. (2003). Just-in-time grammar. *Proceedings of the 2003 International Multiconference in Computer Science and Computer Engineering.*

Bean, D. L., & Riloff, E. (1999). Corpus-based identification of non-anaphoric noun phrases. *Proceedings of the 37th Annual Meeting of the Association for Computational Linguistics* (pp. 373–380). The Association for Computational Linguistics.

Bello, P. (2011). Shared representations of belief and their effects on action selection: A preliminary computational cognitive model. *Proceedings of the 33rd Annual Conference of the Cognitive Science Society* (pp. 2997–3002). Cognitive Science Society.

Bello, P., & Guarini, M. (2010). Introspection and mindreading as mental simulation. In S. Ohlsson & R. Catrambone (Eds.), *Proceedings of the 32nd Annual Conference of the Cognitive Science Society* (pp. 2022–2028). Cognitive Science Society.

Berners-Lee, T., Hendler, J., & Lassila, O. (2001). The Semantic Web. *Scientific American, 284*(5), 34–43.

Besold, T. R., & Uckelman, S. L. (2018) Normative and descriptive rationality: From nature to artifice and back. *Journal of Experimental & Theoretical Artificial Intelligence, 30*(2), 331–344.

Bickerton, D. (1990). *Language and species*. University of Chicago Press.

Bielza, C., Gómez, M., & Shenoy, P. P. (2010). Modeling challenges with influence diagrams: Constructing probability and utility models. *Decision Support Systems, 49*(4), 354–364.

Bies, A., Ferguson, M., Katz, K., & MacIntyre, R. (1995). Bracketing guidelines for Treebank II Style Penn Treebank Project. http://www.cis.upenn.edu/~bies/manuals/root.pdf

Blackburn, P., & Bos, J. (2005). *Representation and inference for natural language: A first course in computational semantics*. Center for the Study of Language and Information.

Bontcheva, K., Tablan, V., Maynard, D., & Cunningham, H. (2004). Evolving GATE to meet new challenges in language engineering. *Natural Language Engineering, 10*(3–4), 349–373.

Bos, J., & Spenader, J. (2011). An annotated corpus for the analysis of VP ellipsis. *Language Resources and Evaluation, 45*, 463–494.

Bowdle, B., & Gentner, D. (2005). The career of metaphor. *Psychological Review, 112*, 193–216.

Bowman, S. R., Angeli, G., Potts, C., & Manning, C. D. (2015). A large annotated corpus for learning natural language inference. *Proceedings of the 2015 Conference on Empirical Methods in Natural Language Processing* (pp. 632–642). The Association for Computational Linguistics.

Boyd, A., Gegg-Harrison, W., & Byron, D. (2005). Identifying non-referential *it*: A machine learning approach incorporating linguistically motivated patterns. *Proceedings of the ACL Workshop on Feature Engineering for Machine Learning in Natural Language Processing* (pp. 40–47). The Association for Computational Linguistics.

Brants, T. (2000). TnT—a statistical part-of-speech tagger. *Proceedings of the Sixth Conference on Applied Natural Language Processing.* The Association for Computational Linguistics.

Bratman, M. E. (1987). *Intentions, plans, and practical reason.* Harvard University Press.

Brick, T., & Scheutz, M. (2007). Incremental natural language processing for HRI. *Proceedings of the ACM/IEEE International Conference on Human-Robot Interaction* (pp. 263–270). Association for Computing Machinery.

Brooks, R. (2015). Mistaking performance for competence. In J. Brockman (Ed.), *What to think about machines that think* (pp. 108–111). Harper Perennial.

Buitelaar, P. P. (2000). Reducing lexical semantic complexity with systematic polysemous classes and underspecification. In A. Bagga, J. Pustejovsky, & W. Zadrozny (Eds.), *Proceedings of the 2000 NAACL-ANLP*

Workshop on Syntactic and Semantic Complexity in Natural Language Processing Systems (pp. 14–19). The Association for Computational Linguistics.

Byron, D. (2004). Resolving pronominal reference to abstract entities [Unpublished doctoral dissertation]. (Technical Report 815). University of Rochester.

Cacciari, C., & Tabossi, P. (Eds). (1993). *Idioms: Processing, structure and interpretation.* Erlbaum.

Callison-Burch, C., Osborne, M., & Koehn, P. (2006). Re-evaluating the role of BLEU in machine translation research. *Proceedings of the 11th Conference of the European Chapter of the Association for Computational Linguistics* (pp. 249–256). The Association for Computational Linguistics.

Cantrell, R., Schermerhorn, P., & Scheutz, M. (2011). Learning actions from human-robot dialogues. *Proceedings of the 2011 IEEE Symposium on Robot and Human Interactive Communication* (pp. 125–130). IEEE.

Carbonell, J. G., & Brown, R. D. (1988). Anaphora resolution: A multi-strategy approach. *Proceedings of the Twelfth International Conference on Computational Linguistics* (pp. 96–101). The Association for Computational Linguistics.

Carbonell, J. G., & Hayes, P. J. (1983). Recovery strategies for parsing extragrammatical language. *American Journal of Computational Linguistics*, *9*(3–4), 123–146.

Carlson, L., Marcu, D., & Okurowski, M. E. (2003). Building a discourse-tagged corpus in the framework of Rhetorical Structure Theory. In J. van Kuppevelt & R. W. Smith (Eds.), *Current and new directions in discourse and dialogue* (pp. 85–112). Kluwer.

Carruthers, P. (2009). How we know our own minds: The relationship between mindreading and metacognition. *Behavioral and Brain Sciences 32*(2), 121–138.

Cartwright, N. (1983). *How the Laws of Physics Lie.* Oxford University Press.

Chambers, C. G., Tanenhaus, M. K., Eberhard, K. M., Filip, H., & Carlson, G. N. (2002). Circumscribing referential domains during real-time language comprehension. *Journal of Memory & Language*, *47*, 30–49.

Chambers, C. G., Tanenhaus, M. K., & Magnuson, J. S. (2004). Actions and affordances in syntactic ambiguity resolution. *Journal of Experimental Psychology*, *30*(3), 687–696.

Charniak, E. (1972). Toward a model of children's story comprehension. [Unpublished doctoral dissertation]. Massachusetts Institute of Technology.

Chinchor, N. (1997). MUC-7 named entity recognition task definition (Version 3.5, September 17). *Proceedings of the Seventh Message Understanding Conference.* http://www-nlpir.nist.gov/related_projects/muc/proceedings/ne_task.html

Chomsky, N. (1957). *Syntactic structures.* Mouton.

Chomsky, N. (1995). *The minimalist program.* MIT Press.

Church, K. (2011). A pendulum swung too far. *Linguistic Issues in Language Technology*, *6*, 1–27.

Church, K., & Hovy, E. (1993). Good applications for crummy machine translation. *Machine Translation*, *8*, 239–258.

Cimiano, P., Unger, C., & McCrae, J. (2014). *Ontology-based interpretation of natural language.* Morgan & Claypool.

Cinková, S. (2009). Semantic representation of non-sentential utterances in dialog. *Proceedings of the EACL 2009 Workshop on Semantic Representation of Spoken Language* (pp. 26–33). The Association for Computational Linguistics.

Clark, A., Fox, C., & Lappin, S. (Eds.). (2010). *The handbook of computational linguistics and natural language processing.* Wiley-Blackwell.

Clark, H. H., & Schaefer, E. F. (1989). Contributing to discourse. *Cognitive Science*, *13*, 259–294.

Clark, H. H. & Wilkes-Gibbs, D. (1986). Referring as a collaborative process. *Cognition*, *22*, 1–39.

Clark, P., Murray, W. R., Harrison, P., & Thompson, J. (2009). Naturalness vs. predictability: A key debate in controlled languages. *Proceedings of the 2009 Conference on Controlled Natural Language* (pp. 65–81). Springer.

Clark, S. (2015). Vector space models of lexical meaning. In S. Lappin & C. Fox (Eds.), *The handbook of contemporary semantic theory* (2nd ed., pp. 493–522). Wiley.

Clegg, A., & Shepherd, A. (2007). Benchmarking natural-language parsers for biological applications using dependency graphs. *BMC Bioinformatics*, *8*(24). https://doi.org/10.1186/1471-2105-8-24

Cohen, K. B., Palmer, M., & Hunter, L. (2008). Nominalization and alternations in biomedical language. *PLOS ONE*, *3*(9), e3158.

Cohen, P., Chaudhri, V., Pease, A., & Schrag, R. (1999). Does prior knowledge facilitate the development of knowledge-based systems? *Proceedings of the 16th National Conference on Artificial Intelligence* (pp. 221–226). American Association for Artificial Intelligence.

Cohen, P. R., & Levesque, H. J. (1990). Rational interaction as the basis for communication. In P. R. Cohen, J. Morgan, & M. Pollack (Eds.), *Intentions in communication* (pp. 221–256). Morgan Kaufmann.

Cohn-Gordon, R., Goodman, N., & Potts, C. (2019). An incremental iterated response model of pragmatics. *Proceedings of the Society for Computation in Linguistics* (Vol. 2, Article 10). https://doi.org/10.7275/cprc-8x17 dissertation

Comrie, B., & Smith, N. (1977). Lingua descriptive questionnaire. *Lingua*, *42*, 1–72.

Cooke, N. J. (n.d.). Knowledge elicitation. Cognitive Engineering Research Institute. http://www.cerici.org/documents/Publications/Durso%20chapter%20on%20KE.pdf

Copestake, A., Flickinger, D., Sag, I., & Pollard, C. (2005). Minimal recursion semantics: An introduction. *Journal of Research on Language & Computation*, *3*(2–3), 281–332.

Coradeschi, S., & Saffiotti, A. (Eds.) (2003). Perceptual anchoring: Anchoring symbols to sensor data in single and multiple robot systems [Special issue]. *Robotics & Autonomous Systems*, *43*(2–3), 83–200.

Core, M., & Allen, J. (1997). Coding dialogs with the DAMSL annotation scheme. *Working Notes of the AAAI Fall Symposium on Communicative Action in Humans and Machines* (pp. 28–35). The AAAI Press.

Crible, L., Abuczki, Á., Burkšaitienė, N., Furkó, P., Nedoluzhko, A., Rackevičienėm, S., Oleškevičienė, G. V., & Zikánová, Š. (2019). Functions and translations of discourse markers in TED Talks: A parallel corpus study of underspecification in five languages. *Journal of Pragmatics*, *142*, 139–155.

Crocker, M. W. (1996). *Computational psycholinguistics: an interdisciplinary approach to the study of language*. Springer.

Davies, M. (2008–). The Corpus of Contemporary American English (COCA): One billion words, 1990–2019. https://www.english-corpora.org/coca/

Davis, E., & Marcus, G. (2015). Commonsense reasoning and commonsense knowledge in artificial intelligence. *Communications of the ACM*, *58*(9), 92–103.

de Marneffe, M., MacCartney, B., & Manning, C. D. (2006). Generating typed dependency parses from phrase structure parses. *Proceedings of the 5th International Conference on Language Resources and Evaluation* (pp. 449–454). European Language Resources Association.

de Marneffe, M., & Potts, C. (2017). Developing linguistic theories using annotated corpora. In N. Ide & J. Pustejovsky (Eds.), *The handbook of linguistic annotation* (pp. 411–438). Springer.

Demberg, V., Keller, F., & Koller, A. (2013). Incremental, predictive parsing with psycholinguistically motivated tree-adjoining grammar. *Computational Linguistics*, *39*(4), 1025–1066.

Denber, M. (1998, June 30). *Automatic resolution of anaphora in English* (Technical Report). Imaging Science Division, Eastman Kodak Co. https://citeseerx.ist.psu.edu/viewdoc/download?doi=10.1.1.33.904&rep=rep1&type=pdf

DeVault, D., Sagae, K., & Traum, D. (2009). Can I finish? Learning when to respond to incremental interpretation results in interactive dialogue. In P. Healey, R. Pieraccini, D. Byron, S. Young, & M. Purver (Eds.), *Proceedings of the 10th Annual Meeting of the Special Interest Group on Discourse and Dialogue, SIGDIAL 2009* (pp. 11–20). The Association for Computational Linguistics.

DeVault, D., Sagae, K., & Traum, D. (2011). Detecting the status of a predictive incremental speech understanding model for real-time decision-making in a spoken dialogue system. *Proceedings of the 12th Annual Conference of the International Speech Communication Association, INTERSPEECH 2011* (pp. 1028–1031). International Speech Communication Association.

Dijkstra, T. (1996). *Computational psycholinguistics: AI and connectionist models of human language processing*. Taylor & Francis.

DiMarco, C., Hirst, G., & Stede, M. (1993). The semantic and stylistic differentiation of synonyms and near-synonyms. *Papers from the AAAI Spring Symposium on Building Lexicons for Machine Translation* (pp. 114–121). (Technical Report SS-93-02). The AAAI Press.

Doddington, G., Mitchell, A., Przybocki, M., Ramshaw, L., Strassel, S., & Weischedel, R. (2004). The automatic content extraction (ACE) program—tasks, data and evaluation. In M. T. Lino, M. F. Xavier, F. Ferreira, R. Costa, & R. Silva (Eds.), *Proceedings of the Fourth International Conference on Language Resources and Evaluation* (pp. 837–840). European Language Resources Association.

Dorr, B. J., Passonneau, R. J., Farwell, D., Green, R., Habash, N., Helmreich, S., Hovy, E., Levin, L., Miller, K. J., Mitamura, T., Rambow, O., & Siddharthan, A. (2010). Interlingual annotation of parallel text corpora: A new framework for annotation and evaluation. *Natural Language Engineering*, *16*(3), 197–243.

Downing, P. (1977). On the creation and use of English compound nouns. *Language*, *53*(4), 810–842.

DuBois, J. W., Chafe, W. L., Meyer, C., Thompson S. A., Englebretson, R., & Martey, N. (2000–2005). *Santa Barbara Corpus of Spoken American English* (parts 1–4). Linguistic Data Consortium.

Eco, U. (1979). *The role of the reader: Explorations in the semiotics of texts*. Indiana University Press.

Elsner, M., & Charniak, E. (2010). The same-head heuristic for coreference. *Proceedings of the 48th Annual Meeting of the Association for Computational Linguistics* (pp. 33–37). The Association for Computational Linguistics.

English, J., & Nirenburg, S. (2007). Ontology learning from text using automatic ontological-semantic text annotation and the Web as the corpus. *Papers from the AAAI 2007 Spring Symposium on Machine Reading* (pp. 43–48). (Technical Report SS-07-06). The AAAI Press.

English, J., & Nirenburg, S. (2010). Striking a balance: Human and computer contributions to learning through semantic analysis. *Proceedings of the IEEE Fourth International Conference on Semantic Computing* (pp. 16–23). IEEE.

Erol, K., Hendler, J., & Nau, D. S. (1994). HTN planning: Complexity and expressivity. *Proceedings of the Twelfth National Conference on Artificial Intelligence* (pp. 1123–1128). The AAAI Press.

Evans, R. (2001). Applying machine learning toward an automatic classification of *it*. *Literary & Linguistic Computing*, *16*(1), 45–57.

Evens, M., & Michael, J. (2006). *One-on-one tutoring by humans and computers*. Erlbaum.

Evsyutina, Y. V., Trukhmanov, A. S., & Ivashkin, V. T. (2014). Family case of achalasia cardia: Case report and review of literature. *World Journal of Gastroenterology*, *20*(4), 1114–1118.

Fass, D. (1997). *Processing metonymy and metaphor*. Ablex.

Feldman, J. (2006). *From molecule to metaphor: A neural theory of language*. MIT Press.

Feldman, J., Dodge, E., & Bryant, J. (2009). A neural theory of language and embodied construction grammar. In B. Heine & H. Narrog (Eds.), *The Oxford handbook of linguistic analysis* (pp. 111–138). Oxford University Press.

Feldman, J. & Narayanan, S. (2004). Embodied meaning in a neural theory of language. *Brain & Language*, *89*, 385–392.

Ferguson, G., & Allen, J. (1998). TRIPS: An integrated intelligent problem-solving assistant. *Proceedings of the Fifteenth National Conference on Artificial Intelligence* (pp. 567–573). The AAAI Press.

Fernández, R., Ginzburg, J., & Lappin, S. 2007. Classifying non-sentential utterances in dialogue: A machine learning approach. *Computational Linguistics*, *33*(3), 397–427.

Fiengo, R., & May, R. 1994. *Indices and identity*. MIT Press.

Fillmore, C. J. (1985). Frames and the semantics of understanding. *Quaderni di Semantica*, *6*(2), 222–254.

Fillmore, C. J., & Baker, C. F. (2009). A frames approach to semantic analysis. In B. Heine & H. Narrog (Eds.), *The Oxford handbook of linguistic analysis* (pp. 313–340). Oxford University Press.

Finin, T. (1980). The semantic interpretation of compound nominals [Unpublished doctoral dissertation]. University of Illinois.

Finlayson, M. A. (2016). Inferring Propp's functions from semantically-annotated text. *Journal of American Folklore*, *129*, 55–77.

Firth, J. R. (1957). A synopsis of linguistic theory, 1930–1955. In J. R. Firth (Ed)., *Studies in linguistic analysis* (pp. 1–32). Blackwell. (Reprinted in *Selected papers of J. R. Firth 1952–1959*, by F. R. Palmer, Ed., 1968, Longman).

Forbus, K. D., Riesbeck, C., Birnbaum, L., Livingston, K., Sharma, A., & Ureel, L. (2007). Integrating natural language, knowledge representation and reasoning, and analogical processing to learn by reading. *Proceedings of the Twenty-Second AAAI Conference on Artificial Intelligence* (pp. 1542–1547). The AAAI Press.

Forbus, K. D. (2018). *Qualitative representations*. MIT Press.

Ford, D. N., & Sterman, J. D. (1998). Expert knowledge elicitation to improve formal and mental models. *System Dynamics Review, 14*, 309–340.

Frigg, R., & Hartman, S. (2020). Models in Science. In E. N. Zalta (Ed.), *The Stanford Encyclopedia of Philosophy* (Spring 2020 Edition). https://plato.stanford.edu/archives/spr2020/entries/models-science

Fuchs, N. E., Kaljurand, K., & Schneider, G. (2006). Attempto controlled English meets the challenges of knowledge representation, reasoning, interoperability and user interfaces. *Proceedings of the Nineteenth International Florida Artificial Intelligence Research Society Conference* (pp. 664–669). The AAAI Press.

Gagné, C. L., & Spalding, T. L. (2006). Using conceptual combination research to better understand novel compound words. *SKASE Journal of Theoretical Linguistics, 3*, 9–16.

Gawande, A. (2009). *The checklist manifesto*. Henry Holt.

Gentner, D., & Maravilla, F. (2018). Analogical reasoning. In L. J. Ball & V. A. Thompson (Eds.), *International handbook of thinking and reasoning* (pp. 186–203). Psychology Press.

Gentner, D., & Smith, L. A. (2013). Analogical learning and reasoning. In D. Reisberg (Ed.), *The Oxford handbook of cognitive psychology* (pp. 668–681). Oxford University Press.

Gibbs, R. W., Jr. (1984). Literal meaning and psychological theory. *Cognitive Science, 8*(3), 275–304.

Gildea, D., & Jurafsky, D. (2002). Automatic labeling of semantic roles. *Computational Linguistics, 28*(3), 245–288.

Ginzburg, J., & Sag, I. A. (2001). *Interrogative investigations: The form, meaning, and use of English interrogatives*. Center for the Study of Language and Information.

Girju, R., Moldovan, D., Tatu, M., & Antohe, D. (2005). On the semantics of noun compounds. *Journal of Computer Speech & Language, 19*(4), 479–496.

Glucksberg, S. (2003). The psycholinguistics of metaphor. *Trends in Cognitive Sciences, 7*(2), 92–96.

Godfrey, J., Holliman, E., & McDaniel, J. (1992). Switchboard: Telephone speech corpus for research and development. *Proceedings of the 1992 IEEE International Conference on Acoustics, Speech and Signal Processing* (Vol. 1, pp. 517–520). IEEE.

Goldstein, A., Arzouan, Y., & Faust, M. (2012). Killing a novel metaphor and reviving a dead one: ERP correlates of metaphor conventionalization. *Brain & Language, 123*, 137–142.

Gonzalo, J., Verdejo, F., Chugur, I., & Cigarran, J. (1998). Indexing with WordNet synsets can improve text retrieval. *Proceedings of the Workshop Usage of WordNet in Natural Language Processing Systems @ ACL/COLING* (pp. 38–44). The Association for Computational Linguistics.

Gorniak, P., & Roy, D. (2005). Probabilistic grounding of situated speech using plan recognition and reference resolution. *Proceedings of the Seventh International Conference on Multimodal Interfaces* (pp. 138–143). Association for Computing Machinery.

Goyal, K., Jauhar, S. K., Li, H., Sachan, M., Srivastava, S., & Hovy, E. (2013). A structured distributional semantic model: Integrating structure with semantics. *Proceedings of the Workshop on Continuous Vector Space Models and Their Compositionality* (pp. 20–29). The Association for Computational Linguistics.

Graff, D., & Cieri, C. (2003). *English Gigaword* (LDC2003T05). Linguistic Data Consortium. https://catalog.ldc.upenn.edu/LDC2003T05

Griffiths, T. L. (2009). Connecting human and machine learning via probabilistic models of cognition. *Proceedings of the 10th Annual Conference of the International Speech Communication Association* (pp. 9–12). ISCA.

Grishman, R., & Sundheim, B. (1996). Message Understanding Conference—6: A brief history. *Proceedings of the 16th International Conference on Computational Linguistics* (pp. 466–471). The Association for Computational Linguistics.

Grosz, B., Joshi, A. K., & Weinstein, S. (1995). Centering: A framework for modelling the local coherence of discourse. *Computational Linguistics, 2*(21), 203–225.

Grove, W. M., & Lloyd, M. (2006). Meehl's contribution to clinical versus statistical prediction. *Journal of Abnormal Psychology, 115*(2), 192–194.

Guarino, N. (1998). Formal ontology in information systems. In N. Guarino (Ed.), *Formal ontology in information systems* (pp. 3–15). IOS Press.

Gunzelmann, G., Gross, J. B., Gluck, K. A., & Dinges, D. F. (2009). Sleep deprivation and sustained attention performance: Integrating mathematical and cognitive modeling. *Cognitive Science, 33*(5), 880–910.

Hahn, U., Romacker, M., & Schulz, S. (1999). How knowledge drives understanding—Matching medical ontologies with the needs of medical language processing. *Artificial Intelligence in Medicine, 15*(1), 25–51.

Hajič, J., Hajičová, E., Mikulová, M., Mírovský, J., Panevová, J., & Zeman, D. (2015). Deletions and node reconstructions in a dependency-based multilevel annotation scheme. *Proceedings of the 16th International Conference on Computational Linguistics and Intelligent Text Processing* (pp.17–31). Springer.

Hajič, J., Hajičová, E., Mírovský, J., & Panevová, J. (2016). Linguistically annotated corpus as an invaluable resource for advancements in linguistic research: A case study. *Prague Bulletin of Mathematical Linguistics, #100*, 69–124.

Hardt, D. (1997). An empirical approach to VP ellipsis. *Computational Linguistics, 23*(4), 525–541.

Harris, D. W. (in press). Semantics without semantic content. *Mind and Language.*

Hasler, L., Orasan, C., & Naumann, K. (2006). NPs for events: Experiments in coreference annotation. *Proceedings of the 5th Edition of the International Conference on Language Resources and Evaluation* (pp. 1167–1172). European Language Resources Association.

Hayes, P. J. (1979). The naive physics manifesto. In D. Michie (Ed.), *Expert systems in the micro-electronic age* (pp. 242–270). Edinburgh University Press.

Hendrickx, I., Kozareva, Z., Nakov, P., Ó Séaghdha, D., Szpakowicz, S., & Veale, T. (2013). SemEval-2013 Task 4: Free paraphrases of noun compounds. *Proceedings of the Seventh International Workshop on Semantic Evaluation* (pp. 138–143). The Association for Computational Linguistics.

Heuer, R. J., Jr. (1999). *Psychology of intelligence analysis.* Central Intelligence Agency Center for the Study of Intelligence. https://www.cia.gov/library/center-for-the-study-of-intelligence/csi-publications/books-and-monographs/psychology-of-intelligence-analysis/

Hirschman, L., & Chinchor, N. (1997). MUC-7 coreference task definition (Version 3.0). *Proceedings of the Seventh Message Understanding Conference.* The Association for Computational Linguistics.

Hirst, G. (1995). Near-synonymy and the structure of lexical knowledge. *Proceedings of the AAAI Symposium on Representation and Acquisition of Lexical Knowledge: Polysemy, Ambiguity, and Generativity* (pp. 51–56). The AAAI Press.

Hobbs, J. R. (2004). Some notes on performance evaluation for natural language systems. University of Southern California Information Sciences Institute. https://www.isi.edu/~hobbs/performance-evaluation.pdf

Hobbs, J. R. (1992). Metaphor and abduction. In A. Ortony, J. Slack, & O. Stock (Eds.), *Communication from an artificial intelligence perspective: Theoretical and applied issues* (pp. 35–58). Springer.

Hobbs, J. R. (2004). Abduction in natural language understanding. In L. Horn & G. Ward (Eds.), *Handbook of pragmatics* (pp. 724–741). Blackwell.

Hoffman, T., & Trousdale, G. (Eds.). (2013). *The Oxford handbook of construction grammar.* Oxford University Press.

Hovy, E., Mitamura, T., Verdejo, F., del Rosal, J., Araki, J., & Philpot, A. (2013). Events are not simple: Identity, non-identity, and quasi-identity. *Proceedings of the First Workshop on EVENTS: Definition, Detection, Coreference, and Representation* (pp. 21–28). The Association for Computational Linguistics.

Howard, R. A., & Matheson, J. E. (2005). Influence Diagrams. *Decision Analysis, 2*(3), 127–143.

Hutchins, W. J. (1986). *Machine translation: Past, present, future.* Longman Higher Education.

Ibrahim, A., Katz, B., & Lin, J. (2003). Extracting structural paraphrases from aligned monolingual corpora. *Proceedings of the Second International Workshop on Paraphrasing* (pp. 57–64). The Association for Computational Linguistics.

Ide, N., & Pustejovsky, J. (Eds.) (2017). *The handbook of linguistic annotation.* Springer.

Ide, N., & Véronis, J. (1993). Extracting knowledge bases from machine-readable dictionaries: Have we wasted our time? *Proceedings of the Workshop from the 1st Conference and Workshop on Building and Sharing of Very Large-Scale Knowledge Bases* (pp. 257–266). AI Communications.

Ide, N., & Wilks, Y. (2006). Making sense about sense. In E. Agirre & P. Edmonds (Eds.), *Word sense disambiguation: Algorithms and applications* (pp. 47–73). Springer.

Inkpen, D., & Hirst, G. (2006). Building and using a lexical knowledge-base of near-synonym differences. *Computational Linguistics, 32*(2), 223–262.

Jackendoff. R. (2002). *Foundations of language: Brain, meaning, grammar, evolution.* Oxford University Press.

Jackendoff, R. (2007). A whole lot of challenges for linguistics. *Journal of English Linguistics, 35*, 253–262.

Jackendoff, R., & Wittenberg, E. (2014). What you can say without syntax: A hierarchy of grammatical complexity. In F. Newmeyer & L. Preston (Eds.), *Measuring grammatical complexity* (pp. 65–82). Oxford University Press.

Jackendoff, R., & Wittenberg, E. (2017). Linear grammar as a possible stepping-stone in the evolution of language. *Psychonomic Bulletin & Review, 24*, 219–224.

Jeong, M., & Lee, G. G. (2006). Jointly predicting dialog act and named entity for spoken language understanding. *Proceedings of the IEEE Spoken Language Technology Workshop* (pp. 66–69). IEEE.

Johnson, K. (2001). What VP ellipsis can do, what it can't, but not why. In M. Baltin & C. Collins (Eds.), *The handbook of contemporary syntactic theory* (pp. 439–479). Blackwell.

Jones, R. M., Wray, R. E., III., & van Lent, M. (2012). Practical evaluation of integrated cognitive systems. *Advances in Cognitive Systems, 1*, 83–92.

Jurafsky, D. (2003). Probabilistic modeling in psycholinguistics: Linguistic comprehension and production. In R. Bod, J. Hay, & S. Jannedy, (Eds.), *Probabilistic linguistics* (pp. 39–96). MIT Press.

Jurafsky, D., & Martin, J. H. (2009). *Speech and language processing: An introduction to natural language processing, speech recognition, and computational linguistics* (2nd ed.). Prentice-Hall.

Kahneman, D. (2011). *Thinking: Fast and slow.* Farrar, Straus and Giroux.

Kahneman, D., & Klein, G. (2009). Conditions for intuitive expertise: A failure to disagree. *American Psychologist, 64*(6), 515–526.

Karlsson, F. (1995). Designing a parser for unrestricted text. In F. Karlsson, A. Voutilainen, J. Heikkilä, & A. Anttila (Eds.), *Constraint grammar: A language-independent framework for parsing unrestricted text* (pp. 1–40). Mouton de Gruyter.

Kempson, R., Meyer-Viol, W., & Gabbay, D. (2001) *Dynamic syntax: The flow of language understanding.* Blackwell.

Kendall, E. F., & McGuinness, D. L. (2019). *Ontology engineering.* Morgan and Claypool.

Kim, S. N., & Nakov, P. (2011). Large-scale noun compound interpretation using bootstrapping and the web as a corpus. *Proceedings of the 2011 Conference on Empirical Methods in Natural Language Processing* (pp. 648–658). The Association for Computational Linguistics.

King, G. W. (1956). Stochastic methods of mechanical translation. *Mechanical Translation, 3*(2), 38–39.

Kingsbury, P., & Palmer, M. (2002). From Treebank to PropBank. *Proceedings of the 3rd International Conference on Language Resources and Evaluation* (pp. 1989–1993). European Language Resources Association.

Kipper, K., Korhonen, A., Ryant, N., & Palmer, M. (2006). Extending VerbNet with novel verb classes. *Proceedings of the Fifth International Conference on Language Resources and Evaluation* (pp. 1027–1032). European Language Resources Association.

Koedinger, K. R., Anderson, J. R., Hadley, W. H., & Mark, M. A. (1997). Intelligent tutoring goes to school in the big city. *International Journal of Artificial Intelligence in Education, 8*, 30–43.

Köhn, A. (2018). Incremental natural language processing: Challenges, strategies, and evaluation. *Proceedings of the 27th International Conference on Computational Linguistics* (pp. 2990–3003). The Association for Computational Linguistics.

Korte, R. F. (2003). Biases in decision making and implications for human resource development. *Advances in Developing Human Resources, 5*(4), 440–457.

Král, P., & Cerisara, C. (2010). Dialogue act recognition approaches. *Computing & Informatics*, *29*, 227–250.

Krippendorff, K. (2010). Krippendorff's alpha. In N. Salkind (Ed.), *Encyclopedia of research design* (pp. 669–674). SAGE.

Kruijff, G. J. M., Lison, P., Benjamin, T., Jacobsson, H., & Hawes, N. (2007). Incremental, multi-level processing for comprehending situated dialogue in human-robot interaction. *Proceedings from the Symposium Language and Robots*.

Lakoff, G. (1993). The contemporary theory of metaphor. In A. Ortony (Ed.), *Metaphor and thought* (2nd ed., pp. 202–251). Cambridge University Press.

Lakoff, G., & Johnson, M. (1980). *Metaphors we live by*. University of Chicago Press.

Langley, P., Laird, J. E., & Rogers, S. (2009). Cognitive architectures: Research issues and challenges. *Cognitive Systems Research*, *10*, 141–160.

Langley, P., Meadows, B., Gabaldon, A., & Heald, R. (2014). Abductive understanding of dialogues about joint activities. *Interaction Studies*, *15*(3), 426–454.

Langlotz, A. (2006). *Idiomatic creativity: A cognitive-linguistic model of idiom-representation and idiom-variation in English*. John Benjamins.

Lapata, M. (2002). The disambiguation of nominalizations. *Computational Linguistics*, *28*(3), 357–388.

Leafgren, J. (2002). *Degrees of explicitness: Information structuring and the packaging of Bulgarian subjects and objects*. John Benjamins.

Lee, H., Chang, A., Peirsman, Y., Chambers, N., Surdeanu, M., & Jurafsky, D. (2013). Deterministic coreference resolution based on entity-centric, precision-ranked rules. *Computational Linguistics*, *39*(4), 885–916.

Lenat, D. (1995). CYC: A large-scale investment in knowledge infrastructure. *Communications of the ACM*, *38*(11), 32–38.

Lenat, D., Miller, G., & Yokoi, T. (1995). CYC, WordNet, and EDR: Critiques and responses. *Communications of the ACM*, *38*(11), 45–48.

Lenci, A., Bel, N., Busa, F., Calzolari, N., Gola, E., Monachini, M., Ogonowski, A., Peters, I., Peters, W., Ruimy, N., Villegas, M., & Zampolli, A. (2000). SIMPLE: A general framework for the development of multilingual lexicons. *International Journal of Lexicography, 13(4), 249–263.*

Lepore, E., & Stone, M. (2010). Against metaphorical meaning. *Topoi*, *29*(2), 165–180.

Levesque, H., Davis, E., & Morgenstern, L. (2012). The Winograd Schema Challenge. *Proceedings of the Thirteenth International Conference on Principles of Knowledge Representation and Reasoning* (pp. 552–561). The AAAI Press.

Levi, J. N. (1979). The syntax and semantics of complex nominals. *Language*, *55*(2), 396–407.

Levin, B. (1993). *English verb classes and alternations: A preliminary investigation*. University of Chicago Press.

Lewis, R. (1993). *An architecturally-based theory of human sentence comprehension* [Unpublished doctoral dissertation]. CMU-CS-93-226. Carnegie Mellon University.

Li, Y., Musilek, P., Reformat, M., & Wyard-Scott, L. (2009). Identification of pleonastic *it* using the Web. *Journal of Artificial Intelligence Research*, *34*(1), 339–389.

Lieber, R., & Štekauer, P. (2009). *The Oxford handbook of compounding*. Oxford University Press.

Lin, D. (1998). Extracting collocations from text corpora. *Proceedings of the COLING-ACL '98 Workshop on Computational Terminology* (pp. 57–63). The Association for Computational Linguistics.

Lindes, P., & Laird, J. E. (2016). Toward integrating cognitive linguistics and cognitive language processing. In D. Reitter & F. E. Ritter (Eds.), *Proceedings of the 14th International Conference on Cognitive Modeling* (pp. 86–92). Penn State.

Liu, B., Hu, M., & Cheng, J. (2005, May 10–14). Opinion observer: Analyzing and comparing opinions on the web. *WWW '05: Proceedings of the 14th International Conference on World Wide Web* (pp. 342–351). Association for Computing Machinery.

Lombrozo, T. (2006). The structure and function of explanations. *Trends in Cognitive Sciences*, *10*, 464–470.

Lombrozo, T. (2012). Explanation and abductive inference. In K. J. Holyoak & R. G. Morrison (Eds.). *Oxford handbook of thinking and reasoning* (pp. 260–276). Oxford University Press.

Lombrozo, T. (2016). Explanatory preferences shape learning and inference. *Trends in Cognitive Sciences, 20*, 748–759.

Löwe, B., & Müller, T. (2011). Data and phenomena in conceptual modeling. *Synthese, 182*, 131–148.

Lu, J., & Ng, V. (2016). Event coreference resolution with multi-pass sieves. *Proceedings of the 10th Language Resources and Evaluation Conference* (pp. 3996–4003). European Language Resources Association.

Lu, J., & Ng, V. (2018). Event coreference resolution: A survey of two decades of research. *Proceedings of the 27th International Joint Conference on Artificial Intelligence* (pp. 5479–5486). International Joint Conferences on Artificial Intelligence.

Lucas, P. (1996). Knowledge acquisition for decision-theoretic expert systems. *AISB Quarterly, 94*, 23–33.

Magnolini, S. (2014). A survey on paraphrase recognition. In L. Di Caro, C. Dodaro, A. Loreggia, R. Navigli, A. Perotti, & M. Sanguinetti (Eds.), *Proceedings of the 2nd Doctoral Workshop in Artificial Intelligence (DWAI 2014), An official workshop of the 13th Symposium of the Italian Association for Artificial Intelligence "Artificial Intelligence for Society and Economy" (AI*AI 2014)* (pp. 33–41). CEUR-WS.org.

Malle, B. (2010). Intentional action in folk psychology. In T. O'Connor & C. Sandis (Eds.), *A companion to the philosophy of action* (pp. 357–365). Wiley-Blackwell.

Mani, I., Pustejovsky, J., & Gaizauskas, R. (Eds.) (2005). *The language of time: A reader*. Oxford University Press.

Manning, C. D., & Schütze, H. (1999). *Foundations of statistical natural language processing*. MIT Press.

Manning, C. D. (2004). Language learning: Beyond Thunderdome. *Proceedings of the Eighth Conference on Computational Natural Language Learning (CoNLL-2004) at HLT-NAACL 2004* (p. 138). The Association for Computational Linguistics.

Manning, C. D. (2006). Local textual inference: It's hard to circumscribe, but you know it when you see it—and NLP needs it. Unpublished manuscript. Stanford University. http://nlp.stanford.edu/~manning/papers/LocalTextualInference.pdf

Manning, C. D., Surdeanu, M., Bauer, J., Finkel, J., Bethard, S. J., & McClosky, D. (2014). The Stanford CoreNLP natural language processing toolkit. *Proceedings of the 52nd Annual Meeting of the Association for Computational Linguistics: System Demonstrations* (pp. 55–60). The Association for Computational Linguistics.

Marcu, D. (2000). The rhetorical parsing of unrestricted texts: A surface based approach. *Computational Linguistics, 26*(3), 395–448.

Marcus, M., Santorini, B., & Marcinkiewicz, M. A. (1993). Building a large annotated corpus of English: The Penn Treebank. *Computational Linguistics, 19*(2), 313–330.

Mascardi, V., Cordì, V., & Rosso, P. (2007). A comparison of upper ontologies. In M. Baldoni, A. Boccalatte, F. De Paoli, M. Martelli, & V. Mascardi (Eds.), *Proceedings of WOA 2007: Dagli Oggetti agli Agenti. 8th AI*IA/TABOO Joint Workshop "From Objects to Agents": Agents and Industry: Technological Applications of Software Agents* (pp. 55–64). Seneca Edizioni Torino.

McAllester, D. A., and Givan, R. (1992). Natural language syntax and first-order inference. *Artificial Intelligence, 56*(1): 1–20.

McCulloch, W. S., & Pitts, W. H. (1943). A logical calculus of the ideas immanent in nervous activity. *Bulletin of Mathematical Biophysics, 5*, 115–133.

McShane, M. (2000). Hierarchies of parallelism in elliptical Polish structures. *Journal of Slavic Linguistics, 8*, 83–117.

McShane, M. (2005). *A theory of ellipsis*. Oxford University Press.

McShane, M. (2009). Reference resolution challenges for an intelligent agent: The need for knowledge. *IEEE Intelligent Systems, 24*(4), 47–58.

McShane, M. (2015). Expectation-driven treatment of difficult referring expressions. *Proceedings of the Third Annual Conference on Advances in Cognitive Systems* (pp. 1–17). Cognitive Systems Foundation.

McShane, M. (2017a). Choices for semantic analysis in cognitive systems. *Advances in Cognitive Systems, 5*, 25–36.

McShane, M. (2017b). Natural language understanding (NLU, not NLP) in cognitive systems. *AI Magazine, 34*(4), 43–56.

McShane, M., & Babkin, P. (2016a). Automatically resolving difficult referring expressions. *Advances in Cognitive Systems*, *4*, 247–263.

McShane, M., & Babkin, P. (2016b). Detection and resolution of verb phrase ellipsis. *Linguistic Issues in Language Technology*, *13*(1), 1–34.

McShane, M., & Beale, S. (2020). A cognitive model of elliptical and anaphoric event coreference. Manuscript submitted for publication.

McShane, M., Beale, S., & Babkin, P. (2014). Nominal compound interpretation by intelligent agents. *Linguistic Issues in Language Technology*, *10*(1), 1–34.

McShane, M., Beale, S. & Nirenburg, S. (2019). Applying deep language understanding to open text: Lessons learned. In A. K. Goel, C. M. Seifert, & C. Freska (Eds.), *Proceedings of the 41st Annual Meeting of the Cognitive Science Society* (pp. 796–802). Cognitive Science Society.

McShane, M., Beale, S., Nirenburg, S., Jarrell, B., & Fantry, G. (2012). Inconsistency as a diagnostic tool in a society of intelligent agents. *Artificial Intelligence in Medicine*, *55*(3), 137–148.

McShane, M., Fantry, G., Beale, S., Nirenburg, S., & Jarrell, B. (2007). Disease interaction in cognitive simulations for medical training. *Proceedings of the MODSIM World Conference, Medical Track*.

McShane, M., Jarrell, B., Fantry, G., Nirenburg, S., Beale, S., & Johnson, B. (2008). Revealing the conceptual substrate of biomedical cognitive models to the wider community. In J. D. Westwood, R. S. Haluck, H. M. Hoffman, G. T. Mogel, R. Phillips, R. A. Robb, & K. G. Vosburgh (Eds.), *Medicine meets virtual reality 16: Parallel, combinatorial, convergent: NextMed by design* (pp. 281–286). IOS Press.

McShane, M., & Nirenburg, S. (2003). Parameterizing and eliciting text elements across languages. *Machine Translation*, *18*(2), 129–165.

McShane, M., & Nirenburg, S. (2012). A knowledge representation language for natural language processing, simulation and reasoning. *International Journal of Semantic Computing*, *6*(1), 3–23.

McShane, M., Nirenburg, S., & Babkin, P. (2015). Sentence trimming in service of verb phrase ellipsis resolution. In G. Airenti, B. G. Bara, & G. Sandini (Eds.), *Proceedings of the EuroAsianPacific Joint Conference on Cognitive Science (EAPCogSci 2015)* (Vol. 1419 of CEUR Workshop Proceedings, pp. 228–233). CEUR-WS.org.

McShane, M., Nirenburg, S., & Beale, S. (2004). OntoSem and SIMPLE: Two multi-lingual world views. In G. Hirst & S. Nirenburg (Eds.), *Proceedings of the Second Workshop on Text Meaning and Interpretation, held in cooperation with the 42nd Annual Meeting of the Association for Computational Linguistics* (pp. 25–32). The Association for Computational Linguistics.

McShane, M., Nirenburg, S., & Beale, S. (2005a). An NLP lexicon as a largely language independent resource. *Machine Translation*, *19*(2), 139–173.

McShane, M., Nirenburg, S., & Beale, S. (2005b). Semantics-based resolution of fragments and underspecified structures. *Traitement Automatique des Langues*, *46*(1), 163–184.

McShane, M., Nirenburg, S., & Beale, S. (2008). Ontology, lexicon and fact repository as leveraged to interpret events of change. In C. Huang, N. Calzolari, A. Gangemi, A. Lenci, A. Oltramari, & L. Prevot (Eds.), *Ontology and the lexicon: A natural language processing perspective* (pp. 98–121). Cambridge University Press.

McShane, M., Nirenburg, S., & Beale, S. (2015). The Ontological Semantic treatment of multiword expressions. *Lingvisticae Investigationes*, *38*(1): 73–110.

McShane, M., Nirenburg, S., & Beale, S. (2016). Language understanding with Ontological Semantics. *Advances in Cognitive Systems*, *4*, 35–55.

McShane, M., Nirenburg, S., Beale, S., Jarrell, B., & Fantry, G. (2007). Knowledge-based modeling and simulation of diseases with highly differentiated clinical manifestations. In R. Bellazzi, A. Abu-Hanna, & J. Hunter (Eds.), *Artificial Intelligence in Medicine: Proceedings of the 11th Conference on Artificial Intelligence in Medicine* (pp. 34–43). Springer.

McShane, M., Nirenburg, S., Beale, S., Jarrell, B., Fantry, G., & Mallott, D. (2013). Mind-, body- and emotion-reading. *Proceedings of the Annual Meeting of the International Association for Computing and Philosophy*.

McShane, M., Nirenburg, S., Beale, S., & O'Hara, T. (2005c). Semantically rich human-aided machine annotation. *Proceedings of the Workshop on Frontiers in Corpus Annotation II: Pie in the Sky, at the 43rd Annual Meeting of the Association for Computational Linguistics* (pp. 68–75). The Association for Computational Linguistics.

McShane, M., Nirenburg, S., Cowie, J., & Zacharski, R. (2002). Embedding knowledge elicitation and MT systems within a single architecture. *Machine Translation*, *17*(4), 271–305.

McShane, M., Nirenburg, S., & Jarrell, B. (2013). Modeling decision-making biases. *Biologically-Inspired Cognitive Architectures*, *3*, 39–50.

McShane, M., Nirenburg, S., Jarrell, B., & Fantry, G. (2015). Learning components of computational models from texts. In M. A. Finlayson, B. Miller, A. Lieto, & R. Ronfard (Eds.), *Proceedings of the 6th Workshop on Computational Models of Narrative* (pp. 108–123). Dagstuhl.

McWhorter, J. H. (2016). *The language hoax*. Oxford University Press.

Meehl, P. E. (1996). *Clinical vs. statistical predictions: A theoretical analysis and a review of the evidence*. Jason Aronson. (Original work published 1954)

Mikulová, M. (2011). *Významová reprezentace elipsy* [*The semantic representation of ellipsis*]. Studies in Computational and Theoretical Linguistics.

Mikulová, M. (2014). Semantic representation of ellipsis in the Prague Dependency Treebanks. *Proceedings of the Twenty-Sixth Conference on Computational Linguistics and Speech Processing* (pp. 125–138). The Association for Computational Linguistics and Chinese Language Processing.

Miller, G. A. (1995). WordNet: A lexical database for English. *Communications of the ACM*, *38*(11), 39–41.

Minsky, M. (1975). A framework for representing knowledge. In P. Winston (Ed.), *The psychology of computer vision*. McGraw-Hill.

Mitkov, R. (2001). Outstanding issues in anaphora resolution. In A. Gelbukh (Ed.), *Computational linguistics and intelligent text processing* (pp. 110–125). Springer.

Mizyed, I., Fass, S. S., & Fass, R. (2009). Review article: Gastro-oesophageal reflux disease and psychological comorbidity. *Alimentary Pharmacology & Therapeutics*, *29*, 351–358.

Moldovan, D., Badulescu, A., Tatu, M., Antohe, D., & Girju, R. (2004). Models for the semantic classification of noun phrases. In D. Moldovan & R. Girju (Eds.), *Proceedings of the Computational Lexical Semantics Workshop at HLT-NAACL 2004* (pp. 60–67). The Association for Computational Linguistics.

Monti, J., Seretan, V., Pastor, G. C., & Mitkov, R. (2018). Multiword units in machine translation and translation technology. In R. Mitkov, J. Monti, G. C. Pastor, & V. Seretan (Eds.), *Multiword units in machine translation and translation technology* (pp. 1–37). John Benjamins.

Mori, M. (2012, June 12). The uncanny valley: The original essay by Masahiro Mori (K. F. MacDorman & N. Kageki, Trans.). *IEEE Spectrum*. https://spectrum.ieee.org/automaton/robotics/humanoids/the-uncanny-valley

Navarretta, C. (2004). Resolving individual and abstract anaphora in texts and dialogues. *Proceedings of the 20th International Conference on Computational Linguistics* (pp. 233–239). The Association for Computational Linguistics.

Navigli, R. (2009). Word sense disambiguation: A survey. *ACM Computing Surveys*, *41*(2), 10:1–10:69.

Navigli, R., Velardi, P., & Faralli, S. (2011). A graph-based algorithm for inducing lexical taxonomies from scratch. In T. Walsh (Ed.), *Proceedings of the 22nd International Joint Conference on Artificial Intelligence* (pp. 1872–1877). The AAAI Press.

Newell, A. (1982). The knowledge level, *Artificial Intelligence*, *18*, 87–127.

Newmeyer, F. J., & Preston, L. B. (2014). *Measuring grammatical complexity*. Oxford University Press.

Nielsen, S. B. (2019). Making a glance an action: Doctors' quick looks at their desk-top computer screens. *Journal of Pragmatics*, *142*, 62–74.

Niles, I., & Pease, A. (2001). Toward a standard upper ontology. *Proceedings of the 2nd International Conference on Formal Ontology in Information Systems* (pp. 2–9). ACM.

Nirenburg, S. 2010. The Maryland Virtual Patient as a task-oriented conversational companion. In Y. Wilks (Ed.), *Close engagements with artificial companions*. John Benjamins.

Nirenburg, S., & McShane, M. (2009). Computational field semantics: Acquiring an Ontological Semantic lexicon for a new language. In S. Nirenburg (Ed.), *Language engineering for lesser-studied languages* (pp. 183–206). IOS Press.

Nirenburg, S., & McShane, M. (2016a). Natural language processing. In S. Chipman (Ed.), *The Oxford handbook of cognitive science* (Vol. 1). Oxford University Press.

Nirenburg, S., & McShane, M. (2016b). Slashing metaphor with Occam's razor. *Proceedings of the Fourth Annual Conference on Advances in Cognitive Systems* (pp. 1–14). Cognitive Systems Foundation.

Nirenburg, S., McShane, M., & Beale, S. (2004). The rationale for building resources expressly for NLP. In M. T. Lino, M. F. Xavier, F. Ferreira, R. Costa, & R. Silva (Eds.), *Proceedings of the Fourth International Conference on Language Resources and Evaluation* (pp. 3–6). European Language Resources Association.

Nirenburg, S., McShane, M., & Beale, S. (2008a). Resolving paraphrases to support modeling language perception in an intelligent agent. In J. Bos & R. Delmonte (Eds.), *Semantics in Text Processing: STEP 2008 Conference Proceedings* (pp. 179–192). College Publications.

Nirenburg, S., McShane, M., & Beale, S. (2008b). A simulated physiological/cognitive "double agent." In J. Beal, P. Bello, N. Cassimatis, M. Coen, & P. Winston (Eds.), *Papers from the Association for the Advancement of Artificial Intelligence Fall Symposium "Naturally Inspired Cognitive Architectures."* The AAAI Press.

Nirenburg, S., McShane, M., & Beale, S. (2010a). Aspects of metacognitive self-awareness in Maryland Virtual Patient. In R. Pirrone, R. Azevedo, & G. Biswas (Eds.), *Cognitive and Metacognitive Educational Systems: Papers from the Association for the Advancement of Artificial Intelligence Fall Symposium* (pp. 69–74). The AAAI Press.

Nirenburg, S., McShane, M., & Beale, S. (2010b). Hybrid methods of knowledge elicitation within a unified representational knowledge scheme. In J. Filipe & J. L. G. Dietz (Eds.), *Proceedings of the International Conference on Knowledge Engineering and Ontology Development* (pp. 177–192). SciTePress.

Nirenburg, S., McShane, M., Beale, S., Wood, P., Scassellati, B., Mangin, O, & Roncone, A. (2018). Toward human-like robot learning. *Natural Language Processing and Information Systems, Proceedings of the 23rd International Conference on Applications of Natural Language to Information Systems (NLDB 2018)* (pp. 73–82). Springer.

Nirenburg, S., Oates, T., & English, J. (2007). Learning by reading by learning to read. *Proceedings of the International Conference on Semantic Computing* (pp. 694–701). IEEE.

Nirenburg, S., & Raskin, V. (2004). *Ontological Semantics*. MIT Press.

Nirenburg, S., Somers, H., & Wilks, Y. (Eds.). (2003). *Readings in machine translation*. MIT Press.

Nirenburg, S., & Wilks, Y. (2001). What's in a symbol: Ontology and the surface of language. *Journal of Experimental & Theoretical AI*, *13*, 9–23.

Nirenburg, S., & Wood, P. (2017) Toward human-style learning in robots. *Proceedings of the AAAI Fall Symposium "Natural Communication for Human-Robot Collaboration."* The AAAI Press.

Norrthon, S. (2019). To stage an overlap—The longitudinal, collaborative and embodied process of staging eight lines in a professional theatre rehearsal process. *Journal of Pragmatics*, *142*, 171–184.

Nouri, E., Artstein, R., Leuski, A., & Traum, D. (2011). Augmenting conversational characters with generated question-answer pairs. *Proceedings of the AAAI Symposium on Question Generation* (pp. 49–52). The AAAI Press.

Noy, N. F., Fergerson, R. W., & Musen, M. A. (2000). The knowledge model of Protégé-2000: Combining interoperability and flexibility. *Proceedings of 12th European Workshop on Knowledge Acquisition, Modeling and Management* (pp. 17–32). Springer.

Nunberg, G. (1987). Poetic and prosaic metaphors. *Proceedings of the 1987 Workshop on Theoretical Issues in Natural Language Processing* (pp. 198–201). The Association for Computational Linguistics.

Nyberg, E. H., & Mitamura, T. (1996). Controlled language and knowledge-based machine translation: Principles and practice. *CLAW 96: Proceedings of the First International Workshop on Controlled Language Applications*. Centre for Computational Linguistics, Katholieke Universiteit Leuven.

Ogden, C. K. (1934). *The system of basic English*. Harcourt, Brace and Company.

O'Hara, T., & Wiebe, J. (2009). Exploiting semantic role resources for preposition disambiguation. *Computational Linguistics*, *35*(2), 151–184.

Olsson, F. (2004). *A survey of machine learning for reference resolution in textual discourse*. Swedish Institute of Computer Science. (SICS Technical Report T2004:02, ISSN 1100–3154, ISRN:SICS-T-2004/02-SE).

Onyshkevych, B. (1997). An ontological semantic framework for text analysis [Unpublished doctoral dissertation]. Carnegie Mellon University.

Palmer, M., Babko-Malaya, O., & Dang, H. T. (2004). Different sense granularities for different applications. *Proceedings of the Second Workshop on Scalable Natural Language Understanding Systems at HLT/NAACL-04*. The Association for Computational Linguistics.

Palmer, M., Gildea, D., & Kingsbury, P. (2005). The proposition bank: An annotated corpus of semantic roles. *Computational Linguistics, 31*(1), 71–105.

Panda, S. C. (2006). Medicine: Science or art? *Mens Sana Monographs*, *4*(1), 127–138.

Panton, K., Matuszek, C., Lenat, D. B., Schneider, D., Witbrock, M., Siegel, N., & Shepard, B. (2006). Common sense reasoning—from Cyc to intelligent assistant. In Y. Cai & J. Abascal (Eds.), *Ambient intelligence in everyday life* (pp. 1–31). Springer.

Papineni, K., Roukos, S., Ward, T., & Zhu, W. (2002). BLEU: A method for automatic evaluation of machine translation. *Proceedings of the 40th Annual Meeting of the Association for Computational Linguistics* (pp. 311–318). The Association for Computational Linguistics.

Paroubek, P., Chaudiron, S., & Hirschman, L. (2007). Principles of evaluation in natural language processing. *Traitement Automatique des Langues*, *48*(1), 7–31.

Pease, A., & Murray, W. (2003). An English to logic translator for ontology-based knowledge representation languages. *Proceedings of the International Conference on Natural Language Processing and Knowledge Engineering* (pp. 777–783). IEEE.

Pereira, F., Tishby, N., & Lee, L. (1993). Distributional clustering of English words. *Proceedings of the 31st Annual Meeting of the Association for Computational Linguistics* (pp. 183–190). The Association for Computational Linguistics.

Perrault, C. R. (1990). An application of default logic to speech act theory. In P. R. Cohen, J. Morgan, & M. E. Pollack (Eds.), *Intentions in communication* (pp. 161–185). MIT Press.

Piantadosi, S. T., Tily, H., & Gibson, E. (2012). The communicative function of ambiguity in language. *Cognition*, *122*, 280–291.

Poesio, M. (2004). Discourse annotation and semantic annotation in the GNOME corpus. *Proceedings of the 2004 ACL Workshop on Discourse Annotation* (pp. 72–79). The Association for Computational Linguistics.

Poesio, M., & Artstein, R. (2005). The reliability of anaphoric annotation, reconsidered: Taking ambiguity into account. *Proceedings of the Workshop on Frontiers in Corpus Annotation II: Pie in the Sky* (pp. 76–83). The Association for Computational Linguistics.

Poesio, M., Mehta, R., Maroudas, A., & Hitzeman, J. (2004). Learning to resolve bridging references. *Proceedings of the 42nd Annual Meeting on Association for Computational Linguistics* (pp. 143–150). The Association for Computational Linguistics.

Poesio, M., Stevenson, R., di Eugenio, B., & Hitzeman, J. (2004). Centering: A parametric theory and its instantiations. *Computational Linguistics*, *30*(3), 309–363.

Poesio, M., Stuckardt, R., Versley, Y. (Eds.) (2016). *Anaphora resolution: Algorithms, resources, and applications*. Springer.

Pulman, S. (1996). Controlled language for knowledge representation. *CLAW 96: Proceedings of the First International Workshop on Controlled Language Applications* (pp. 233–242). Centre for Computational Linguistics, Katholieke Universiteit Leuven.

Purver, M., Eshghi, A., & Hough, J. (2011). Incremental semantic construction in a dialogue system. In J. Bos & S. Pulman (Eds.), *Proceedings of the 9th International Conference on Computational Semantics* (pp. 365–369). The Association for Computational Linguistics.

Pustejovsky, J. (1995). *The generative lexicon*. MIT Press.

Pustejovsky, J., & Batiukova, O. (2019). *The lexicon*. Cambridge University Press.

Pustejovsky, J., Knippen, R., Littman, J., & Saurí, R. (2005). Temporal and event information in natural language text. *Language Resources and Evaluation*, *39*(2–3), 123–164.

Pustejovsky, J., Krishnaswamy, N., Draper, B., Narayana, P., & Bangar, R. (2017). Creating common ground through multimodal simulations. In N. Asher, J. Hunter, & A. Lascarides (Eds.), *Proceedings of the IWCS Workshop on Foundations of Situated and Multimodal Communication*. The Association for Computational Linguistics.

Raskin, V., & Nirenburg, S. (1998). An applied ontological semantic microtheory of adjective meaning for natural language processing. *Machine Translation*, *13*(2–3), 135–227.

Ratinov, L., & Roth, D. (2012). Learning-based multi-sieve co-reference resolution with knowledge. In J. Tsujii, J. Henderson, & M. Paşca (Eds.), *Proceedings of the 2012 Joint Conference on Empirical Methods in Natural Language Processing and Computational Natural Language Learning* (pp. 1234–1244). The Association for Computational Linguistics.

Recasens, M., Martí, M. A., & Orasan, C. (2012). Annotating near-identity from coreference disagreements. In N. Calzolari, K. Choukri, T. Declerck, M. Uğur Doğan, B. Maegaard, J. Mariani, A. Moreno, J. Odijk, S. Piperidis (Eds.), *Proceedings of the Eighth International Conference on Language Resources and Evaluation* (pp. 165–172). European Language Resources Association.

Reiter, E. (2010). Natural language generation. In A. Clark, C. Fox, & S. Lappin (Eds.), *Handbook of computational linguistics and natural language processing* (pp. 574–598). Wiley Blackwell.

Resnik, P., & Lin, J. (2010). Evaluation of NLP systems. In A. Clark, C. Fox, & S. Lappin (Eds.), *Handbook of computational linguistics and natural language processing* (pp. 271–296). Wiley Blackwell.

Roncone, A., Mangin, O., & Scassellati, B. (2017). Transparent role assignment and task allocation in human robot collaboration. *Proceedings of the IEEE International Conference on Robotics and Automation* (pp. 1014–1021). IEEE.

Rosario, B., & Hearst, M. (2001). Classifying the semantic relations in noun compounds via a domain-specific lexical hierarchy. In L. Lee & D. Harman (Eds.), *Proceedings of Empirical Methods in Natural Language Processing* (pp. 82–90). The Association for Computational Linguistics.

Rosenbloom, P. S., Newell, A., & Laird, J. E. (1991). Toward the knowledge level in Soar: The role of the architecture in the use of knowledge. In K. VanLehn (Ed.), *Architectures for Intelligence: The 22nd Carnegie Mellon Symposium on Cognition*. Erlbaum.

Roy, D. (2005). Grounding words in perception and action: Computational insights. *TRENDS in Cognitive Sciences*, *9*(8), 389–396.

Roy, D. K., & Reiter, E. (Eds.) (2005). Connecting language to the world. *Artificial Intelligence*, *167*(1–2).

Sampson, G. (2003). Thoughts on two decades of drawing trees. In A. Abeillé (Ed.), *Treebanks: Building and using parsed corpora* (pp. 23–41). Kluwer.

Schank, R. (1972). Conceptual dependency: A theory of natural language understanding. *Cognitive Psychology*, *3*, 532–631.

Schank, R., & Abelson, R. P. (1977). *Scripts, plans, goals and understanding: An inquiry into human knowledge structures*. Erlbaum.

Scheutz, M., Eberhard, K., & Andronache, V. (2004). A real-time robotic model of human reference resolution using visual constraints. *Connection Science Journal*, *16*(3), 145–167.

Scheutz, M., Harris, J., & Schmemerhorn, P. (2013). Systematic integration of cognitive and robotic architectures. *Advances in Cognitive Systems*, *2*, 277–296.

Scheutz, M., Krause, E., Oosterveld, B., Frasca, T., & Platt, R. (2017). Spoken instruction-based one-shot object and action learning in a cognitive robotic architecture. In S. Das, E. Durfee, K. Larson, W. Winikoff (Eds.), *Proceedings of the Sixteenth International Conference on Autonomous Agents and Multiagent Systems* (pp. 1378–1386). International Foundation for Autonomous Agents and Multiagent Systems.

Schlangen, D., & Lascarides, A. (2003). The interpretation of non-sentential utterances in dialogue. *Proceedings of the 4th SIGdial Workshop on Discourse and Dialogue* (pp. 62–71). The Association for Computational Linguistics.

Sedivy, J. C. (2007). Implicature during real time conversation: A view from language processing research. *Philosophy Compass*, *2*(3), 475–496.

Shadbolt, N., & Burton, M. (1995). Knowledge elicitation: A systematic approach. In E. N. Corlett & J. R. Wilson (Eds.), *Evaluation of human work: A practical ergonomics methodology* (pp. 406–440). CRC Press.

Shannon, C. E., & Weaver, W. (1964). *The mathematical theory of communication*. University of Illinois Press. (Original work published 1949)

Shapiro, S. C., & Ismail, H. O. (2003). Anchoring in a grounded layered architecture with integrated reasoning. *Robotics & Autonomous Systems*, *43*(2–3), 97–108.

Shi, C., Verhagen, M., & Pustejovsky, J. (2014). A conceptual framework of online natural language processing pipeline application. *Proceedings of the Workshop on Open Infrastructures and Analysis Frameworks for HLT* (pp. 53–59). The Association for Computational Linguistics and Dublin City University.

Shirky, C. (2003). The Semantic Web, syllogism, and worldview. (First published November 7, 2003, on the "Networks, Economics, and Culture" mailing list.) https://www.gwern.net/docs/ai/2003-11-07-clayshirky-thesemanticwebsyllogismandworldview.html

Shutova. E. (2015). Design and evaluation of metaphor processing systems. *Computational Linguistics*, *40*, 579–623.

Sidner, C. L. (1981). Focusing for interpretation of pronouns. *Journal of Computational Linguistics*, *7*, 217–231.

Simon, H. (1957). *Models of man, social and rational: Mathematical essays on rational human behavior in a social setting.* Wiley.

Skulsky, H. (1986). Metaphorese. *Noûs*, *20*(3), 351–369.

Sowa, J. F. (2004, February 24). Common logic controlled English. http://www.jfsowa.com/clce/specs.htm

Sparck Jones, K. (2004). What's new about the Semantic Web? Some questions. *ACM SIGIR Form*, *38*(2), 18–23.

Stead, W. W., & Lin, H. S. (Eds.). 2009. *Computational technology for effective health care: Immediate steps and strategic directions.* National Research Council; National Academies Press.

Steels, L. (2008). The symbol grounding problem has been solved, so what's next? In M. De Vega, G. Glenn-berg, & G. Graesser (Eds.), *Symbols, embodiment and meaning* (pp. 223–244). Academic Press.

Steen, G. (2011). From three dimensions to five steps: The value of deliberate metaphor. *Metaphorik.de*, *21*, 83–110.

Steen, G. (2017). Deliberate metaphor theory: Basic assumptions, main tenets, urgent issues. *Intercultural Communication*, *14*(1), 1–24.

Stich, S., & Nichols, S. (2003). Folk psychology. In S. Stitch & T. A. Warfield (Eds.), *The Blackwell Guide to Philosophy of Mind* (pp. 235–255). Basil Blackwell.

Stipp, D. (1995, November 13). 2001 is just around the corner. Where's Hal? *Fortune.*

Stock, O., Slack, J., & Ortony, A. (1993). Building castles in the air: Some computational and theoretical issues in idiom comprehension. In C. Cacciari & P. Tabossi (Eds.), *Idioms: Processing, structure and interpretation* (pp. 229–248). Erlbaum.

Stolcke, A., Ries, K., Coccaro, N., Shriberg, E., Bates, R., Jurafsky, D., Taylor, P., Martin, R., Meteer, M., & Van Ess-Dykema, C. (2000). Dialogue act modeling for automatic tagging and recognition of conversational speech. *Computational Linguistics*, *26*(3), 339–371.

Stork, D. G. (1997). Scientist on the set. An interview with Marvin Minsky. In D. G. Stork (Ed.), *HAL's legacy: 2001's computer as dream and reality.* MIT Press.

Stork, D. G. (2000). *HAL's legacy: 2001's computer as dream and reality.* MIT Press.

Stoyanov, V., Gilbert, N., Cardie, C., & Riloff, E. (2009). Conundrums in noun phrase coreference resolution: Making sense of the state-of-the-art. In K. Su, J. Su, J. Wiebe, & H. Li (Eds.), *Proceedings of the Joint Conference of the 47th Annual Meeting of the Association for Computational Linguistics and the 4th International Joint Conference on Natural Language Processing of the Asian Federation of Natural Language Processing* (pp. 656–664). The Association for Computational Linguistics.

Strube, M. (1998). Never look back: An alternative to centering. *Proceedings of the 36th Annual Meeting of the Association for Computational Linguistics and the 17th International Conference on Computational Linguistics* (Vol. 2, pp. 1251–1257). The Association for Computational Linguistics.

Suhr, A., Lewis, M., Yeh, J., & Artzi, Y. (2017). A corpus of natural language for visual reasoning. *Proceedings of the 55th Annual Meeting of the Association for Computational Linguistics* (Vol. 2, pp. 217–223). The Association for Computational Linguistics.

Sukkarieh, J. Z. (2003). Mind your language! Controlled language for inference purposes. *Proceedings of the Joint Conference Combining the 8th International Workshop of the European Association for Machine Translation and the 4th Controlled Language Applications Workshop.* The European Association for Machine Translation.

Taylor, A., Marcus, M., & Santorini, B. (2003). The Penn Treebank: An overview. In A. Abeillé (Ed.), *Treebanks: Building and using parsed corpora* (pp. 5–22). Kluwer.

ter Stal, W. G., & van der Vet, P. E. (1993). Two-level semantic analysis of compounds: A case study in linguistic engineering. In G. Bouma (Ed.), *Papers from the 4th Meeting on Computational Linguistics in the Netherlands (CLIN 1993)* (pp. 163–178). Rijksuniversiteit Groningen.

Tratz, S., & Hovy, E. (2010). A taxonomy, dataset, and classifier for automatic noun compound interpretation. *Proceedings of the 48th Annual Meeting of the Association for Computational Linguistics* (pp. 678–687). The Association for Computational Linguistics.

Traum, D. R. (1994). A computational theory of grounding in natural language conversation [Unpublished doctoral dissertation]. TR 545. University of Rochester.

Traum, D. R. (1999a). Computational models of grounding in collaborative systems. *Working Notes of AAAI Fall Symposium on Psychological Models of Communication* (pp. 124–131). The AAAI Press.

Traum, D. R. (1999b). Speech acts for dialogue agents. In M. Wooldridge & A. Rao (Eds.), *Foundations and theories of rational agents* (pp. 169–201). Kluwer.

Traum, D. R. (2000). 20 questions for dialogue act taxonomies. *Journal of Semantics*, *17*(1), 7–30.

Tulving, E., & Schacter, D. L. (1990). Priming and human memory systems. *Science*, *247*, 301–306.

Turney, P. D., & Pantel, P. (2010). From frequency to meaning: Vector space models of semantics. *Journal of Artificial Intelligence Research*, *37*, 141–188.

Ungerer, F., & Schmid, H. (2006) *An introduction to cognitive linguistics*. Routledge.

Uschold, M. (2003). Where are the semantics in the Semantic Web? *AI Magazine*, *24*(3), 25–36.

Veale, T. (2012). *Exploding the creativity myth: The computational foundations of linguistic creativity*. Bloomsbury Academic.

Vieira, R., & Poesio, M. (2000). An empirically-based system for processing definite descriptions. *Computational Linguistics*, *26*(4), 525–579.

Wahlster, W. (2000). Mobile speech-to-speech translation of spontaneous dialogs: An overview of the final Verbmobil system. In W. Wahlster (Ed.), *Verbmobil: Foundations of speech-to-speech translation* (pp. 3–21). Springer.

Weaver, W. (1955). Translation. In W. N. Locke & A. D. Booth (Eds.), *Machine translation of languages: Fourteen essays* (pp. 15–23). MIT Press. (Original memorandum composed 1949).

Webber, B. L. (1988). Discourse deixis: Reference to discourse segments. *Proceedings of the Twenty-Sixth Annual Meeting of the Association for Computational Linguistics* (pp. 113–122). The Association for Computational Linguistics.

Webber, B. L. (1990). *Structure and ostension in the interpretation of discourse deixis* (Technical Report MS-CIS-90–58). University of Pennsylvania.

Wiebe, J., Wilson, T., & Cardie, C. (2005). Annotating expressions of opinions and emotions in language. *Language Resources & Evaluations*, *39*(2–3), 165–210.

Wilks, Y. (2000). Is word sense disambiguation just one more NLP task? *Computers & the Humanities*, *34*, 235–243.

Wilks, Y. (2004). Artificial companions. *Proceedings of the First International Conference on Machine Learning for Multimodal Interaction* (pp. 36–45). Springer.

Wilks, Y. (2011). Computational semantics requires computation. In C. Boonthum-Denecke, P. M. McCarthy, & T. Lamkin (Eds.), *Cross-disciplinary advances in applied natural language processing: Issues and approaches* (pp. 1–8). IGI Global.

Wilks, Y. (2009). Ontotherapy, or how to stop worrying about what there is. In N. Nicolov, G. Angelova, & R. Mitkov (Eds.), *Recent Advances in Natural Language Processing V: Selected papers from RANLP 2007* (pp. 1–20). John Benjamins.

Wilks, Y., Catizone, R., Worgan, S., Dingli, A., Moore, R., Field, D., & Cheng, W. (2011). A prototype for a conversational companion for reminiscing about images. *Computer Speech & Language*, *25*(2), 140–157.

Wilks, Y. (1975). Preference semantics. In E. L. Keenan (Ed.), *Formal semantics of natural language: Papers from a colloquium sponsored by the King's College Research Centre* (pp. 321–348). Cambridge University Press.

Wilks, Y. A., Slator, B. M., & Guthrie, L. M. (1996). *Electric words: Dictionaries, computers, and meanings*. MIT Press.

Winograd, T. (1972). *Understanding natural language.* Academic Press.

Winston, P. H. (2012). The right way. *Advances in Cognitive Systems, 1*, 23–36.

Winther, R. G. (2016). The Structure of Scientific Theories. In E. N. Zalta (Ed.), *The Stanford Encyclopedia of Philosophy* (Winter 2016 Edition). https://plato.stanford.edu/archives/win2016/entries/structure-scientific-theories

Wittgenstein, L. (1953). *Philosophical Investigations.* Oxford, UK: Blackwell.

Wong, W., Liu, W., & Bennamoun, M. (2012). Ontology learning from text: A look back and into the future. *ACM Computing Surveys, 44*(4), 20:1–20:36.

Woods, W. A. (1975). What's in a link: Foundations for semantic networks. In D. G. Bobrow & A. M. Collins (Eds.), *Representation and understanding: Studies in cognitive science* (pp. 35–82). Academic Press.

Woodward, J. (2019). Scientific explanation. In E. N. Zalta (Ed.), *The Stanford encyclopedia of philosophy* (Winter 2019 ed.). https://plato.stanford.edu/archives/win2019/entries/scientific-explanation

Yuret, D. (1996, February 13). *The binding roots of symbolic AI: A brief review of the Cyc project.* MIT Artificial Intelligence Laboratory.

Zaenen, A. (2006). Mark-up barking up the wrong tree. *Computational Linguistics, 32*, 577–580.

Zaenen, A., Karttunen, L., & Crouch, R. S. (2005). Local textual inference: Can it be defined or circumscribed? *Proceedings of the ACL 2005 Workshop on Empirical Modeling of Semantic Equivalence and Entailment* (pp. 31–36). The Association for Computational Linguistics.

Zhu, Z., & Hu, H. (2018). Robot learning from demonstration in robotic assembly: A survey. *Robotics, 7*(2), 17. https://doi.org/10.3390/robotics7020017

Zipf, G. K. (1949). *Human behavior and the principle of least effort.* Addison-Wesley.

Zlatev, J. (2010). Phenomenology and cognitive linguistics. In S. Gallagher & D. Schmicking (Eds.), *Handbook of phenomenology and cognitive science* (pp. 415–446). Springer.

Index

Page numbers in italic indicate a figure and page numbers in bold indicate a table on the corresponding page.